Bone Circulation and Vascularization in Normal and Pathological Conditions

NATO ASI Series

Advanced Science Institutes Series

A series presenting the results of activities sponsored by the NATO Science Committee, which aims at the dissemination of advanced scientific and technological knowledge, with a view to strengthening links between scientific communities.

The series is published by an international board of publishers in conjunction with the NATO Scientific Affairs Division

A	**Life Sciences**	Plenum Publishing Corporation
B	**Physics**	New York and London
C	**Mathematical and Physical Sciences**	Kluwer Academic Publishers
D	**Behavioral and Social Sciences**	Dordrecht, Boston, and London
E	**Applied Sciences**	
F	**Computer and Systems Sciences**	Springer-Verlag
G	**Ecological Sciences**	Berlin, Heidelberg, New York, London,
H	**Cell Biology**	Paris, Tokyo, Hong Kong, and Barcelona
I	**Global Environmental Change**	

Recent Volumes in this Series

Volume 243A — Biological Effects and Physics of Solar and Galactic Cosmic Radiation, Part A
edited by Charles E. Swenberg, Gerda Horneck, and E. G. Stassinopoulos

Volume 243B — Biological Effects and Physics of Solar and Galactic Cosmic Radiation, Part B
edited by Charles E. Swenberg, Gerda Horneck, and E. G. Stassinopoulos

Volume 244 — Forest Development in Cold Climates
edited by John Alden, J. Louise Mastrantonio, and Søren Ødum

Volume 245 — Biology of *Salmonella*
edited by Felipe Cabello, Carlos Hormaeche, Pasquale Mastroeni, and Letterio Bonina

Volume 246 — New Developments in Lipid–Protein Interactions and Receptor Function
edited by K. W. A. Wirtz, L. Packer, J. Å. Gustafsson, A. E. Evangelopoulos, and J. P. Changeux

Volume 247 — Bone Circulation and Vascularization in Normal and Pathological Conditions
edited by A. Schoutens, J. Arlet, J. W. M. Gardeniers, and S. P. F. Hughes

Series A: Life Sciences

Bone Circulation and Vascularization in Normal and Pathological Conditions

Edited by

A. Schoutens

Hôpital Erasme
Brussels, Belgium

J. Arlet

Centre Hospitalo-Universitaire Rangueil
Toulouse, France

J. W. M. Gardeniers

Katholieke Universiteit Nijmegen
Nijmegen, The Netherlands

and

S. P. F. Hughes

Hammersmith Hospital
London, United Kingdom

Springer Science+Business Media, LLC

Proceedings of a NATO Advanced Research Workshop on
Bone and Bone Marrow Circulation in Normal and Pathological Conditions,
held September 25–26, 1992,
in Brussels, Belgium

NATO-PCO-DATA BASE

The electronic index to the NATO ASI Series provides full bibliographical references (with keywords and/or abstracts) to more than 30,000 contributions from international scientists published in all sections of the NATO ASI Series. Access to the NATO-PCO-DATA BASE is possible in two ways:

—via online FILE 128 (NATO-PCO-DATA BASE) hosted by ESRIN, Via Galileo Galilei, I-00044 Frascati, Italy

—via CD-ROM "NATO-PCO-DATA BASE" with user-friendly retrieval software in English, French, and German (©WTV GmbH and DATAWARE Technologies, Inc. 1989)

The CD-ROM can be ordered through any member of the Board of Publishers or through NATO-PCO, Overijse, Belgium.

```
             Library of Congress Cataloging-in-Publication Data
─────────────────────────────────────────────────────────────────────
Bone circulation and vascularization in normal and Pathological
  conditions / edited by A. Schoutens ... [et al.].
      p.   cm. -- (NATO ASI series. Series A, Life sciences ; v.
  247)
     "Proceedings of a NATO Advanced Research Workshop on Bone and Bone
  Marrow Circulation in Normal and Pathological Conditions, held
  September 25-26, 1992, in Brussels, Belgium"--T.p. verso.
     Includes bibliographical references and index.
     ISBN 978-0-306-44523-1     ISBN 978-1-4615-2838-8 (eBook)
     DOI 10.1007/978-1-4615-2838-8
     1. Bones--Blood-vessels--Congresses.  2. Bones--Necrosis-
  -Congresses.  3. Bone marrow--Blood-vessels--Congresses.  4. Bones-
  -Pathophysiology--Congresses.   I. Schoutens, A.  II. NATO Advanced
  Research Workshop on Bone and Bone Marrow Circulation in Normal and
  Pathological Conditions (1992 : Brussels, Belgium)  III. Series.
     [DNLM: 1. Bone and Bones--blood supply--congresses.  2. Bone
  Diseases--physiopathology--congresses.  3. Blood Circulation-
  -congresses.   WE 225 B71255 1992]
  QP88.2.B587  1993
  612.7'5--dc20
  DNLM/DLC
  for Library of Congress                                    93-4599
                                                                CIP
─────────────────────────────────────────────────────────────────────
```

ISBN 978-0-306-44523-1

© 1993 by Springer Science+Business Media New York
Originally published by Plenum Press New York in 1993

PREFACE

The Association Internationale de Recherche sur la Circulation Osseuse, A.R.C.O., was founded in London in December 1989 by a small group of doctors, surgeons and researchers in basic sciences who had been involved for many years in the study of bone circulation and its disorders. They had met several times in Toulouse, during the International Symposia on Bone Circulation held there since 1973 and they wished to carry their contacts further.

In founding A.R.C.O., they established as their primary aims the encouragement and furtherance of research, organisation of meetings and promotion of knowledge on the subject.

At the present time, the Association has over a hundred members from more than twenty countries in Europe, America and Asia. All have the conviction that bone tissue and its pathology can only be truly known and studied if one has an understanding of its vascular system and the way its circulation functions. This concept, apparently beyond question, has not yet been adopted by all physicians and scientists who are interested in bone. From time to time, one comes across teaching programmes on bone pathology which make no mention of bone circulation.

For this reason, during our annual meeting, we decided to review present-day knowledge on this subject and to establish what might be called the state of the art on bone circulation in normal and pathological conditions. We asked each authority to present a synthesis of his work and to give his opinion on points open to controversy.

The subiects or themes of the lectures have been chosen for their fundamental interest and practical value, and also according to the particular competence of each of the contributors.

These lectures have been brought together in this book under three headings dealing with :

1. Basic anatomical and physiological concepts and methods of investigation of bone circulation.
2. The circulatory aspect of the various disorders of bone tissue in which the circulation can always be seen to be involved, to a varying degree.
3. Osteonecrosis, and in particular osteonecrosis of the femoral head. We thought it essential that an important part of our work should be devoted to this disease, where circulatory failure, ischaemia, is the main lesional agent and which has such damaging consequences on the joints.

This meeting could not have taken place and this book could not have appeared without the material help of NATO and the dedication of the Belgian members of our association, under the efficient, enthusiastic and warm-hearted guidance of Professor André Schoutens.

Jacques Arlet

CONTENTS

FRACTURE HEALING AND BONE GRAFTS

CIRCULATORY ASPECTS OF BONE DISORDERS

OSTEONECROSIS

1. General Aspects of Osteonecrosis

2. Methods of Diagnosis

3. Treatment

4. Arco Perspective for Staging

BONE CIRCULATION AND
VASCULARIZATION IN
NORMAL AND PATHOLOGICAL CONDITIONS

A PROPOS DE LA VASCULARISATION DES OS LONGS

PAR

R. DE MARNEFFE

(Laboratoire d'Histologie. Faculté de Médecine. Université libre de Bruxelles.)

Au cours de ces dernières années, les études que nous avons poursuivies sur la vascularisation des os longs nous paraissent de nature à éclairer certains aspects de la physiopathologie osseuse. Ces recherches ont porté principalement sur la vascularisation des os longs du rat, du cobaye et du lapin. Mais les quelques constatations que nous avons pu effectuer sur du matériel humain montrent que le schéma général de la vascularisation des os longs s'applique également à l'anatomie humaine.

Nous avons abordé ces recherches par deux groupes de techniques essentiellement différentes. D'une part, les injections vasculaires au moyen d'encre de Chine gélatinée, de substances opaques aux rayons X et surtout de matières plastiques conduisent à une conception assez complète du mode de vascularisation des os longs. D'autre part, des modifications expérimentales apportées à certains réseaux vasculaires permettent de pénétrer l'intimité vasculaire de la compacte diaphysaire et leurs rapports avec la vitalité du tissu osseux lui-même.

Nous allons rapporter ici les constatations essentielles que ces recherches ont permis de faire.

I. — OBSERVATIONS MORPHOLOGIQUES.

Le sang arrive dans la *cavité diaphysaire* par l'artère nourricière principale. Elle ne donne aucune collatérale au tissu osseux lui-même. Elle se divise rapidement en branches ascendantes et descendantes qui se dirigent vers les extrémités osseuses. Chacune de ces branches se ramifie à son tour en branches secondaires (fig. 1). A proximité du cartilage de conjugaison (rappelons que celui-ci persiste très longtemps chez le rat, le cobaye et le lapin), ces branches secondaires se ramifient à nouveau brusquement pour former un véritable réseau sous-chondral. Dépendant de ce réseau sous-chondral, nous avons remarqué l'existence d'expansions digitiformes borgnes qui pénètrent perpendiculairement dans le cartilage conjugal. Il s'agit de vrais « culs-de-sac vasculaires » qui occupent les espaces laissés libres par la lyse des chondrocytes les plus profonds.

Mais l'artère nourricière principale n'est pas la seule voie d'apport sanguin dans la

Bone Circulation and Vascularization in Normal and Pathological Conditions
Edited by A. Schoutens *et al.*, Plenum Press, New York, 1993

cavité diaphysaire. En effet, celle-ci est encore irriguée par des artères nourricières secondaires : pour le fémur, l'artère digitale à son extrémité supérieure et les artères métaphysaires (sus-condyliennes internes et externes) à son extrémité inférieure ; pour le tibia, l'artère diaphysaire antérieure au tiers moyen de l'os et les artères métaphysaires interne, médiane et externe dans la région rétro-tubérositaire de l'extrémité supérieure. Tous ces vaisseaux se perdent rapidement dans le réseau capillaire de la moelle osseuse. Nous avons remarqué que tous les orifices vasculaires comportent en plus du vaisseau artériel, un ou deux vaisseaux veineux ainsi qu'un ou deux éléments nerveux. Le réseau vasculaire de la moelle osseuse est un système parfaitement clos, formé de capillaires plus ou moins ténus qui sillonnent la trame conjonctive.

Pour quitter la cavité médullaire des os longs, le sang emprunte un système veineux extrêmement particulier. La cavité médullaire est parcourue sur toute sa hauteur par un large sinus veineux central. Par de petits vaisseaux transversaux, celui-ci draine le réseau capillaire de la moelle osseuse. A ses extrémités, ce sinus veineux semble se terminer par des expansions borgnes légèrement renflées.

Les voies de sortie de ce sinus comportent deux systèmes veineux distincts. D'une part, des veines qui cheminent parallèlement aux artères : on décrit pour le tibia les veines diaphysaires principale (postérieure) et secondaire (antérieure), pour le fémur la veine nourricière principale. D'autre part, on reconnaît des voies veineuses de sortie, isolées, généralement très larges et aboutissant à des orifices osseux distincts : ce sont pour le tibia la veine diaphysaire inférieure et à l'extrémité proximale les deux gros vaisseaux veineux rétrotubérositaires externe et médian ; pour le fémur, ce sont les deux ou trois grosses veines qui à son extrémité supérieure s'ouvrent au dehors dans la cavité digitale. Par leur calibre important, plusieurs fois supérieur à celui de la veine nourricière principale et par leur nombre, ces gros vaisseaux veineux de l'extrémité proximale du tibia et du fémur assurent le plus gros débit sanguin veineux. Ils constituent, à notre avis, la voie efférente principale de la circulation de retour (fig. 2).

Enfin, signalons encore que le réseau capillaire de la moelle osseuse communique directement avec l'extérieur par les veines métaphysaires. Celles-ci accompagnent les artères du même nom. Ce système veineux est indépendant de celui constitué par le grand sinus veineux central.

L'irrigation des *épiphyses* est simple. Elle est assurée par deux ou plusieurs artères nourricières qui pénètrent dans la cavité épiphysaire, s'y arborisent et se perdent dans un réseau médullaire semblable à celui de la cavité diaphysaire. Le sang se collecte peu à peu dans un réseau veineux et quitte l'épiphyse par des veines épiphysaires accolées aux artères. La plupart des paquets vasculaires destinés aux épiphyses sont accompagnés d'un nerf.

Si nous étudions la vascularisation du *tissu osseux compact*, nous voyons qu'elle est assurée par des capillaires corticaux longitudinaux qui occupent les canaux de Havers. Ces capillaires donnent des branches transversales qui s'anastomosent aux capillaires des canaux de Havers voisins. Chaque canal de Havers contient habituellement un seul capillaire et non pas, comme il a été admis longtemps, un paquet vasculaire formé d'une artère et d'une veine ; exceptionnellement, on peut y rencontrer deux capillaires. La compacte est irriguée par un vaste réseau capillaire ; celui-ci est en continuité parfaite d'une part avec le réseau périosté et d'autre part avec le réseau capillaire de la moelle osseuse. La vascularisation osseuse est donc tributaire de deux systèmes capillaires différents, l'un interne, médullaire, et l'autre externe, périosté. Mais si l'on examine l'importance de ces deux systèmes, on constate qu'elle varie aux divers niveaux de la diaphyse. A la partie proximale de celle-ci, les capillaires corticaux proviennent en majeure partie de la cavité médullaire et sont orientés obliquement vers le bas et le dehors. L'apport périosté y est pratiquement nul. Dans la partie moyenne de l'os, un certain nombre de capillaires périostés participent à l'irrigation de la compacte, l'apport médullaire est moins marqué. Enfin, à l'extrémité inférieure de la diaphyse, l'apport d'origine médullaire a disparu. Les capillaires corticaux, dirigés vers le bas et le dedans, dépendent uniquement du réseau périosté.

II. — Observations expérimentales.

Sans vouloir entrer dans le détail de ces expériences qui ont porté sur le fémur du lapin, signalons que nous avons modifié la vascularisation d'un os long en supprimant les divers apports sanguins destinés à la compacte diaphysaire. Ces expériences sont intéressantes par elles-mêmes et par leurs différentes combinaisons. Nous avons supprimé l'apport vasculaire interne (médullaire) en curettant la cavité diaphysaire et en la remplissant de paraffine stérile. La suppression de l'apport vasculaire externe a été réalisée en décollant le périoste de la diaphyse et en isolant celle-ci au moyen d'une gaine en tissu plastique. Enfin, nous avons privé la compacte diaphysaire des apports sanguins dépendant des canaux de Havers ou des extrémités osseuses par divers traits de scie. L'étude histologique a porté sur l'intégrité tissulaire de la compacte osseuse dans le segment diaphysaire intéressé par la dévascularisation.

Dans l'ensemble, ces expériences mettent l'accent sur l'importance de l'apport vasculaire périosté pour assurer l'intégrité du tissu osseux compact de la diaphyse. Au contraire, les artères nourricières diaphysaires desservent uniquement le réseau capillaire de la moelle osseuse. Elles n'interviennent que très peu dans l'irrigation de la corticale diaphysaire.

Quand les apports vasculaires externe (périosté) et interne (médullaire) sont supprimés, la vascularisation qui arrive par les canaux de Havers dans la région opérée est insuffisante pour assurer la vitalité du tissu osseux.

Enfin, signalons qu'au cours d'expériences où nous avions supprimé simultanément les circulations extrapériostée et médullaire nous avons eu la surprise d'observer l'apparition d'une ostéite fibreuse tout à fait typique caractérisée par des remaniements osseux importants dans lesquels destruction et apposition se mêlent d'une façon anarchique et s'accompagnent de fibrose médullaire. Cette ostéite fibreuse siège uniquement dans le territoire diaphysaire dévascularisé (fig. 3).

Conclusions.

Les constatations morphologiques et expérimentales que nous venons de rapporter, ouvrent un jour n'ouveau sur certains faits cliniques bien connus.

La grosse capacité du système veineux et surtout l'importance et la multiplicité de ses voies efférentes sont de nature à éviter les fluctuations de la pression sanguine intraosseuse. C'est le large débit de ces voies de drainage qui assure l'écoulement facile des transfusions intramédullaires.

La vascularisation particulièrement précaire de certaines épiphyses rend parfaitement compte des nécroses avasculaires posttraumatiques observées dans la tête fémorale ou dans l'épiphyse tibiale inférieure. Elle interviendrait également dans la pathogénie des luxations congénitales de la hanche et de leurs complications.

Les détails de l'irrigation que nous avons observés au niveau de la compacte expliquent la difficulté avec laquelle on obtient parfois la guérision de certaines fractures diaphysaires situées au tiers inférieur du tibia et du fémur. Au contraire, les régions métaphysaires distale du fémur et proximale du tibia, particulièrement bien vascularisées, guérissent facilement. Chez l'enfant, cette vascularisation est probablement en rapport avec la fertilité du cartilage de conjugaison de ces deux régions.

Enfin, les lésions d'ostéite fibreuse observées après suppression combinée des apports vasculaires externe et interne sont à rapprocher de celles trouvées dans les cals hypertrophiques des fractures traitées par ostéosynthèse centro-médullaire. Elles sont également comparables aux lésions microscopiques de la dysplasie fibreuse, de l'ostéome ostéoïde et de certains kystes osseux localisés. Il existerait, à la base de ces différentes affections, un trouble étiologique vasculaire localisé.

Ces recherches conduisent donc à une meilleure compréhension de la physiologie circulatoire des os longs et à une thérapeutique plus judicieuse.

RÉSUMÉ.

La circulation diaphysaire est assurée par des vaisseaux nourriciers principaux et secondaires principalement destinés à la moelle osseuse. La circulation épiphysaire dépend de deux ou plusieurs paquets vasculaires épiphysaires. Les voies veineuses de retour comportent deux systèmes veineux distincts : des veines nourricières principales et secondaires qui accompagnent les artères, de grosses veines isolées en rapport avec le sinus veineux central et qui s'ouvrent au dehors à l'extrémité proximale de l'os. Ces voies veineuses assurent le drainage du réseau capillaire médullaire. La suppression de l'apport vasculaire externe (périosté) altère la structure de la compacte diaphysaire. La suppression simultanée des apports vasculaires externe (périosté) et interne (médullaire) provoque l'apparition de lésions d'ostéite fibreuse. La connaissance approfondie de la circulation osseuse diaphysaire et épiphysaire rend parfaitement compte de nombreux faits cliniques bien connus.

THE CIRCULATION IN LONG BONES

R. DE MARNEFFE

Clinique d'Orthopédie, Hôpital Universitaire Brugmann (Emeritus)
Brussels, Belgium

This research work carried out long ago and published in 1951 is still considered today to be a valuable reference. This is why an up-to-date summary is presented as an historical introduction to this Course on bone circulation.

MATERIAL AND METHODS

The animals used were the rat, the guinea-pig and the rabbit. The femur and the tibia were studied in order to visualize the blood vessels. Intravascular injections were given into the hind limbs, through the abdominal aorta using gelatinized Indian ink, Novobaryum and Neoprene Latex 572, a plastic solution familiar from Trueta's excellent work on "The Renal Circulation" (1947). The observations were made macroscopically, by minute and progressive dissection after resorption of bone tissue, and by histological analysis.

RESULTS

General Morphology of Bone Circulatoin

Dry Bones Examination. A great number of holes were found in the rat, dog and human femora and tibiae.

What was surprising was that, in all species including man, the holes were all symetric and corresponded mathematically. As these discoveries are covered by the following studies I refer the reader to the original paper.

The Arterial Supplies (afferent). Neoprene Latex only fills the afferent arteries and stops at the capillaries.

Femur: the diaphysis has two main arterial supplies: the principal nutrient artery and accessory arteries which are located proximally at the trochanteric fossa, and distally at the metaphyseal region. The upper epiphysis (head and great trochanter) and the lower epiphysis receive multiple supplies.

Bone Circulation and Vascularization in Normal and Pathological Conditions
Edited by A. Schoutens *et al.*, Plenum Press, New York, 1993

7

Tibia: the diaphysis has three arterial supplies: the principal nutrient artery and two accessory arteries, mid-shaft anterioles and, proximally, at the back of the tibial tuberosities. The upper epiphysis presents multiple arterial supplies surrounding the tibial spinal processes; and the lower epiphysis, surrounding both malleoli.

All arteries, wether diaphyseal or epiphyseal, after passing through cortical bone without collaterals reach the medullary cavities where they ramify. All arteries are accompanied by veins which run in parallel, thus, all these vascular structures are in fact "arterio-venous bunches". Finally, the perfect symmetry of the vascular distribution in these two long bones should be noted.

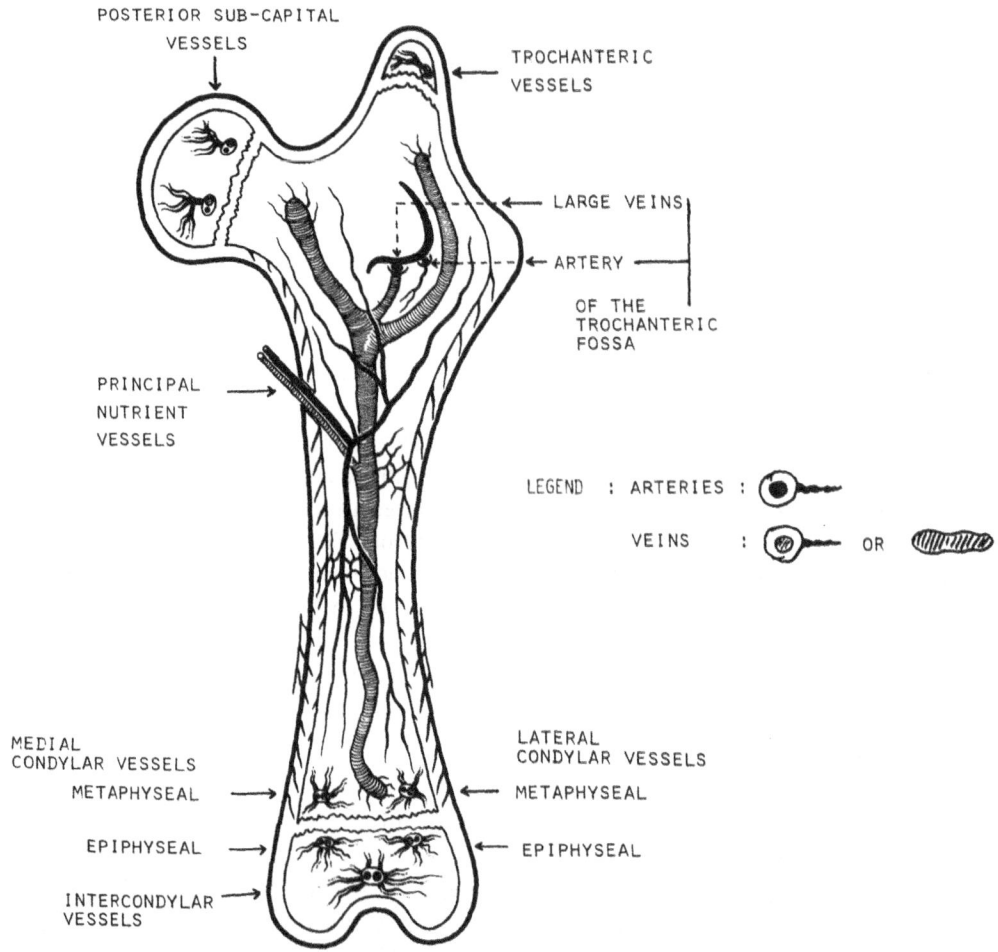

Figure 1. Summary of the general morphology of bone circulation in the rabbit's femur.

The Medullary Vascular Structures. Indian ink completely fills the vascular bed of the bone, from the arterial to the venous side, including the capillaries. In no instance was Indian ink found outside the vessels, thus, bone circulation is, at all levels, a perfectly closed system.

Bone Marrow (Medullary Cavity) is best seen in the vertebra of the rat's tail, when injected with Indian ink (De Marneffe, 1948). It shows the transcortical nutrient artery and its progressive ramification as far as the capillaries.

Sinuses: When opening the medullary cavity of the femur or of the tibia, an enormous sinus is found running all along the medullary canal, with two or three digitations proximally in the femur. These vascular sinuses are completely surrounded by a thin cellular wall. They are a kind of "collector".

The Venous Pathways (efferent). At the proximal femoral extremity, in the trochanteric fossa, there is a wide venous opening communicating directly with one of the

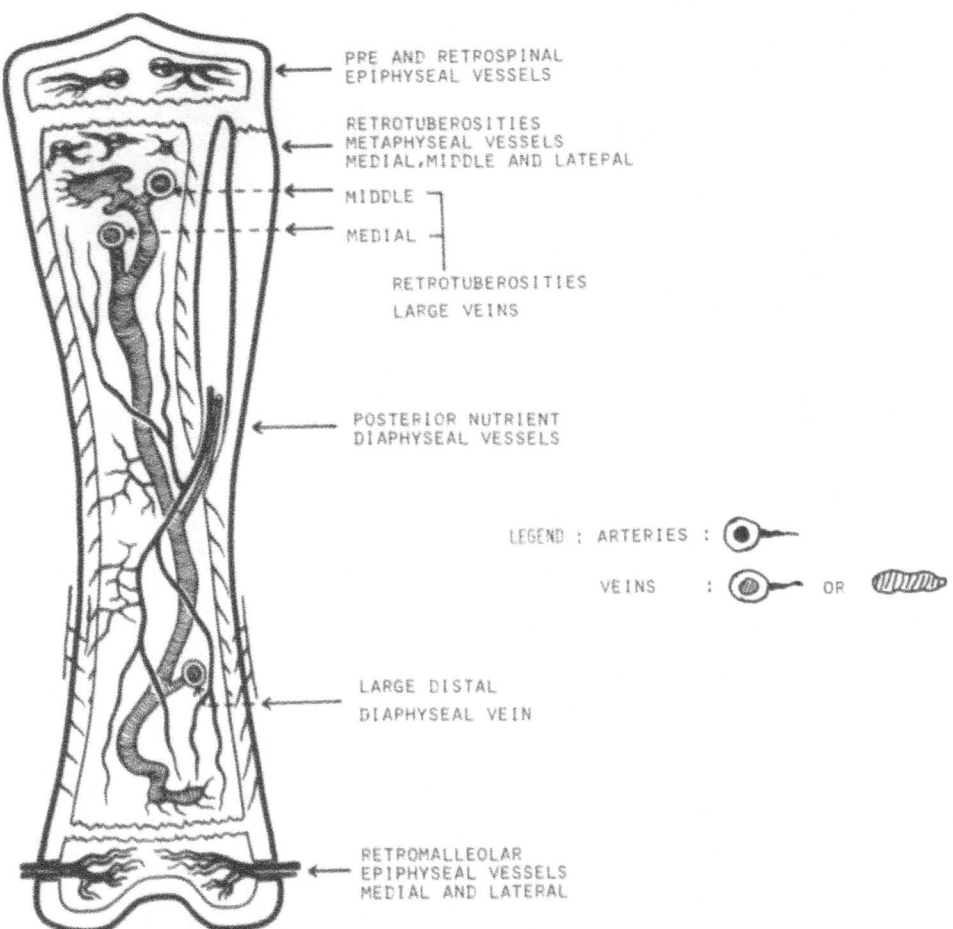

PRE AND RETROSPINAL
EPIPHYSEAL VESSELS

RETROTUBEROSITIES
METAPHYSEAL VESSELS
MEDIAL, MIDDLE AND LATERAL

MIDDLE

MEDIAL

RETROTUBEROSITIES
LARGE VEINS

POSTERIOR NUTRIENT
DIAPHYSEAL VESSELS

LEGEND : ARTERIES :

VEINS : OR

LARGE DISTAL
DIAPHYSEAL VEIN

RETROMALLEOLAR
EPIPHYSEAL VESSELS
MEDIAL AND LATERAL

Figure 2. Summary of the general morphology of bone circulation in the rabbit's tibia.

medullary sinuses. At the tibia, three wide venous openings were found: 2 proximally, at the back of the tibial tuberosities and one distally, below the principal nutrient bunch.

As mentioned above, all the afferent arteries are accompanied by parallel efferent structures which are veins.

Cortical Bone Irrigation. As many authors have described, cortical irrigation is in perfect continuity with the vessels of the periosteum outside and the capillaries of the medullary cavity inside. It is accepted that irrigation within the cortical bone itself, with the osteocytes as the final destination, is ensured by the Haversian Canals. The Haversian Canal is generally occupied by only one vessel. Very exceptionnally two have been found, one being thinner than the other. The orientation of the capillaries have been examined running in the diaphyseal cortical bone of the femur and the tibia and it has been found that, anatomically, at the proximal end of the diaphysis, the cortex seems to be irrigated preferentially by the medullary cavity, at the mid-shaft the irrigation is equally mixed and at the distal end it seems that the prevalence is periosteal .

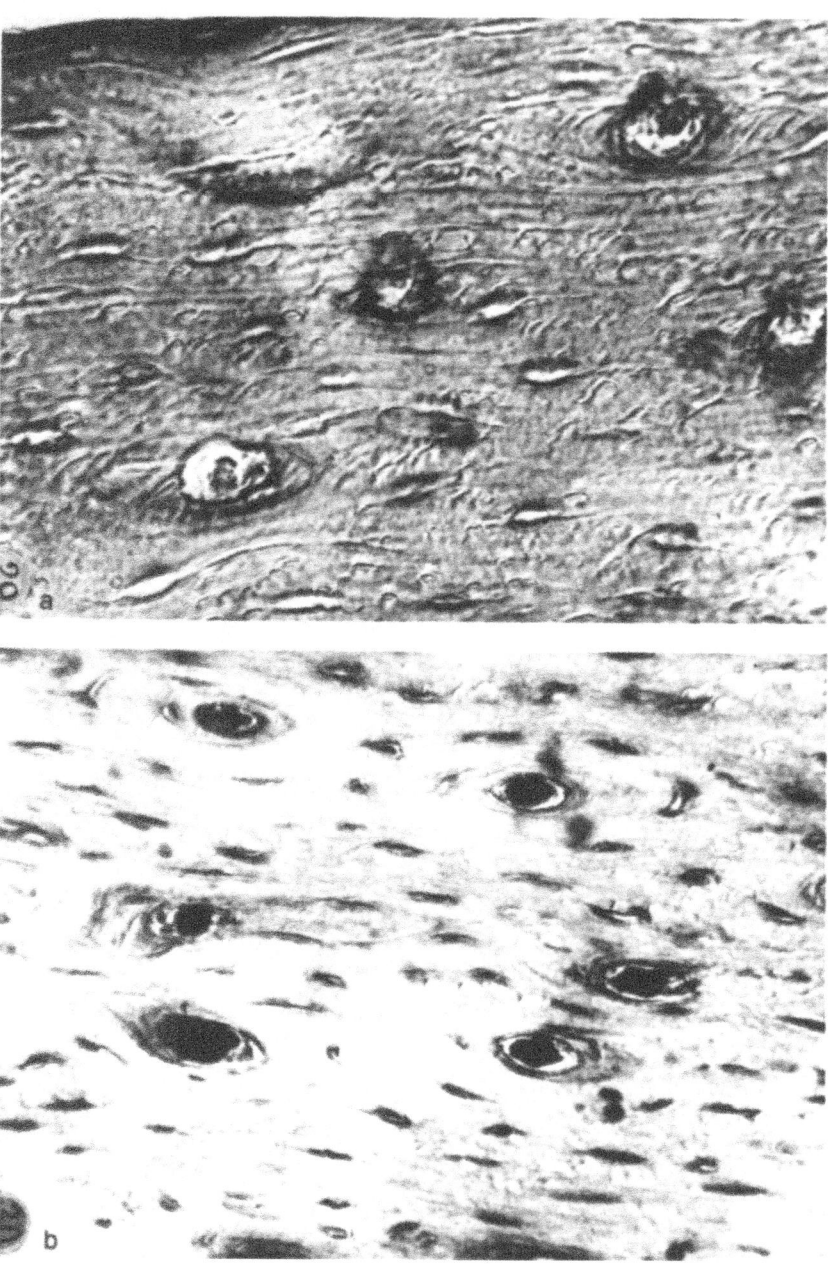

The Role of Each Vascular Supply to the Cortex

The last point to be understand was the importance of each blood supply in cortical bone irrigation and also to observe, when one source is interrupted, what possibilities exist of compensation by other systems. In other words, the chances of survival of bone tissue in different conditions. This was felt to be very important particularly in clinical situations.

Experimental Procedures. The various blood supplies to the cortical bone were interrupted by surgical procedures: - Interruption of the blood supply from the medullary cavity by curetage of the cavity and infilling with paraffin (M.C.); - Ligature of the principal nutrient artery (Lg.Pr.N.A.); - Separation of the periosteal supply from the diaphyseal cortex by a plastic sheet wrapping (Pl.S.); - Periosteal and medullary supplies interrupted by an association of plastic sheet wrapping and medullary curetage; - The Haversian irrigation interrupted by a quadrangular cortical osteotomy, combined with a plastic sheet wrapping and medullary curetage (Q.O.); - A double diaphyseal osteotomy (one proximal and one distal) was performed to evaluate the specific role of the principal nutrient artery. The nutrient artery was kept intact while all other cortical irrigations were destroyed (D.O.); - An extra-periosteal separation using a plastic sheet wrapping associated with a principal nutrient artery ligature completed my experiments.

Observations. The vitality of bone tissue and its resistance to the experimental vascular ischemia were examined by histology to study - the state of osteocytes: normal, altered or absent, leaving empty osteoplasts; - the quality of the ground substance; - the presence or absence of Indian ink.
Interruption of the blood supply from the medullary cavity or from the principal nutrient artery only affects the diaphyseal cortical bone slightly or indeed not at all. The role of the periosteal vessels is mainly to the outer part of the cortex. When medullary and periosteal supplies are interrupted simultaneously, the Haversian canals are incapable of ensuring cortex survival alone. When the principal nutrient artery is the only means of irrigation left, bone death ensures. The combined interruption of the periosteal supply and of the principal nutrient artery produces a complex picture composed of bone resorption, bone apposition and medullary fibrosis similar to "Osteitis Fibrosa" .

CONCLUS I ONS

In this pioneer work using experimental surgical procedures, the influence of the blood supply on cortical bone vitality has been evaluated and capacities for compensating and to ensuring the survival of bone tissue has been examined. The periosteal supply is shown to be particularly important, and by a combined vascular interruption, a peculiar bone reaction similar to "Osteitis Fibrosa" has been produced.

Figure 3. (A) Rabbit's femur cortical bone, two weeks after periosteal blood supply interruption: alteration of the fundamental substance and of the osteocytes, osteoplasts, no Indian ink. (B) Non operated femur of same animal for comparison: normal bone tissue and Indian ink.

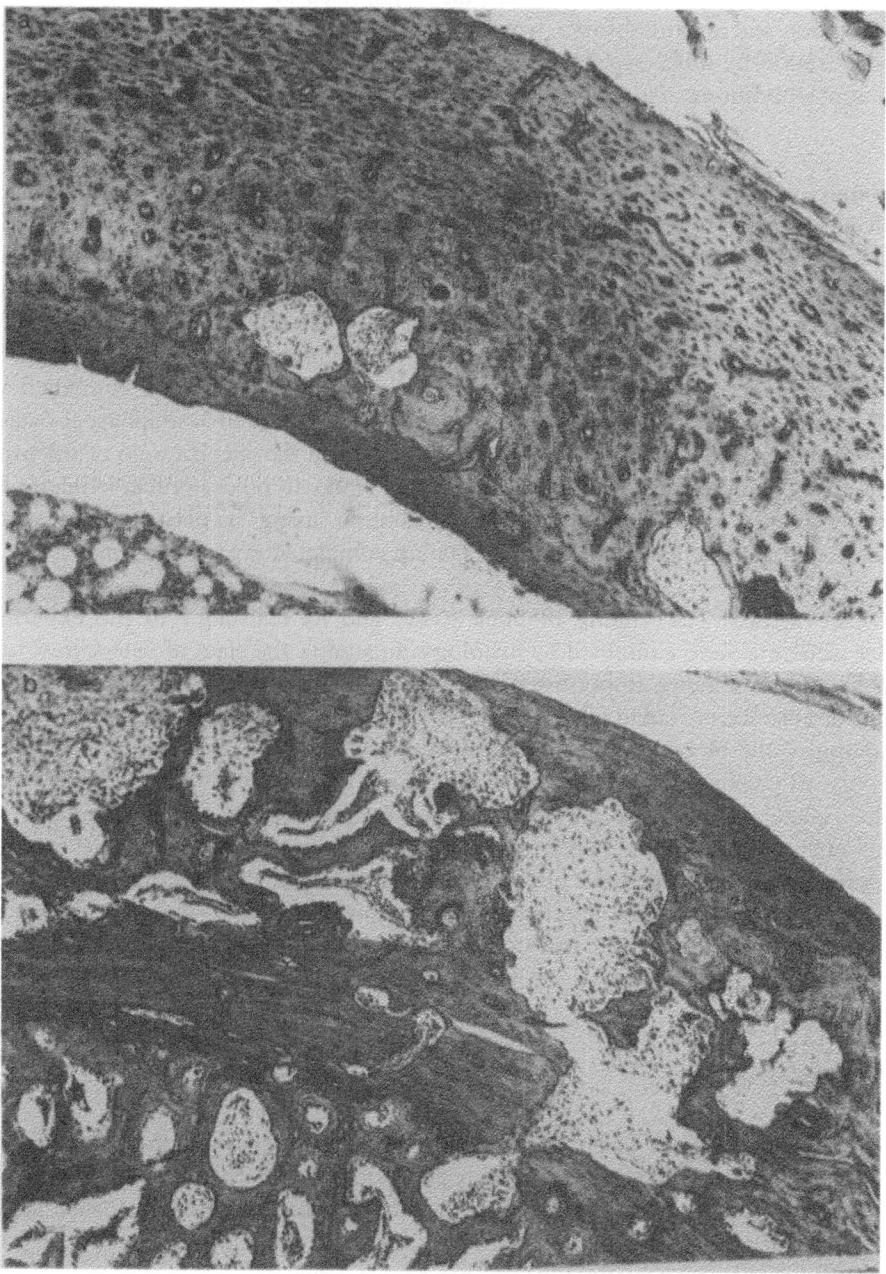

Figure 4. (A) Rabbit's femur, two weeks after a combined vascular interruption: experimental "osteitis fibrosa", first stage lesion. (B) Similar experiment: second stage lesion showing apposition, resorption and medullar fibrositis. (C) Same animal: cortical bone outside the operated zone. Practically normal structures. (D) Same animal: cortical bone of the non operated femur. Normal bone structures.

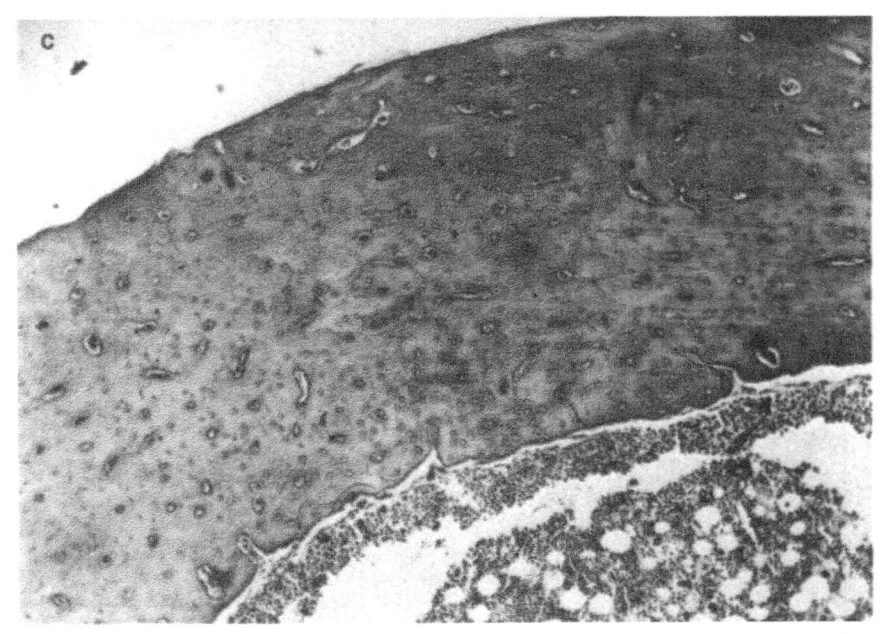

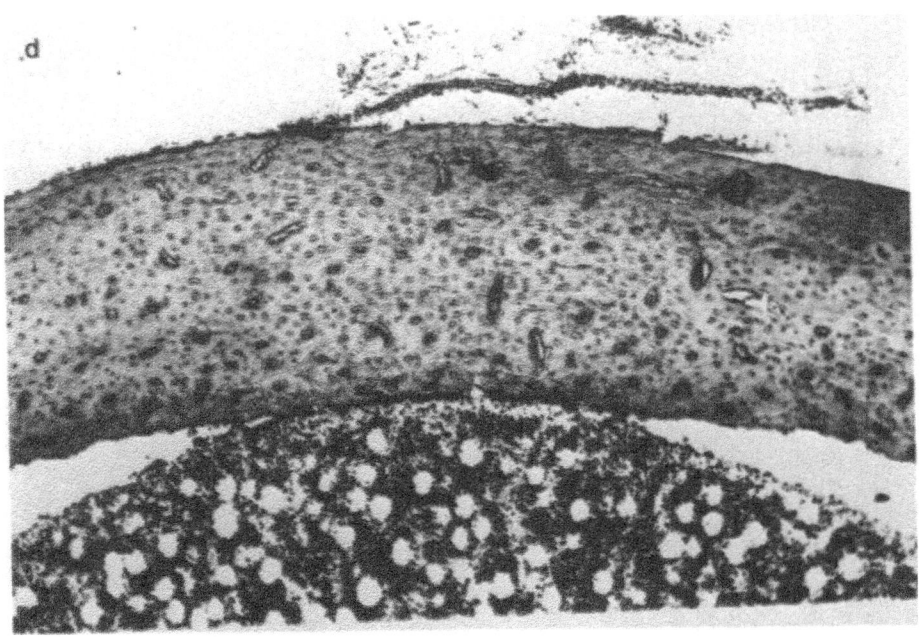

BLOOD SUPPLIES INTERRUPTIONS	SURGERY LG, M,C,	PR,N,A,	PL,S,	Q,O,	D,O,	OBSERVATIONS (BONE TISSUE REACTIONS)
MEDULLARY	X					SUB-NONE
PRINCIPAL NUTRIENT ARTERY		X				SUB-NONE
PERIOSTEAL			X			CORTEX EXTERNAL HALF PART
MEDULLARY AND PERIOSTEAL	X		X			SUB-TOTAL NECROSIS
MEDULLARY, PERIOSTEAL AND HARVERSIAN CANALS	X		X	X		TOTAL NECROSIS
ALL SUPPLIES EXCEPT PRINCIPAL NUTRIENT ARTERY			X		X	PROGRESSIVE NECROSIS FROM THE EXTERNAL TO THE INTERNAL CORTEX
EXTRA-PERIOSTEAL AND PRINCIPAL NUTRIENT ARTERY		X	X			RESORPTION, APPOSITION AND MEDULLARY FIBROSIS ("OSTEITIS FIBROSA")

Figure 5. Summary of the experimental surgical procedures, their combinations and results on the cortical bone vitality.

REFERENCES

de Marneffe, R.: A propos de la vascularisation de la vertèbre caudale du rat. C.R. Soc Biol. 142: 1093-1095 (1948).

de Marneffe, R.: Recherches morphologiques et expérimentales sur la vascularisation osseuse. Acta Chir. Belg. 50:1-80 (1951).

de Marneffe, R.: A propos de la vascularisation osseuse des os longs. Rev. Chir. Orthop. 38: 64-69 (1952).

de Marneffe, R.: Données actuelles concernant la vascularisation osseuse dans la maladie de Paget. Acta Cardiol. Belg. 8: 181-188 (953).

de Marneffe, R.: Production expérimentale d'ostéoporose et d'ostéite fibreuse chez le lapin. Acta Chir. Belg. et Acta Orthop. Belg., Suppl. I: 194-197 (1956).

Trueta, J., A. Barclay, P. Daniel, K. Franklin, M. Prichard: Studies of the renal circulation. Blackwell Scientific Publication, Oxford 1947.

ANATOMY AND PHYSIOLOGY

MORPHOLOGY AND DISTRIBUTION OF BLOOD VESSELS AND BLOOD FLOW IN BONE

M. BROOKES

Academic Orthopaedic Research Unit
United Medical Schools, St.Thomas' Hospital
London, Great Britain

METHODS OF INVESTIGATION

The principal methods of investigating the blood supply of bone include

Bone Angiography. A human limb or entire laboratory animal is perfused intravascularly with a radiopaque 45 % barium sulphate suspension and fixed in a 10% formalin solution in normal saline. Subsequently the bony parts are decalcified in 5 % nitric acid solution, washed in tap water and stored in formalin. Sections of bone are prepared for microfocal X-ray analysis, often with some form of magnification, to display blood vessels in bone (Brookes, 1990).

Indian Ink Perfusion and **Light Microscopy** are especially useful for revealing details of bone vascularization in small laboratory animals (Brookes, 1971).

Bone Blood Flow Measurement can be carried out in the laboratory by the hydrogen washout method (Brookes, 1987). It can be modified to yield a continuous record of blood flow rate in bone marrow extending over weeks, but electrode poisoning is a serious hazard (Revell and Heatley, 1987). Regional bone blood flow rates in compact and cancellous bone and bone marrow are more conveniently measured by the method of arteriolar blockade (Brookes, 1970) and its variant techniques. Radioisotope uptake methods deployed with mathematical modelling are available for human bone (Brookes, 1987; Charkes, et al., 1979, 1980).

BLOOD VESSELS OF THE DIAPHYSIS

The branches of the parent nutrient artery ramify in bone marrow and are distributive in function. Medullary arterioles, however, control the blood flow rate and are

Bone Circulation and Vascularization in Normal and Pathological Conditions
Edited by A. Schoutens *et al.*, Plenum Press, New York, 1993

19

the resistance vessels of the osseous circulation. They diminish in calibre from 100 μm to the 5 μm metarterioles. The latter are equipped with precapillary sphincters, and feed into the cortical capillaries and marrow sinusoids, the capacitance or reservoir vessels of the cortex and marrow. They regulate the volume of blood in bone.

Young Cortex: Medullary Supply. In the long bones of rats, rabbits and dogs (Brookes, 1986), as well as human fetal and adolescent material (Brookes, 1990), intra-arterial perfusion of barium sulphate shows that the larger branches of the nutrient artery lie generally in a subcortical location. Histologically they are "small arteries" with much elastic tissue in the tunica media. They give rise to arterioles whose finest branches, the metarterioles, terminate in the marrow sinusoids, or penetrate the bone endosteally to end into the cortical capillary reservoir (Brookes, 1971).
When periosteum and surrounding muscle are well perfused, a similar penetration of the external surface of the cortex by periosteal arterioles is not seen. A minor exception to this is that diaphyseal zygapophyses such as the linea aspera, gluteal tuberosity and soleal line, are pierced by a few periosteal arterioles. Hence, bone angiography suggests that the arterial blood input to the cortex of long bones, in youth, comes overwhelmingly from the marrow, a principle first emphasised by Brookes and Harrison (1957). It is interesting to note that shortly after Wilhelm von Roentgen's discovery of X-rays (1895), Soulié (1904) used bone angiography in dogs, and pointed out the absence of an anastomosis between medullary arteries in bone cortex and the arteries of the periosteum.

Centrifugal Blood Flow in Bone Cortex. The Haversian, cortical capillaries freely communicate with medullary sinusoids internally, and with periosteal capillaries externally. Cortical capillaries pass directly into the periosteal capillary network in regions lacking muscle attachment and are continuous with intramuscular venules, where muscles have a fleshy attachment. Hence, the anatomical evidence suggests that there is a centrifugal flow of blood from the endosteal to the periosteal surface of the cortex (Brookes, 1971), the blood current draining ultimately into periosteal and intramuscular veins. These provide a mechanism of venous escape from the cortex.
Centrifugal blood flow has been directly observed in living human femoral bone cortex after the injection of Evans Blue during amputation (Lamas, et al., 1946). Centrifugal uptake of Disulphine Blue has also been demonstrated in the cortex of rabbit long bones (Gunst, 1980). The use of ferritin has confirmed the presence of centrifugal flow in intact chick bone; and not only of blood, but also of extravascular bone water in the tibial cortex (Dillaman, 1984).

Old Cortex: Reduced Medullary, Increased Periosteal Supply. In the laboratory rat, cortical flow normally declines exponentially with age; so that in the aged rat, bone blood flow is reduced to a tenth of what it was in youth (Brookes, 1971).
Experimentally, nutrient (Brookes, 1960) or systemic (Marneffe, 1951) artery obstruction in rabbit long bones, causes osteoporosis and the development of a periosteal arterial supply to the now ischaemic bone.
According to Ramseier (1962), the arteries of human bone marrow are commonly subject to arteriosclerosis, increasing in severity with advancing years. He reported that grade for grade, arterial disease appears at least 10 years earlier in femoral marrow than in the arteries of the brain, heart, or striated muscle. Similarly, in human tibiae from limbs am-

putated for peripheral gangrene (Brookes, 1960b), the marrow was poorly vascularized and a periosteal blood supply to the cortex was prominent.

The combined results suggest that in human bone, ageing is accompanied by marrow ischaemia, and that an ever increasing proportion of the total blood supply to the cortex, comes from the periosteal vessels.

Blood Supply to Human Bone. Recently, the author has perfused intravascularly with barium sulphate suspension entire human lower limbs post mortem, and the femoral diaphyses of twenty-five subjects have been angiographed (Brookes, 1990). The age at death ranged from 21-91 years, and death was by violence or supradiaphragmatic disease. The results suggest that below the age of 30 years, the diaphyseal cortex is arterialized from the marrow alone (Figure 1). After this time, the arterial supply to the cortex is increasingly periosteal (Figure 2), and the vascularity of the mid shaft marrow diminishes. In advanced years (70 plus) the periosteal supply is dominant (Figure 3). In a male aged 56 years, the upper femur had been reamed and the stem of a hip prosthesis implanted using acrylic cement, 4 years before death. Angiography showed that the femoral cortex in life was wholly sustained by the periosteum (Figure 4).

In youthful bone, the flow of blood in the cortical capillaries is centrifugal, from the marrow to the periosteum. This might well apply to ageing bone also. Even with the marrow cavity packed with cement, and the cortex supplied wholly by periosteal arterioles, blood flow in the cortical capillaries was still centrifugal, to allow venous escape at the periosteal surface.

Biomechanics of the Osseous Circulation. The following haemodynamic factors influence blood flow in the cortex.

Driving pressure. Pressure within the marrow cavity is considerably higher (45 - 60 mmHg) than the 12-15 mmHg in extraosseous capillaries. The vascular driving pressure is therefore centrifugal across the cortex, from marrow to periosteum.

Pulse pressure. This is of the order of 8-10 mmHg in bone marrow, confined in an unyielding cortical container (Stolk, 1987). Thus the pulse pressure promotes centrifugal flow at each heart beat.

Muscle pump. Systemic venous valves are plentiful in the veins lying in intermuscular spaces. Muscle contraction empties the veins. The venous valves however, prevent retrograde flow. It follows that muscle activity pumps the blood from the bone towards the heart (Langer, 1876).

Impact force transmission. This has been demonstrated photographically during human locomotion as a wave of deformation passing up the limb (Light, 1987). In a simulated laboratory exercise utilising rabbits, impact forces have been demonstrated manometrically to pass along the bone marrow. These add to the pre-existing medullary pressures, and wring out the blood from the marrow as if it were a sponge.

Quantification of Bone Flow. In round figures, 20 % of the cardiac output is delivered to the skeleton in the resting adult human. The skeletal perfusion rate as an overall average is about 10ml/100g/min (Charkes, et al., 1980). In rats the corresponding values are 10 % of the cardiac output (Brookes, 1978) and 20ml/100g/min (Brookes, 1970).

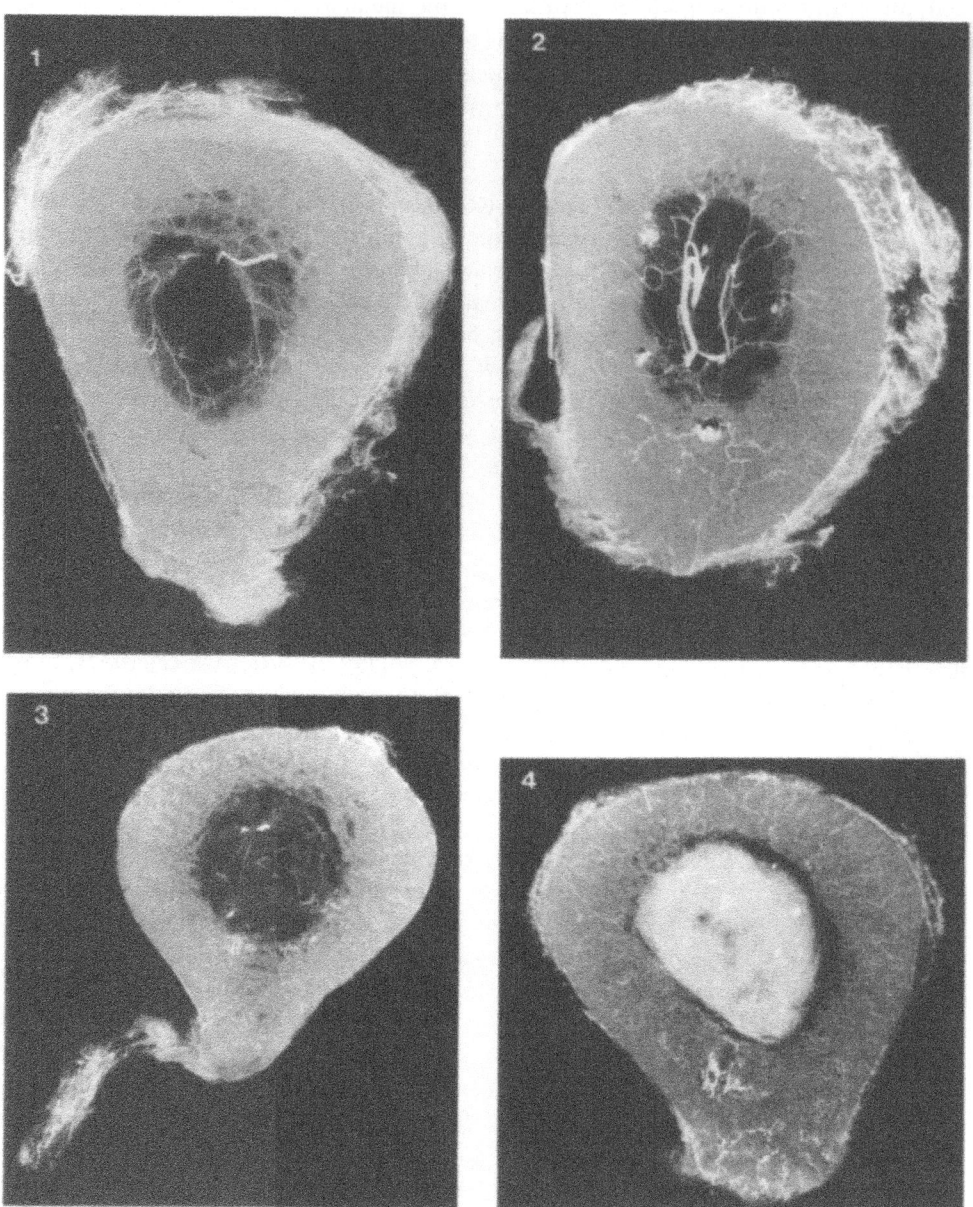

Figure 1. Angiograph of a 1 cm thick section of the mid-femoral diaphysis of a 21-year-old youth who died by violence. The limb perfused 3 days later shows the cortex supplied overwhelmingly from the marrow. The latter was itself necrotic and poorly perfused. Note periosteal supply to the linea aspera.

Figure 2. Angiograph of a 1 cm thick section of the mid-femoral diaphysis of a 41-year-old man. The marrow shows some peripheral ischaemia, but the cortex is still largely supplied by medullary arterioles. Note the nutrient artery in its canal, supplying related cortex.

Figure 3. Angiograph of a 72-year-old woman. The marrow is severely ischaemic, and gross endosteal osteoporosis is present. The cortex is supplied almost entirely by periosteal arterioles.

Figure 4. Angiograph from the mid-femoral diaphysis of a 56-year-old man. A hip prosthesis had been implanted 3 years previously. The section passes through the acrylic cement filling the marrow cavity below the stem of the prosthesis. The angiograph demonstrates that the cortex can be sustained entirely by periosteal arterioles.

Obstruction of the Periosteal Supply. The periosteum is a potential source of blood supply to the cortex, and always a route of venous escape for the cortical blood flow. It follows that obstruction of the surface with an orthopaedic plate, might disturb the underlying osseous circulation. Two nylon plates made of sterile Cerclene (How-medica) bands were applied to 14 rabbit femora. One surface of the plate was flat and

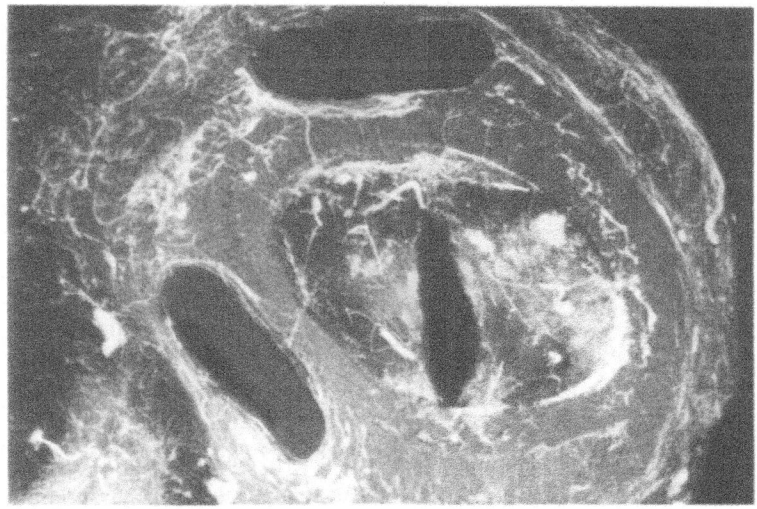

Figure 5. Microangiograph of mid-shaft femoral section, 12 weeks after the application of 2 nylon plates. Vessels have grown into the interface between the flat plates and the cortical surface. The cortex is severely hyperaemic. Periosteal and medullary supply to ischaemic bone.

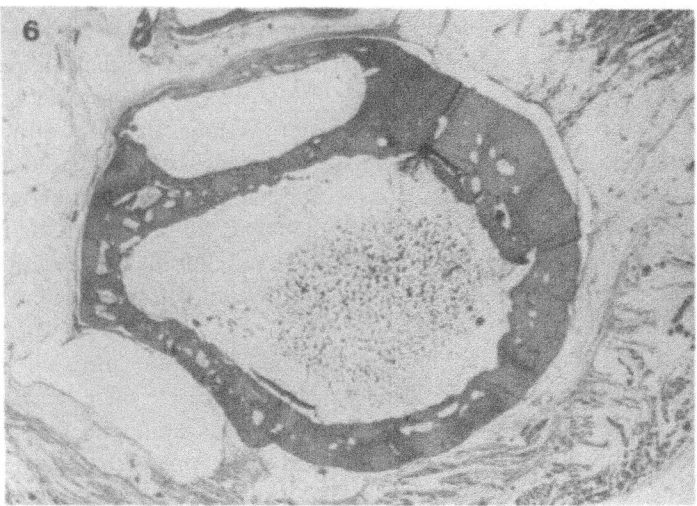

Figure 6. Photomicrograph of paraffin section taken from specimen shown in Figure 5. Severe osteoporosis; the cortex is much reduced in thickness and riddled with large spaces full of bone marrow (cortical medullization). Haematoxylin and eosin.

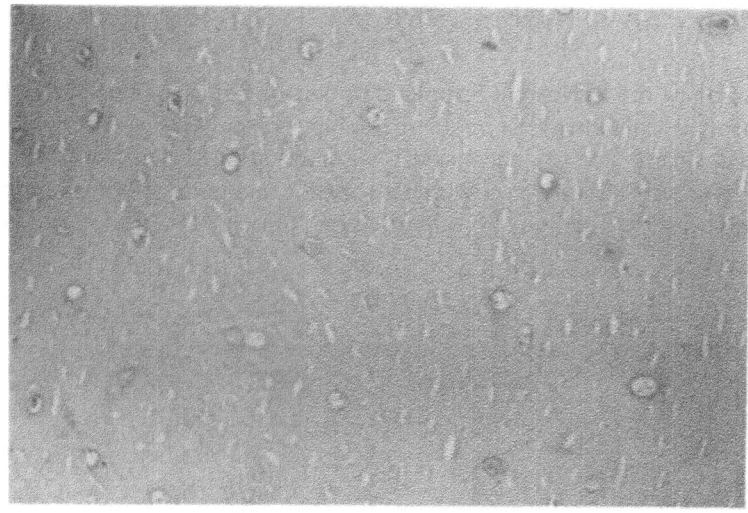

Figure 7. Photomicrograph of a transverse section through the femoral shaft showing the general appearance of dead osteocytes in the cortex, one week after plating. Haematoxylin and eosin.

made immediate contact with the femoral bone surface. The other side presented a trough lying between the edges of the plate, and made only linear contact with the bone surface. The plates were held in place by 4 nylon ties. In 12 rabbits the flat surface of the plates was applied to the femoral surface. The plates could not transmit mechanical force. The bones were not fractured. The rabbits were sacrificed serially up to 12 weeks postoperatively by intra-arterial barium sulphate perfusion. Angiographic and histological examination followed. In 2 rabbits the other, excavated, side of the plates was applied. In these mock experiments, the plates were elevated above and did not obstruct, the bone surface. They were still held in place by 4 nylon ties. The bones were examined 6 weeks postoperatively.

At 4 weeks post-operatively, a marked hypervascularity affected the inner two thirds of cortex beneath the plates. Subsequently, erosion cavities were much in evidence. Severe bone removal from the periosteal aspect was also encountered, resulting in thinning of the plated cortex, sometimes in association with endosteal buttressing deep to the plate. Florid angiographic hyperaemia was present in the osteoporotic cortex, which showed no sign of abatement 12 weeks later (Figures 5, 6). In the rabbits examined at one week post-operatively, bone loss was not in evidence, but histology of the cortex showed severe osteocyte pyknosis and absence of cell nuclei in many bone lacunae. Obstructing the periosteal surface had killed the osteocytes (Figure 7). In the mock operated rabbits where the plates had been inverted and rendered non-obstructive, the cortex below the plates 6 weeks postoperatively was not hyperaemic nor reduced in thickness. Osteoporosis was minimal.

Increased bone fragility following internal fixation of fractures utilising bone plates, has often been attributed to the fracture itself causing increased bone remodelling (Schenk and Willeneger, 1964); or to the rigidity of the plates (Uhtoff and Dubuc, 1971); or to stress protection (Tonino et al., 1976) whereby bridging apparatus diverts applied mechanical forces away from the reparative site. The results here using flexible nylon plates on unfractured bone, suggest that these factors do not have the significance attributed to them. On the contrary, the consequence of preventing the normal centrifugal blood flow by the application of an obstructive plate is to cause a profound ischaemia in the related cortex and much cell death, followed by a sustained vascular congestion with removal of dead bone at the periosteal surface, as well as in the cortex generally, accounting for the observed severe osteoporosis.

Obstruction of the Medullary Supply. The response of the unfractured tibia to reaming alone has also been investigated, and compared with the results of reaming followed by simulated nailing. The medullary cavity of a tibia was reamed in 14 rabbits, from a proximal point on the tibial plateau behind the ligamentum patellae. In another group of 14, reaming was followed by simulated nailing of the marrow cavity. A flexible polyethylene tube was inserted in the reamed medulla to represent a steel intramedullary nail. The bone was not fractured. The investigation was followed at serial intervals up to 12 weeks post-operatively. The animals were sacrificed by intra-arterial perfusion of barium sulphate. Segments of the tibiae were then studied angiographically and histologically.

Reaming alone destroyed medullary structure completely. The medullary supply to the cortex was suspended for a week. At this time vascular regeneration was present in the marrow cavity. A pronounced deposit of external callus was observed up to 8 weeks post-operatively. The callus was supplied by periosteal blood vessels. Histologically, hard callus was still detectable at 12 weeks on the surface of the unbroken bone (Figure 8). Reaming and simulated nailing of rabbit tibiae caused the inner cortex to be ischaemic up to 2 weeks post-operatively. The indwelling tube was tightly packed into the marrow cavity. Blood vessels proliferated in the confined space between the tube and endocortex after 2 weeks. Again, there was a marked periosteal deposit of living new bone on the external surface of the bone (Figure 9).

Reaming, or reaming and simulated nailing, bring about marked periosteal new bone formation (external callus). This follows the destruction of the medullary circulation in both instances. The periosteal circulation however is spared, permitting external callus to form and maintaining the cortex alive by its auxiliary supply function. In the

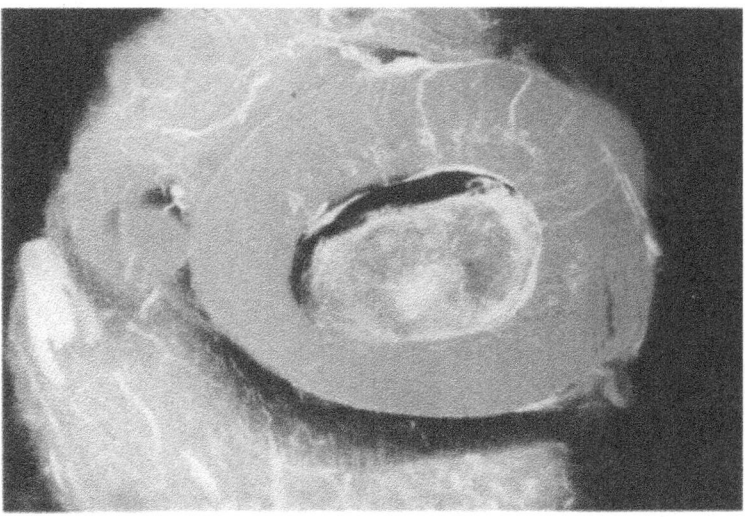

Figure 8. Microangiograph of a rabbit tibial section taken 12 weeks after intramedullary reaming. The marrow is fully restored and the cortex is slightly hyperaemic. Quasi normal appearances.

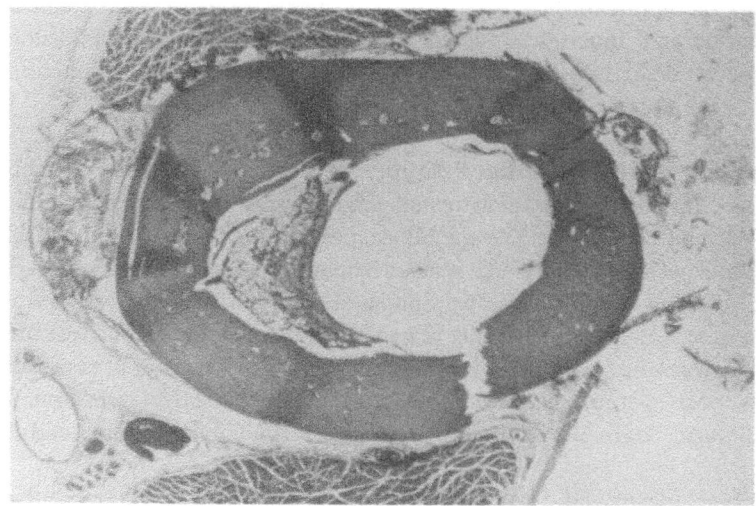

Figure 9. Photomicrograph of paraffin section taken from mid shaft tibia, 12 weeks after combined reaming and nailing. The nylon tube occupied the central marrow space. The thick cortex shows some Haversian enlargement; osteoporosis is minimal. Haematoxylin and eosin.

treatment of difficult fractures by intramedullary nailing, the promotion of external callus may be important in achieving firm bony union and avoiding nonunions.

Frank osteoporosis, reduction in cortical thickness, and medullization of the cortex were seen in ischaemic bone deep to obstructive plates. These features were absent in the reamed, or reamed and nailed specimens. The present results suggest therefore that the damaging effect of reaming on the cortical circulation is neither severe nor sustained. It is noted that resistance to both angulatory and rotational deformation is made possible by a locked and slotted intramedullary nail without recourse to a tightly fitting implant. Finally, two tibiae were reamed and perfused and examined the next day. It has

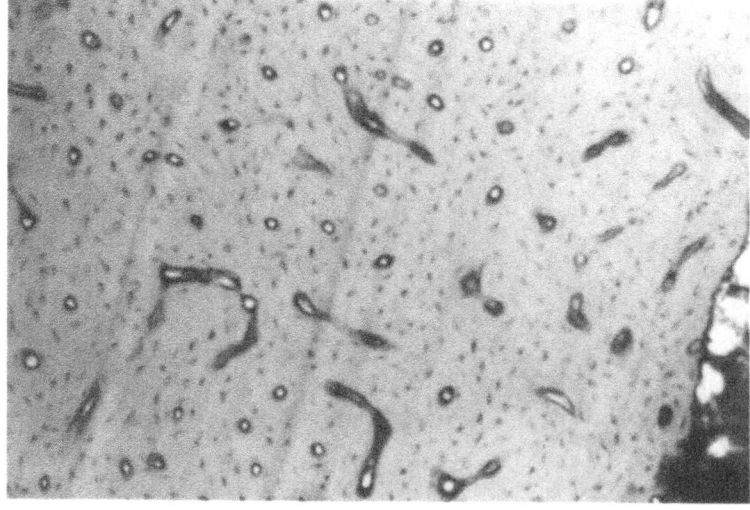

Figure 10. Photomicrograph of a transverse section of a rabbit tibial shaft, following reaming of the medullary cavity; showing the normal cortical appearances. Haematoxylin and eosin.

been suggested that the external callus observed after intramedullary nailing, is the inductive consequence of marrow debris being pushed into the periosteum by medullary reaming. The hypothesis has not been sustained here. Histologically, the cortex appeared to be alive and entirely normal. Of course, in fractured bone, medullary reaming/nailing may fill the fracture gap with marrow and thereby promote union.

SUMMARY

Normal Flow. In youth, medullary arterioles supply the diaphyseal cortex of long bones. As age advances, the marrow becomes less and less vascular; the cortical vascular capacitance decreases and an auxiliary periosteal arterial supply to the cortex develops.
Direction of flow in the cortical capillaries is normally from the marrow to the periosteum. Centrifugal flow is promoted by a variety of biomechanical forces.

Periosteal Obstruction. Plating the intact rabbit bone with flexible nylon strips obstructs the normal venous escape of blood from the cortex. This results in massive osteocyte death and a sustained osteoporosis. Non-obstructive plating however, does not cause bone loss.

Medullary Destruction. Intramedullary reaming and nailing do not cause osteoporosis. The periosteal vessels are spared. The marrow vessels regenerate in less than 2 weeks. External callus deposits are extensive in both reamed and nailed, intact bone.

REFERENCES

Brookes, M.: Sequelae of experimental partial ischaemia in long bones of the rabbit. J. Anat. (London). 94: 552-561 (1960a).

Brookes, M. : The vascular reaction of tubular bone to ischaemia in peripheral occlusive vascular disease. J. Bone Joint Surg. 42B: 110-125 (1960b).

Brookes, M.: Arteriolar blockade: a method of measuring blood flow rates in the skeleton. J. Anat. (London) 106: 557-563 (1970).

Brookes, M.: The Blood Supply of Bone. London: Butterworth (1971).

Brookes, M.: Blood flow measurements by fractional distribution compared with ^{18}F uptake in the skeleton. In: J. Arlet and P. Ficat (eds): Circulation Osseuse (2nd SIC0). pp. 141-147. Université Paul Sabatier Publ., Toulouse (1978).

Brookes, M.: An anatomy of the osseous circulation. Bone. 3: 32-35 (1986).

Brookes, M.: Bone blood flow measurement, Part 2. Bone. 4: 33-3 (1987).

Brookes, M.: Blood Flow in the Diaphysis of Long Bones. ARC0 Newsletter. (Toulouse). 2: 75-85 (1990).

Brookes, M. and R.G. Harrison: The vascularization of the rabbit femur and tibiofibula. J. Anat. (London). 91: 61-72 (1957).

Charkes, N.D., M. Brookes and P.T. Jr. Makler: Studies of skeletal tracer kinetics: II. Evaluation of a five compartment model of [18F] fluoride kinetics in rats. J. Nucl. Med. 20: 1150 - 1157 (1979).

Charkes, N.D., M. Brookes and P.T. Jr. Makler: Radiofluoride kinetics. In: L.G. Colombetti (eds): Principles of Radiopharmacology. pp. 225-242. CRC Press. Boca Ratton, 1980.

Dillaman, R.M.: Movement of ferritin in the 2-day-old chick femur. Anat. Rec. 209: 445-543 (1984).

Gunst, M.A.: Interference with bone blood supply through plating of intact bone. In: H.K. Uhthoff (eds): Current Concepts of Internal Fixation of Fractures. pp. 268-276, Springer Verlag , Berlin 1980.

Lamas, A., D. Amado and J.C. da Costa: La circulation du sang dans l'os. Presse med. 54: 862-863 (1946).

Langer, K.: Uber das Gefässsystem der Röhrenknochen, mit Beiträgen zur Kenntnis des Baues und der Entwicklung des Knochengewebes. Denkschr. K.K. Akad. Wiss. Wien. 36: 1-40 (1876).

Light, H.: Personal communication. MRC Research Unit, Stanmore, Middlesex, London (1987).

Marneffe, R. de: Recherches morphologiques et expérimentales sur la vascularisation osseuse. Acta Med. Belge (1951).

Ramseier, E.: Untersuchungen über arteriosklerotische Veränderungen der Knochenarterien. Virchows Arch. Path. Anat. 336: 77-86 (1962).

Revell, W.J. and F.W. Heatley. Long term sequential measurements of bone blood flow within a single animal by the hydrogen washout method. In: J. Arlet, B. Mazières and D. Hungerford (eds): La Circulation Osseuse (4th SICO, Toulouse). Springer Verlag, Berlin 1987.

Soulié, A.: Sur les applications de la radiographie stéréoscopique à l'étude des artères des os. (Note technique) C.R. Ass. Anat. 6: 172-174 (1904).

Schenk, R. and H. Willenegger: Zur Histologie der primären Knochenheilung. Langenbecks Arch. klin. Chir. 308: 440-452 (1964).

Stolk, P.W.T.: Personal communication. Veterinary Institute, University of Utrecht, Netherlands (1987).

Tonino, A.J, C.L. Davidson, P.J. Klopper and L.A. Linclau: Protection from stress in bone and its effects. J. Bone Joint Surg. 58B: 107-113 (1976).

Uhthoff, H.K. and F.L. Dubuc: Bone structure in the dog under rigid internal fixation. Clin. Orthop. 81: 165-170 (1971).

MICROVASCULARIZATION, OSTEOGENESIS, AND MYELOPOIESIS IN NORMAL AND PATHOLOGICAL CONDITIONS

Ebbe Stender HANSEN

Department of Orthopedics
Institute of Experimental Clinical Research
University of Aarhus
Aarhus N, Denmark

INTRODUCTION

The microvasculature of bone plays a crucial role in skeletal ontogeny, remodelling, and repair, and is involved in most pathological states in bone and bone marrow. The review suggested here summarizes recent reports on interactions between the intraosseous vascular system and the processes of osteogenesis and myelopoiesis. Emphasis will be placed on the active biological role of vascular endothelium and on a growing appreciation in the literature that autonomic and sensory nerves play an integrating role in both vascular and cellular functions in bone.

MICROVASCULARIZATION AND OSTEOGENESIS

Modes of Osteogenesis

The most common mode of osteogenesis is endochondral bone formation, where ossification takes place in cartilaginous tissue. It occurs in diaphyses, metaphyses, and epiphyses of tubular bones during embryogenesis, in fracture healing, and during heterotopic bone formation induced by demineralized bone matrix (DBM) (Carrington and Reddi, 1991). Endochondral bone formation represents a highly ordered sequence of events that begins with proliferation of pluripotent mesenchymal cells, their differentiation into chondrocytes, and secretion of cartilage specific extracellular matrix, which calcifies by a vitamin D_3 dependent mechanism. Chondrocytes pass through an apparently predetermined set of discrete developmental stages from proliferation through maturity to the hypertrophic stage before degeneration (Caplan, 1991). These developmental stages are represented morphologically in the growth plate strata, but occur in less orderly arrange-

Bone Circulation and Vascularization in Normal and Pathological Conditions
Edited by A. Schoutens *et al.*, Plenum Press, New York, 1993

29

ment during fracture healing and development of DBM-induced ossicles as well. Ossification commences with vascular invasion of the calcified cartilage. Differentiation and proliferation of osteoblast precursors is restricted to the immediate vicinity of the invading vessel sprout (Shapiro et al., 1988). During osteoid secretion osteoblasts become arranged as a monolayer of highly polar secretory cells with base along the invading vessel endothelium and secretory face toward the osteoid front, where a bone strut forms. The osteoblasts gradually mature and become embedded in the secreted bone matrix as osteocytes, while a new layer of osteoid is being secreted by new osteoblasts adjacent to the vessel. This spatial sandwich-like arrangement of a monolayer of polar secretory osteoblasts with a vascular endothelial lining on one side and an osteoid seam on the other is highly characteristic and occurs wherever bone formation takes place. The final step in osteogenesis is establishment of a red marrow by invasion of the marrow space between struts with myelopoietic cells. This also occurs in DBM-induced ossicles (Carrington and Reddi, 1991).

Intramembranous bone formation occurs during embryonal development of flat bones of the skull. The process begins with avascular mesenchymal condensation. Differentiation of the central cells into first preosteoblasts and then into secretory osteoblasts coincides with vascular invasion of the bone anlage, indicating a close interplay between differentiation of osteoblasts and angiogenesis (Thompson et al., 1989).

Bone remodelling takes place throughout life by coupled osteoclastic resorption and bone formation in the so-called basic multicellular unit (BMU) containing osteoblasts, osteoclasts, blood vessels, and perivascular stromal tissue. The cortical BMU is an osteon, which is a Haversian vessel loop capped with osteoclasts tunnelling through old cortical bone (the cutting cone) flanked with secretory osteoblasts elaborating concentric shells of new bone matrix (the closing cone). Analogous processses take place in the trabecular bone BMU, where blood vessels also can be observed in close contact with bone (Draenert and Draenert, 1980). Dilated venous vessels with high linear velocity of blood in direct apposition to bone have been linked to the process of bone resorption during remodelling (McClugage and McCuskey, 1973).

Vascular Invasion Precedes Osteogenesis

By vital microscopy of bone ingrowth into an implanted bone chamber it has recently been directly observed that angiogenesis precedes osteogenesis (Winet et al., 1990). The classic experiment demonstrating a pivotal role of vascular invasion for endochondral osteogenesis was partial ablation of the growth plate blood supply on either the metaphyseal or the epiphyseal aspects of the growth cartilage in rabbits. Performed on the epiphyseal side this led to focal osteonecrosis with complete disruption of the growth plate structure, indicating that nutrition of proliferating chondrocytes in the growth plate cartilage depends on the epiphyseal blood supply. However, ablation of vessels on the metaphyseal aspect of the growth plate removed the possibility for vascular invasion of mineralized matrix and resulted in deficient osteogenesis with increased width of the hypertrophic chondrocyte stratum, resembling the histologic appearance in rickets (Trueta, 1963). A recent investigation on vascular morphology in experimental rickets in rats (Hunter et al., 1991) casts further light on this process. In vitamin D_3 depletion mature arterioles, venules, and capillaries with full basal membrane and pericyte lining formed a vascular mesh without any definite orientation and with no apparent growth front throughout the metaphysis including the region immediately adjacent to hypertrophic chondrocytes in the growth cartilage. Vitamin D_3 repletion immediately led to a burst of vascular sprouting and invasion of cartilage by immature vessels consisting exclusively of attenuated endothelial cells with no pericyte lining or basal membrane. Invasion had particular predilection for regions with hypertrophic chondrocytes rather than for cartilage matrix. Within 4 days, the normal vascular morphology of growth plates had been reestablished completely. The notion that the metaphyseal vasculature is involved in rickets is not new. The interesting point here, however, is that vitamin D_3 depletion did not distort or disrupt the metaphyseal circulation. Rather, metaphyseal vessels matured and established a fully developed closed nutritional type of circulation, resulting in deficient osteogenesis. Endothelial invasivity and, consequently, osteogenesis were immediately and completely restored with vitamin D_3 repletion. The precise mode of action of vitamin D_3 in this process has not been elucidated, but it seems clear from this experiment that vascular invasivity is critical to osteogenesis.

Initiation of Angiogenesis

In the vessel, angiogenesis begins with local degradation of the basement membrane of the parent venule, followed by migration of endothelial cells and cell division in a zone a few cells behind the leading sprout tip. The invading vascular bud consists of a monolayer of attenuated squamous fenestrated endothelium and lacks both the surrounding layer of pericytes and the basal membrane. Continuity of the endothelium during growth is maintained by formation of tight junctions between daughter cells even before they separate (Hunter and Arsenault, 1990). Circulation is established after formation of a lumen and fusion of adjacent sprouts, which thus form a labyrinth of interconnected sinusoidal complexes with bulbous extensions invading chondrocyte cavities in the zone of ossification (Stanka et al., 1991).

Initiation of angiogenesis and vascular invasion rests on an intricate interplay between messenger molecules. Cartilage is kept avascular by elaboration of inhibitors of angiogenesis and vascular invasion by chondrocytes. One of these recently identified factors has been shown to inhibit endothelial cell proliferation and migration and to neutralize a collagenase produced and used by endothelial cells during tissue invasion (Moses et al., 1990). Chondrocytes also produce basic fibroblast growth factor (bFGF), which stimulates both proliferation and migration of endothelial cells (Klagsbrun and Smith, 1980). Furthermore, a family of heparin binding growth factors have been identified, including acidic fibroblast growth factor (aFGF) and endothelial cell growth factor (ECGF), which all exhibit mitogenicity and migration stimulating effects on endothelial cells (Folkman, 1985), also in bone (Streeten and Brandi, 1990). The signal for vascular invasion into growth plate cartilage appears to be a shift in the balance between angiogenesis inhibitors and stimulators in favor of the latter, maybe initiated by release of large amounts of chondrocyte bFGF in the zones of cartilage hypertrophy and degeneration or by the event of cartilage matrix calcification, which might alter the properties of the cartilage angiogenesis inhibitors. Protease inhibitor activity is maximal during mesenchymal cell proliferation and chondrogenesis and decreases coincident with an increase in protease activity during cartilage calcification (Reddi and Kuettner, 1981). Recently, plasminogen activator (PA), which converts plasminogen to the potent peptidase plasmin, has been shown to peak at the time of vascular invasion, suggesting a role in the degradation of calcified cartilage matrix during vascular invasion (Desimone and Reddi, 1992). The cellular source of this burst of PA activity has not been established, but may be either hypertrophic chondrocytes or the vascular endothelium.

Vascularization of Secondary Ossification Centers

Vascularization of secondary centers of ossification differ among species. In smaller animals such as the mouse, the distal femoral epiphyseal cartilage is completely avascular until the fifth postnatal day, when hypertrophic chondrocytes appear centrally; from that day progressive invasion of vascular buds occurs from the periphery, whereas the first sign of ossification can be detected histologically on the eighth day (Floyd et al., 1987). Thus, the sequence of events leading to ossification of the chondroepiphysis is dependent upon the appearance of hypertrophic chondrocytes, which definitely precedes both matrix calcification and peripheral vascular invasion in the murine distal femur.

In larger mammals, including rabbits, pigs, and humans, the chondroepiphyses are extensively vascularized by cartilage canals containing blood vessels and mesenchymal tissue even before chondrocyte hypertrophy begins to take place. These canals presumably serve a nutritive purpose when the epiphyseal cartilage volume is too large for nutrition to take place by diffusion alone. Vessels of the cartilage canals are not transformed to invasive endothelium, however, without the appearance of hypertrophic chondrocytes (Ganey et al., 1992). The mechanisms by which vessels of the cartilage canals are kept unreactive or how they have become accepted in the otherwise avascular cartilage environment are not completely understood. Chondrocyte hypertrophy appears first in the region most densely equipped with cartilage canals, followed by transition of the vascular endothelium, vascular invasion, and ossification.

BONE ENDOTHELIUM

Bone Endothelial Cell Cultures

More precise exploration of biological signals activating bone vascular endothelium has been attempted in an endothelial cell culture grown from the bovine sternum (Streeten and Brandi, 1990). In this system, a pronounced mitogenic response (^{3}H-thymidine incorporation) was found when cultures were stimulated with ECGF-alpha. Other substances with mitogenic effects were ascorbic acid, bFGF, platelet-derived growth factor (PDGF), insulin, insulin-like growth factors 1 and 2, and progesterone. Inhibition of ^{3}H-thymidine incorporation was observed with transforming growth factor beta and high doses of heparin. No response was found when cultures were exposed to other circulating or local hormones or growth factors with known effects on bone metabolism, including calcitonin, PTH, vitamin D_3, estradiol, testosterone, epidermal growth factor, dexamethasone, thrombin, or transferrin (Streeten and Brandi, 1990).

Genetic Relation between Bone Endothelium and Osteoblasts

Elucidation of the genealogy of bone cells has been a classic problem in bone biology. Current opinion holds that two distinct cell lines supply osteoblast and osteoclast precursers, osteoblasts being derived from stromal cells, and osteoclasts from the monocyte-macrophage cell line. The obligatory intimate relationship between bone vessels and osteoblasts has raised the theory that bone endothelial cells or pericytes might be osteoblast precursors (Trueta, 1963). At the ultrastructural level similarities and transitional forms between preosteoblasts and sinusoid endothelial cells have been observed (Burkhardt et al., 1984). Furthermore, bone endothelial cells in culture react with increased cyclic adenosine mono-phosphate (cAMP) production when stimulated with PTH, just as osteoblasts do. This is a unique feature of bone endothelial cells, which has not been observed in other endothelial cell systems, and suggests a genetic relationship between endothelial cells and osteoblasts (Streeten and Brandi, 1990). Recent observations of ultrastructural transitions in the very early marrow callus after closed rib fracture in the rabbit has provided new evidence on this point (Brighton and Hunt, 1991): The marrow callus 12 hours after fracture comprised three distinct zones. The fracture was sealed with a fibrin clot. Adjacent to the clot there was a narrow zone of high cellular density, and between this and normal marrow a wider zone of low cellular density. Within the region of high cellular density there was disruption of normal marrow architecture, disappearance of blood vessels, extravasated red blood cells, degenerating hematopoietic cells, and numerous polymorphic mesenchymal cells. The neighbouring region of low cellular density had altered marrow architecture, but blood vessels were preserved, and endothelial cells were seen to undergo transformation, enlargement, and extravasation. The first signs of bone formation were observed at 24 hours in the region of low cellular density adjacent to blood vessels. A consistent finding was close proximity between transformed endothelial cells, polymorphic mesenchymal cells, osteoblasts, and immature osseous foci. A probable interpretation of these findings is that blood vessels disappeared in the region of high cellular density because their endothelial cells were transformed into polymorphic mesenchymal cells and osteoblast precursors, and that endothelial cell transformation in the region of low cellular density represented both mesenchymal and osteoblastic differentiation and vessel sprouting in the process of osteogenesis during formation of a marrow callus. Further support for this view has been provided by the recent demonstration of basement membrane components (laminin and heparin sulphate) within cartilaginous tissue undergoing differentiation 3 days after fracture (Hulth et al., 1990).

Evidence for an Endothelial Control of Bone Cells

The close spatial relationship between cells of the microcirculation and bone forming cells suggests that endothelial cells or pericytes may exert direct influence over the activity of bone forming cells. Indeed, endothelial cells and pericytes have been shown to synthesize a soluble substance with mitogenic effects on bone cells (Guenther et al., 1986; Jones et al., 1992), which indicates that vessels in addition to providing osteoblast precursors may act to increase the number of these cells at an osteogenic site. In addition, ECGF, which is the most potent mitogen for endothelial cells, exerts a mitogenic effect on osteoblasts as well. The effect is enhanced by the presence of heparin, and heparin containing mast cells accumulate in fracture callus. ECGF is present in large amounts in bone matrix and is released by skeletal injury or during osteoclastic resorption, suggesting that this growth factor might play a central role in the synchroneous vascular and osteoblastic response in bone repair and remodelling (Canalis et al., 1987).
Osteoclast activity in bone is regulated by its main circulating inhibitor calcitonin in association with locally produced modulators. Nitric oxide (NO) is a strong vasodilating substance produced by vascular endothelium upon cholinergic stimulation and is involved in both neurotransmission and regulation of blood flow, also in bone (Brinker et al., 1990; Davis and Wood, 1992; Roorda et al., 1992). Interestingly, NO has recently been shown to be a potent inhibitor of osteoclast function (MacIntyre et al., 1991). The abundance of NO producing endothelial cells in bone marrow and their proximity to osteoclasts suggests that marrow endothelial cells may take part in the regulation of osteoclast activity in bone.

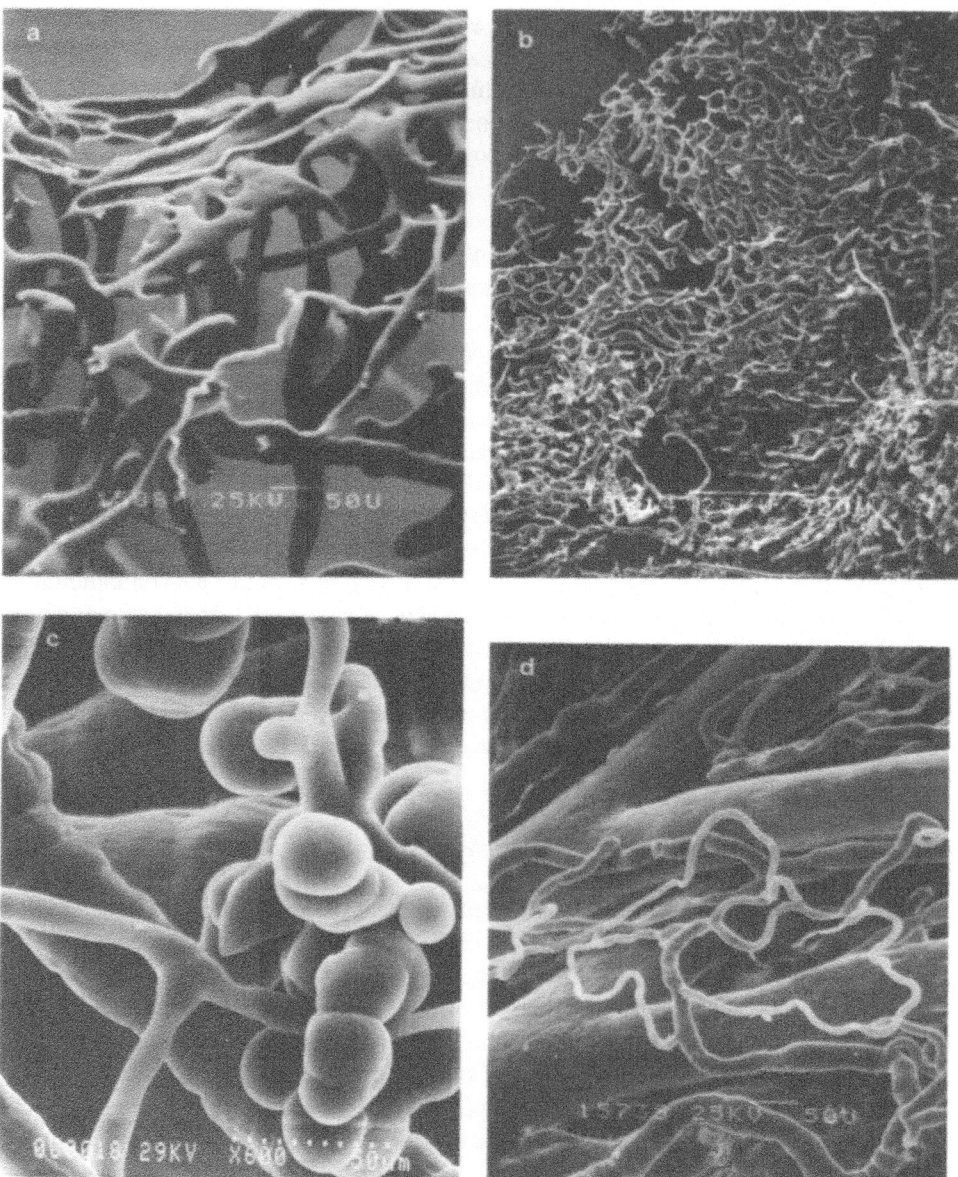

Figure 1. Microvascular morphology of bone by scanning electron microscopy of polymethylmetha-crylate corrosion casts of the rabbit tibia. A) Bone marrow with endosteal venous sinusoidal plexus drai-ning centraipetally (down). Note presence of sinusoids, capillaries, and arterioles with circular constriction rings from the action of vascular smooth muscle cells. B) Extensive widely interconnecting sinusoid net-work typical of hemopoietic tissue. From metaphyseal cancellous bone of proximal tibia. C) In the fore-ground a branching medullary capillary and several bulbous venous sinusoids. The background consists of casts from larger resistance arteries (diameter greater than 100 μm) and one obliquely crossing arteriolar structure (25-30 μm) with circular constriction rings. D) Branching intraosseous artery (top) and vein (bot-tom) among anastomosing capillaries of the proximal tibial epiphysis, where the sinusoid vascular structure is not as prominent as in the metaphysis. Note the shape of the impressions from endothelial cells, which may be used for distinction between arteries (oblong impressions) and veins (rounded impressions).

MARROW MICROVASCULARIZATION

Sinusoids

Structurally the red marrow microcirculation is arranged in a closed and widely anastomosing three dimensional hexagonal lattice of thin walled sinusoids. The sinusoid walls consist of a single continuous layer of endothelial cells interconnected by tight junctions, with no complete basal membrane and only occasionally a complete adventitial layer of perivascular cells (Wickramasinghe, 1991). Functional temporary slits or pores in the endothelial lining allow newly formed blood cells to enter the circulation by diapedesis. Bone marrow also contains long and slender conventional capillaries, which prevail in yellow marrow. Blood flow of bone marrow decreases during conversion from red hematopoietic marrow to yellow fatty marrow. This shift, which is an early event in osteonecrosis of the hip, occurs gradually with age and is associated with decreasing blood flow, decreased density of marrow vessels (Burkhardt et al., 1987), and a gradual shift in vascular structure from the sinusoidal arrangement to a pure capillary circuit. Thus, the sinusoidal vascular arrangement is specific for myelopoiesis.

Marrow Vascularity and Myelopoiesis

Major drives for hematopoiesis are circulating erythropoietin from the renal parenchyme and the arterial oxygen tension. Furthermore, its regulation rests on complex interactions among cell-derived bioactive molecules and their target cells, i.e. hematopoietic stem and progenitor cells and accessory cells. Both stimulating factors (colony stimulating factors, (CSFs)) and suppressors (prostaglandins of the E series (PGE), ferritins, transferrin, the interferons, and tumor necrosis factor) have been characterized (Broxmeier and Williams, 1988), and their delicate interaction in normal and pathological hematopoiesis is a current subject of considerable interest. Vascular endothelial cells undoubtedly play a central role in the initiation and regulation of myelopoiesis: The monocyte-macrophage appears to be centrally placed in myeloid regulation through its production of interleukin-1 (IL-1), which works through induction of endothelial cells and stromal fibroblasts to produce the myeloid growth factors, i.e. the Colony Stimulating Factor (CSFs). However, the same CSFs and IL-1 also stimulate monocytes-macrophages, fibroblasts, and endothelial cells to synthesize prostaglandin E (PGE), which is an important downregulator of myelopoiesis in vivo (Pelus, 1990). Furthermore, circulating myeloid stem cells have been shown to require interaction with vascular endothelial cells to proliferate and form colonies with CSF stimulation in vitro (Monroy et al., 1992). At the ultrastructural level direct cell to cell contact exists between vascular endothelial cells and myeloid cells, but this contact is lost in early fracture healing (Brighton and Hunt, 1991), evidencing myelopoietic downregulation during osteogenesis. Such an effect might be mediated by the generation of PGE_2 at the fracture site. Bone marrow hypervascularity with increased blood flow, increased number of sinusoids, and endothelial cell hyperplasia has been implicated in polycythemia vera and myelofibrosis, maybe as a primary step in the pathogenesis of these disorders (Baglin et al., 1991; Thiele et al., 1992; Martiat et al., 1987). Conversely, decreased density of marrow vessels has been observed in bone biopsies from patients with aplastic anemia (Burkhardt et al., 1987). By morphometry of bone marrow in granulocytic leukemia and myelofibrosis it has been suggested that marrow blood flow in these states depends on marrow cellularity (Lahtinen et al., 1982), although a claim has been made to the contrary (Martiat et al., 1987).

35

Table 1. Considerable heterogeneity exists in intraosseous vascularity, hemodynamics, and bone metabolism. Regional blood flow (RBF), vascular volume (VV), mean blood transit time (TT), tissue hematocrit (tHct), and uptake of bone seeking diphosphonate (^{99m}Tc-DPD) in selected functionally and anatomically different regions of the femur in growing dogs (mean, SEM, n=8)

	RBF[a]	VV[b]	TT[c]	tHct[d]	[tHct-I][e]	^{99m}Tc-DPD[f]
Midshaft diaphysis, cortex	4.4 *0.5*	1.2 *0.1*	18 *3*	0.25 *0.02*	[0.73 *0.03*]	17 *3*
Midshaft diaphysis, marrow	14.9 *1.9*	11.5 *0.9*	54 *7*	0.19 *0.02*	[0.55 *0.03*]	-
Distal metaphysis, cortex,	7.9 *1.2*	3.1 *0.2*	29 *5*	0.22 *0.02*	[0.64 *0.03*]	33 *7*
Distal metaphysis spongiosa, average	9.8 *1.4*	10.8 *0.5*	77 *9*	0.18 *0.02*	[0.53 *0.02*]	30 *7*
adjacent to marrow	19.3 *1.9*	11.8 *0.6*	41 *5*	0.19 *0.02*	[0.55 *0.03*]	32 *7*
intermediate	9.4 *2.1*	12.6 *0.9*	101 *17*	0.18 *0.02*	[0.52 *0.02*]	31 *6*
adjacent to region of growth	2.7 *0.6*	9.2 *0.7*	301 *64*	0.17 *0.01*	[0.48 *0.03*]	28 *7*
region of growth	31.1 *7.2*	5.4 *0.4*	13 *2*	0.15 *0.01*	[0.44 *0.03*]	119 *23*
Distal epiphysis, central spongiosa	5.4 *0.6*	5.0 *0.3*	66 *12*	0.21 *0.02*	[0.61 *0.03*]	23 *5*
Distal epiphysis, subchondral bone	10.0 *1.8*	3.2 *0.2*	27 *6*	0.22 *0.02*	[0.63 *0.03*]	30 *6*

a   RBF, regional blood flow (ml/min/100g tissue), measured with 15μm radioactive microspheres

b   VV, vascular volume (ml/100 g tissue) determined as the sum of distribution spaces of ^{125}I-fibrinogen (plasma space) and ^{51}Cr-erythrocytes (red cell space).

c   TT, mean transit time of blood (seconds) determined by VV divided by RBF.

d   tHct, tissue hematocrit, determined by the distribution space of ^{51}Cr-erythrocytes (red cell space) divided by the vascular volume.

e   tHct-I, tissue Hct index is tHct expressed as a fraction of central arterial Hct (average 35 3 in this experiment).

f   99mTc-DPD, uptake of ^{99m}Technetium-diphosphonate, tissue-to-blood ratio 2 hours after i.v. injection.

Marrow blood flow is increased in hypovolemic as well as hemolytic anemia (Chen et al., 1986; Iversen et al., 1992; Schoutens et al., 1990), but uncertainty prevails about the cause. One study found that erythropoietin stimulation caused increased blood flow and oxygen consumption of bone marrow, suggesting a link between stimulated erythropoiesis and marrow blood flow (Iversen et al., 1992). Another study found normalization of marrow flow if chronic anemia was acutely compensated, indicating that increased marrow flow in anemia is due to altered blood rheology rather than to stimulated erythropoiesis, and could not detect any influence of erythropoietin supplementation or systemic hypoxia on marrow blood flow rates (Schoutens et al., 1990).

NEURAL CONTROL OF BONE PHYSIOLOGY

Nervous Control of Bone Blood Flow

It has long been known that autonomic nerves are present in bone tissue with effect on blood flow regulation. Both alpha 1 and 2 adrenergic receptors are involved, while beta receptors only appear to play a minor role (Dean et al., 1990). Sympathectomy has been shown to increase resting bone blood flow, but normalization occurs after 5-6 weeks (Davis et al., 1987), evidencing other adaptive phenomena active in bone blood flow regulation. Recently, screening of bone tissue for the presence of immunoreactivity to a large number of potential neurotransmitters demonstrated the presence of nerve terminals containing tyrosine hydroxylase (TH), the enzyme responsible for conversion of tyrosine to noradrenaline, neuropeptide Y (NPY), a widespread potent vasoconstrictor, calcitonin gene-related peptide (CGRP), a potent vasodilator produced by the calcitonin gene along with calcitonine), vasoactive intestinal peptide (VIP), a vasodilating peptide, and substance P (SP), in normal rat bone (Bjurholm, 1989; Hill and Elde, 1991). All these neuropeptides also occur in developing DBM induced heterotopic bone with the first fibres appearing at the chondroblastic stage, i.e. before ossification begins (Bjurholm, 1989). Both noradrenaline and NPY exert profound vasoconstriction in bone in an in situ and in vivo perfused porcine tibia (Lindblad et al., 1993), while CGRP, VIP and SP have been shown to induce relaxation of resistance arteries isolated from cancellous bone (Lundgaard et al., 1993).

Nervous Control of Bone Metabolism

There is evidence to suggest that the four neuroendocrine peptides identified in bone take part in the regulation of bone cell physiology. Thus, VIP stimulates bone resorption in culture (Hohman et al., 1983), while CGRP inhibits bone resorption, albeit only at concentrations thousandfold greater than those of calcitonin (Zaidi et al., 1987). It is uncertain whether this effect of CGRP has any physiologic significance, but it is noteworthy that calcitonin is a circulating hormone, while CGRP is released locally at nerve terminals and thus may be present locally in much higher concentrations. Osteoblasts in culture express receptors for noradrenaline, NPY, CGRP, VIP, (and PTH), but not SP, utilizing cAMP as secondary messenger, and NPY inhibits the increase in cAMP generation produced by noradrenaline and PTH (Bjurholm, 1989). This inhibition by NPY of PTH induced cAMP formation represents the first example of interaction between a circulating calciotropic hormone and a local neuropeptide (Bjurholm, 1989).

DISCUSSION

It is apparent that the vascular system in bone is not just a system for nutrition and solute exchange, but is intimately linked to the processes of osteogenesis and myelopoiesis at the ultrastructural level. Correspondingly, the structural, metabolic, and functional heterogeneity, which exists among different parts of the skeleton as well as within each single bone, is associated with considerable topographic variation in vascularity and hemodynamics (see table). Thus, cortical bone blood flow has been shown to vary directly with number of remodelling sites (McCarthy and Lang, 1992) as well as with the corrected mineral apposition rate, which is an index of the work rate of mineralization in each cortical BMU (Reeve et al., 1988). Furthermore, a relationship exists between the regional distribution of bone hemodynamics and uptake of bone seeking tracers such as [85]Sr (McCarthy and Lang, 1992; McInnis et al. 1980) and ^{99m}Tc-diphosphonate (Hansen et al., 1991; McKinstry et al., 1982), which is of value clinically in radionuclide imaging of the skeleton. Accelerated bone remodelling with negative balance and a resultant bone loss is associated with increased bone blood flow. This occurs in e.g. limb immobilization or disuse (Semb, 1969) and in juxtaarticular bone in arthritis (Bünger, 1987; Hansen et al., 1991). Daily physical exercise, which influences the balance between bone formation and resorption positively, also is associated with increased bone blood flow (Jurvelin et al. 1988).

Although the mechanisms controlling the vascular system and blood flow rates to bone and those regulating osteogenesis and myelopoietic activity are interrelated, they are certainly not identical and far from completely elucidated. This has been demonstrated elegantly in an experiment, in which the bone loss associated with spinal cord injury was partially prevented by indomethacin, whereas the concomitant increase in bone blood flow was not (Schoutens et al., 1988). Direct electric currents stimulate both osteogenesis and bone vascularization (Nanmark et al., 1988). An intriguing current theory is that the link between mechanical loading and bone metabolism is electric streaming potentials generated by movement of solutes through bone driven by mechanical stresses in bone and perhaps also by hydrostatic stresses of the circulation (Kelly and Bronk, 1990).

Thorough discussion of interactions between the vascular system and bone metabolism in bone pathology is beyond the scope of this review. However, interesting examples are decreased bone and marrow vascularity in patients with osteoporosis (Burkhardt et al., 1987), reversibly reduced bone blood flow in streptozotocin-induced experimental diabetes in rats (Lucas, 1987), progressive bone loss in insulin dependent diabetics with microvascular complications, but not in patients without microangiopathy (Mathiasen et al., 1990), and association between arteriosclerosis and osteoporosis (Laroche et al., 1992).

The demonstration of complex autonomic and sensory innervation in bone appearing before or simultaneous with vascularization during osteogenesis raises the possibility that vascular, osteogenic, myelopoietic, and immunologic functions in bone all may be under nervous control. Future research on the involvement of nerves in bone development, repair, and pathology must receive high priority.

ABSTRACT

The vasculature of bone plays a major role in the initiation and scaffolding of osteogenesis during skeletal ontogeny, remodelling, and repair. Active biological roles of vascular endothelial cells include vascular invasion of cartilage prior to osteogenesis, extra-vascular migration and transformation into osteoprogenitor cells, secretion of mitogens for osteoblasts, and secretion of vasodilating nitrous oxide and prostaglandins with modulating influences on bone vascular tone as well as bone cell physiology. The specialized sinusoidal vascular structure of bone marrow reflects active hematopoiesis, in which endothelial cells supply growth factors and facilitates cell diapedesis from marrow to the blood stream. Adrenergic and peptidergic neurons influence both vascular physiology and metabolic processes in bone and marrow and may serve to integrate physiological processes in bone. Consequently, bone vascularity and blood flow appear to exhibit covariation with osteogenic and hematopoietic activity in normal bone as well as in important examples of bone and bone marrow pathology.

ACKNOWLEDGEMENTS

The scanning electron micrographs of polymethyl methacrylate vascular casts were kindly provided by Dr Shu-Zheng He, M.D., M.S., Department of Orthopedics, University Hospital of Aarhus, Denmark.

REFERENCES

Baglin, T.P., J. Crocker, A. Timmins, S. Chandler, B.J. Boughton: Bone marrow hypervascularity in patients with myelofibrosis identified by infra-red thermography. Clin. Lab. Haemat. 13: 341-348 (1991).

Bjurholm, A.: Neuroendocrine peptides in bone. Thesis. ISBN 91-7900-875-5, Stockholm 1989.

Brighton, C.T., R.M. Hunt: Early histological and ultrastructural changes in medullary fracture callus. J. Bone Joint Surg. 73A: 832-847 (1991).

Brinker, M.R., H.L. Lippton, S.D. Cook, A.L. Hyman: Pharmacological regulation of the circulation of bone. J. Bone Joint Surg. 72A: 964-975 (1990).

Broxmeyer, H.E., D.E. Williams: The production of myeloid blood cells and their regulation during health and disease. Crit. Rev. Oncol. Hematol. 8: 173-226 (1988).

Bünger, C. Hemodynamics of the juvenile knee. Joint effusion and synovial inflammation studied in dogs. Acta Orthop. Scand. 58 (suppl 222): 1-104 (1987).

Burkhardt, B., R. Bartl, B. Frisch, K. Jäger, C. Mahl, W. Hill, G. Kettner: The structural relationship of bone forming and endothelial cells of the bone marrow. In: Arlet J, Ficat RP, Hungerford DS (eds): Bone circulation, pp 2-14. Williams and Wilkins, Baltimore 1984.

Burkhardt, B., G. Kettner, W. Bohm, M. Schmidmeier, R. Schlag, B. Frisch, B. Mallman, W. Eisenmenger, T.H. Gilg: Changes in trabecular bone, hematopoiesis and bone marrow vessels in aplastic anemia, primary osteoporosis, and old age: a comparative histomorphometric study. Bone 8: 157-164 (1987).

Canalis, E., J. Lorenzo, W.H. Burgess, T. Maciag: Effects of endothelial cell growth factor on bone remodelling in vitro. J. Clin. Invest. 79: 52-58 (1987).

Caplan, A.I.: Mesenchymal stem cells. J. Orthop. Res. 9: 641-650 (1991).

Carrington, J.L., A.H. Reddi: Parallels between development of embryonic and matrix-induced endochondral bone. Bioessays 13: 403-408 (1991).

Chen, L.T., M.F. Chen, V.L. Porter: Increased bone marrow blood flow in rabbits with acute hemolytic anemia. Am. J. Hematol. 22: 35-41 (1986).

Davis, R.F., L.C. Jones, D.S. Hungerford: The effect of sympathectomy on blood flow in bone. Regional distribution and effect over time. J. Bone Joint Surg. 69: 1384-1390 (1987).

Davis, T.R, M.B. Wood: Endothelial control of long bone vascular resistance. J. Orthop. Res. 10: 344-349 (1992).

Dean, M.T., M.B. Wood, P.M. Vanhoutte: Antagonist drugs and bone vascular smooth muscle. J. Orthop. Res. 10: 104-111 (1992).

DeSimone, D.P., A.H. Reddi: Vascularization and endochondral bone development: Changes in plasminogen activator activity. J. Orthop. Res. 10: 320-324 (1992).

Draenert, K., Y. Draenert: The vascular system of bone marrow. Scanning Electron Microscopy IV: 113-122 (1980).

Floyd, W.E., D.J. Zaleske, A.L. Schiller, C. Trahan, H.J. Mankin: Vascular events associated with the appearance of the secondary center of ossification in the murine distal femoral epiphysis. J. Bone Joint Surg. 69A: 185-190 (1987).

Folkman, J.: Toward an understanding of angiogenesis: search and discovery. Perspectives in Biology and Medicine 29: 10-36 (1985).

Ganey, T.M., S.M. Love, J.A. Ogden: Development of vascularization in the chondroepiphysis of the rabbit. J. Orthop. Res. 10: 496-510 (1992).

Guenther H.L., H. Fleisch, N. Sorgente: Endothelial cells in culture synthesize a potent bone cell active mitogen. Endocrinology 119: 193-201 (1986).

Hansen, E.S., K. Søballe, T.B. Henriksen, V.E. Hjortdal, C. Bünger: ^{99m}Tc-diophosphonate uptake and hemodynamics in arthritis of the immature dog knee. J. Orthop. Res. 9:191-202 (1991).

Hill, E.L., R. Elde: Distribution of CGRP-0, VIP-, D beta H-, SP-, and NPY-immunoreactive nerves in the periosteum of the rat. Cell Tissue Res. 264: 469-480 (1991).

Hohmann, E.L., L. Levine, A.H. Tashjian Jr: Vasoactive intestinal peptide stimulates bone resorption via a cyclic adenosine 3',5' monophosphate dependent mechanism. Endocrinology 112: 1233-1239 (1983).

Hulth, A., O. Johnell, L. Lindberg: Demonstration of blood-vessellike structures in cartilaginous callus by antilaminin and antiheparin sulphate proteoglycan antibodies. Clin. Orthop. 254: 289-293 (1990).

Hunter, W.L., A.L. Arsenault: Endothelial cell division in metaphyseal capillaries during endochondral bone formation in rats. Anat. Rec. 227: 351-358 (1990).

Hunter, W.L., A.L. Arsenault, A.B. Hodsman: Rearrangement of the metaphyseal vasculature of the rat growth plate in rickets and rachitic reversal: a model of vascular arrest and angiogenesis renewed. Anat. Rec. 229: 453-461 (1991).

Iversen, P.O., G. Nicholaysen, H.B. Benestad: Blood flow to bone marrow during development of anemia or polycythemia in the rat. Blood 79: 594-601 (1992).

Jones, A.R., C.C. Clark, C.T. Brighton: The effects of rat microvascular cell products on cultured rat calvarial bone cells. Trans. Orthop. Res. Soc. 17: 188 (1992).

Jurvelin, J., T. Lahtinen, I. Kiviranta, I. Arnala, R. Lappalainen, M. Tammi, H.J. Helminen: Blood flow, histomorphology, and elemental composition of the canine femur after physical training or immobilization. Acta Physiol. Scand. 132: 385-389 (1988).

Kelly, P.J., J.T. Bronk: Venous pressure and bone formation. Microvasc. Res. 39: 364-375 (1990).

Klagsbrun, M., S. Smith: Purification of a cartilage-derived growth factor. J. Biol. Chem. 255: 10859-10866 (1980).

Lahtinen, R., T. Lahtinen, T. Romppanen: Bone and bone-marrow blood flow in chronic granulocytic leukemia and primary myelofibrosis. J. Nucl. Med. 23: 218-224 (1982).

Laroche, M., C. Ribot, J.L. Puech, J.M. Pouilles, J. Arlet, H. Boccalon, P. Puel, B. Mazieres: Arteriopathy of the lower limbs and osteoporosis in men. Ass. Res. Circ. Osseous News Letter 4: 30-36 (1992).

Lindblad, B.E., L.B. Nielsen, A. Bjurholm, E.S. Hansen: Vasoconstrictive action of neuropeptide Y in bone. Trans. Orthop. Res. Soc. 18 (1993).

Lucas, P.D.: Reversible reduction in bone blood flow in streptozotocin-diabetic rats. Experientia 43: 894-895 (1987)

Lundgaard, A., C. Aalkjœr, M.J. Mulvany, A. Bjurholm, E.S. Hansen: Effects of calcitonin gene-related peptide and vasoactive intestinal peptide on resistance vessels isolated from cancellous bone. Trans. Orthop. Res. Soc. 18 (1993).

Macintyre, I., M. Zaidi, A.S. Alam, H.K. Datta, B.S. Moonga, P.S. Lidbury, M. Hecker, J.R. Vane: Osteoclastic inhibition - an action of nitric oxide not mediated by cyclic GMP. Proc. Natl. Acad. Sci. USA 88: 2936-2940 (1991).

Martiat, P., A. Ferrant, M. Cogneau, A. Bol, C. Michel, J. Rodhain, J.L. Michaux, G. Sokal: Assessment of bone marrow blood flow using positron emission tomography: no relationship with bone marrow cellularity. Br. J. Haematol 66: 307-310 (1987).

Mathiassen, B., S. Nielsen, J.S. Johansen, D. Hartwell, J. Ditzel, P. Rodbro, C. Christiansen: Long-term bone loss in insulin-dependent diabetic patients with microvascular complications. J. Diabet. Complic. 4: 145-149 (1990).

McCarthy, I.D., Y. Lang: The relationship between bone blood flow, strontium clearance, and bone turnover in the tibial cortex. Trans. Orthop. Res. Soc. 17: 11 (1992).

McClugage, S.J., R.S. McCuskey: Relationship of the microvascular system to bone resorption and growth in situ. Microvasc. Res. 6: 132-134 (1973).

McInnis, J.C., R.A. Robb, P.J. Kelly: The relationship of bone blood flow, bone tracer deposition, and endosteal new bone formation. J. Lab. Clin. Med. 96: 511-522 (1980).

McKinstry, P., J.E. Schnitzer, T. Light, J.A. Ogden, P. Hoffer: Relationship of ^{99m}Tc-MDP uptake to regional osseous circulation in skeletal immature and mature dogs. Skeletal Radiol. 8: 115-121 (1982).

Monroy, R.L., T.A. Davis, T.B. Nielsen, A.J. Staton: Peripheral blood hematopoietic progenitor/stem cells proliferate to form colonies in liquid culture but require contact with vascular endothelial cells and GM-CSF. Int. J. Cell Cloning 10: 105-115 (1992).

Moses, M.A., J. Sudhalter, R. Langer: Identification of an inhibitor of neovascularization from cartilage. Science 248: 1408-1410 (1990).

Nanmark, U., F. Buch, T. Albrechtsson: Influence of direct currents on bone vascular supply. Scand. J. Plast. Reconstr. Surg. 22: 113-115 (1988).

Pelus, L.M.: Modulation of myelopoiesis by prostaglandin E_2: Demonstration of a novel mechanism of action in vivo. Immunol. Res. 8: 176-184 (1989).

Reddi, A.H., K.E. Kuettner: Vascular invasion of cartilage: correlation of morphology with lysozome, glycosaminoglycans, protease, and protease-inhibitor activity during endochondral bone development. Dev. Biol. 82: 217-223 (1981).

Reeve, J., M. Arlot, R. Wootton, C. Edouard, M. Tellez, R. Hesp, J.R. Green, P.J. Meunier: Skeletal blood flow, iliac histomorphometry, and strontium kinetics in osteoporosis: A relationship between blood flow and corrected apposition rate. J. Endocrin. Metab. 66: 1124-1131 (1988).

Roorda, J., C.S. Han, M.B. Wood: In vivo confirmation of bone blood flow control by endothelial derived relaxation factor in the canine tibia. Trans. Orthop. Res. Soc. 17: 8 (1992).

Schoutens, A., M. Verhas, N. Dourov, P. Bergmann, F. Caulin, A. Verschaeren, M. Mone, A. Heilporn: Bone loss and bone blood flow in paraplegic rats treated with calcitonin, diphosphonate, and indomethacin. Calcif. Tissue Int. 42: 136-143 (1988).

Schoutens, A., M. Verhas, N. Dourov, A. Verschaeren, M. Mone, A. Heilporn: Anemia and marrow blood flow in the rat. Br. J. Haematol. 74: 514-518 (1990).

Semb, H.: Experimental limb disuse and bone blood flow. Acta Orthop. Scand. 40: 552-562 (1969).

Shapiro, I.M., E.E. Golub, B. Chance, C. Piddington, O. Oshima, O.C. Tuncay, J.C. Haselgrove: Linkage between energy status of perivascular cells and mineralization of the chick growth cartilage. Dev. Biol. 129: 372-379 (1988).

Stanka, P., U. Bellack, A. Lindner: On the morphology of the terminal microvasculature during endochondral ossification in rats. Bone Miner. 13: 93-101 (1991).

Streeten, E.A., M.L. Brandi: Biology of bone endothelial cells. Bone Miner. 10: 85-94 (1990).

Thiele, J., V. Rompcik, S. Wagner, R. Fisher: Vascular architecture and collagen type IV in primary myelofibrosis and polycythemia vera: an immunomorphometric study on trephine biopsies of the bone marrow. Br. J. Hematol 80: 227-34 (1992).

Thompson, T.J., P.D. Owens, D.J. Wilson: Intramembranous osteogenesis and angiogenesis in the chick embryo. J. Anat. 166: 55-65 (1989).

Trueta, J.: The role of vessels in osteogenesis. J. Bone Joint Surg 45B: 402-418 (1963).

Wickramasinghe, SN.: Observations on the ultrastructure of sinusoids and reticular cells in human bone marrow. Clin. Lab. Haemat. 13: 263-278 (1991).

Winet, H., J.Y. Bao, R. Moffat: A control model for tibial cortex neovascularization in the bone chamber. J. Bone Miner. Res. 5: 19-30 (1990).

Zaidi, M., K. Fuller, P.J.R. Bevis, R.E. Gainesdas, T.J. Chambers, I. MacIntyre: Calcitonin Gene-Related Peptide inhibits osteoclastic bone resorption: A comparative study. Calcif. Tiss. Int. 40 : 149-154 (1987).

ENDOTHELIAL CELLS AND BONE CELLS

Olusola O.A. ONI, S. DEARING and S. PRINGLE

University of Leicester Medical School
Glenfiled General Hospital
Leicester, United Kingdom

INTRODUCTION

There is a long, and distinguished tradition of belief in the theory that endothelial cells give rise to bone cells. Haller (1763), Hunter (1835) and Keith (1927) have all proposed that bone originated from the blood vessels. Figure 1 shows a schema of the progeny of bone cells proposed by Trueta (1963) who envisaged a syncytium of cells originating from the endothelial cell and terminating in the osteocyte. According to Trueta (1963), dying osteocytes and endothelial cells produce a vascular stimulating factor which attracts vessels to the fracture site. As a consequence, in the fracture callus, vessels migrate towards the ischaemic area at the centre of the fracture. Bone deposition radiates from, and is moulded onto, the vascular pattern. Trueta's schema is based on the concept of contiguity of cells. Recently, Brighton and Hunt (1991) provided electron microscopic evidence in support of the theory. This evidence is only indirect, however. The fact that cells are lying next to one another is not a conclusive proof that they are related. On the other hand, Hulth et al. (1990) using immunocytochemical methods which localised laminin and other basement membrane components have demonstrated vessel-like structures in the callus cartilage. These findings have not yet been corroborated.

Immunocytochemical methods have now become available which may be used to exclusively label endothelial cells and their progeny. Lectins such as Ulex europeus I-peroxidase (UEP) are used by pathologists to distinguish tumours of vascular origin from other tumours (Holthofer et al., 1982; Walker 1985). There are available monoclonal antibodies specifically raised against endothelial cell proteins (Pringle and de Bono, 1988). These methods have been used in this study to investigate the origins of the fracture callus.

Bone Circulation and Vascularization in Normal and Pathological Conditions
Edited by A. Schoutens *et al.*, Plenum Press, New York, 1993

43

MATERIALS AND METHODS

Lectin Binding Study

Samples of the fracture callus were obtained from four 3-10-day old adult human tibial diaphyseal fractures, three 1-2 week old osteotomies of the adult rabbit tibia treated by external fixation and three 4-14-day old closed fractures of the adult rabbit tibia. All the samples were fixed in formalin, decalcified, routinely processed and embedded in paraffin wax. 6 μm sections were obtained from each sample and stained with haematoxylin and eosin (HE).

The methods of lectin binding have been previously described by Walker (1985): four 6 μm sections were obtained from each specimen and two of these were trypsin digested at 37° for 30 minutes. Next, all the sections were washed and then exposed to 0.3% hydrogen peroxide for 30 minutes to block the endogenous peroxidase. One trypsinised and one non-trypsinised sections from each specimen were incubated at room temperature for 1 hour with either 10 μg/ml UEP (Sigma) or with labelled lectin with 0.1 m L-fucose. After washing, the peroxidase was localised using the diaminobenzidene-hydrogen peroxide reaction.

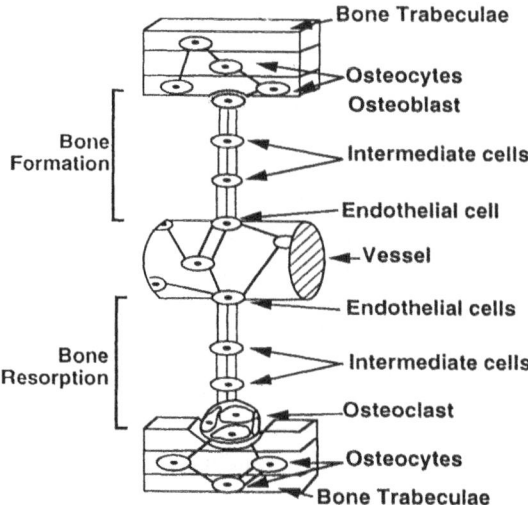

Figure 1. Trueta's syncytium theory.

Monoclonal Antibody Study

Samples of the periosteum close to the fracture site were obtained from eight adults undergoing open reduction and internal fixation 3-10 days following a tibial diaphyseal fracture.

Six specimens were routinely fixed, decalcified, processed and embedded in paraffin wax. 6 μm thick sections were obtained and comparative sections from each specimen were stained with HE or using an immunocytochemical method similar to that previously described by Stafford et al. (1992). Four different monoclonal antibodies; namely, D6G11, A3E10, D8G8 and P14G11, and previously described by Pringle and de Bono (1988), were used. Antibodies D6G11, A3E10 and D8G8 were raised against bovine aortic endothelium. All cross-react with human umbilical vein endothelium (HUVEC) in gluteraldehyde fixed tissue cultures and stain endothelium to varying degrees in acetone fixed cryostat sections. D6G11 and A3E10 appear to have the same epitope specificity of a 45kD protein as shown by Western blot analysis of endothelial cell (EC) lysates. D8G8 shows specificity for a triplet band protein of 15-18kD molecular weight.

Antibody P14G11 was raised against porcine aortic endothelium and cross-reacts with HUVEC on non-fixed as well as fixed cells. It is specific for a 75kD epitope of EC lysate proteins. All the monoclonals stain the cells along the edge of the denuded area of a scratch-damaged cell monolayer. These cells are injured sub-lethally and appear to be involved in the initial stages of repair of the damaged monolayer.

Frozen sections were obtained from the two remaining specimens, fixed in acetone for 10 minutes and then air dried. The endogenous peroxidase activity was blocked by incubation at room temperature for 30 minutes in 0.3 % H_2O_2. The sections were washed in running tap water for 5 minutes and then equilibrated in TRIS buffered saline (TBS) (pH 7.5) for 5 minutes. Non specific binding sites were blocked by incubation in rabbit serum (1/5 dilution) for 10 minutes. Next, the sections were incubated with antibody for 60 minutes at room temperature. The monoclonal antibodies were in the form of cell culture supernatant and used as neat solutions. The sections were washed in TBS for 5 minutes, incubated with peroxidase conjugated anti-mouse antibody for 30 minutes at room temperature (1:100 dilution, Dako) and then rewashed before the substrate (diaminobenzidine - 1 mg/ml in TBS + 100 µl/10 ml 3 % H_2O_2) was added for 2-3 minutes. The sections were washed again before being counterstained in Mayer's haematoxylin.

In a parallel study, an attempt was made to localise urokinase in the callus tissue using a monoclonal anti-urokinase antibody (Serotec) and an indirect immunoperoxidase method. It has been suggested that migrating endothelial cells, in the process of angiogenesis, utilise this enzyme to digest connective tissue matrices to aid invasion of the tissue to be vascularised (Pepper et al., 1988).

RESULTS

In the HE sections, areas of endochondral ossification were observed adjacent to areas of intramembranous ossification. The reasons for the divergent differentiation pathways were not apparent.

Lectin Binding Study

Positive binding was observed only in the human specimens stained with UEP. In these specimens, lectin activity was restricted to the vascular structures and to red blood cells. No other cell types were stained. Clear and consistent results were obtained with no background staining.

In the areas of intra-membranous ossification, positively stained vessels were observed within the new bone trabeculae (Figure 2) and in the intervening soft tissues. In the areas of endochondral ossification, positively stained vessels were also observed within the new bone trabeculae (Figure 3) and within the callus cartilage (Figure 4).

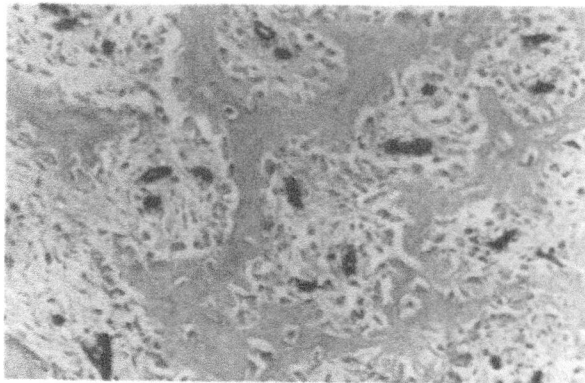

Figure 2. Photomicrograph of a section of the human fracture callus showing intramembranous ossification and lectin-binding of only the endothelial cells.

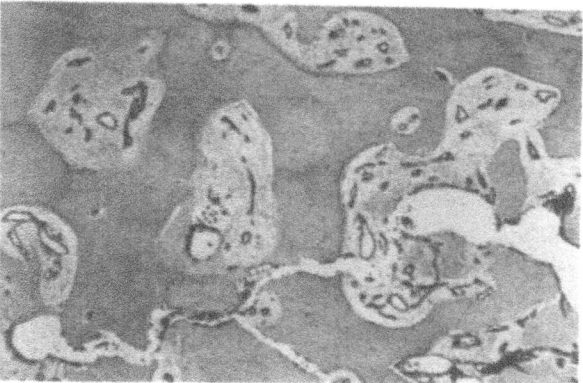

Figure 3. Photomicrograph of a section of the human fracture callus showing endochondral ossification and lectin-binding of only the endothelial cells.

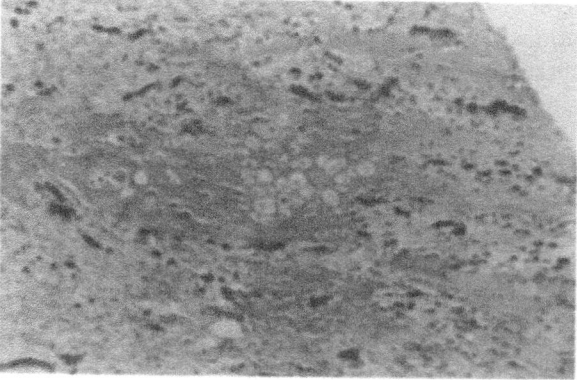

Figure 4. Photomicrograph of a section of the human fracture callus showing newly formed cartilage and lectin-binding of only the vascular elements.

Monoclonal Antibody Study

In the sections stained for endothelial cell proteins, positive binding was observed only in the vascular structures within newly formed bone trabeculae (Figure 5) and in the intervening soft tissues. The bone and cartilage cells were not stained in all specimens. The results from the two frozen sections were similar to those obtained from the six fixed specimens. The sections stained for urokinase also gave a negative result with regards to staining of the bone and cartilage cells.

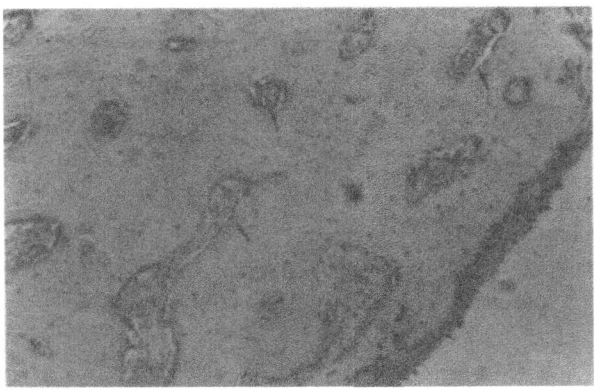

Figure 5. Photomicrograph of a section of the human fracture callus showing immunoreactive positive endothelial cells (arrows) (D8G8 antibody).

DISCUSSION

The methods used in this study are specific with regards to identifying endothelial cells and their progeny. The monoclonal antibodies have been shown by Pringle and de Bono (1988), in an in vitro model of endothelial damage and regrowth, to result in reproducible immunoperoxidase staining. These antibodies have been further characterised by Western blotting and I^{125} labelling. Because EC proteins are more specific to blood vessels than the basement membrane components investigated by Hulth et al. (1990), immunolocalisation of endothelial cell proteins provides a better indication of the vascularity of the fracture callus.

Lectins are plant or animal proteins which bind saccharide moieties in glycoproteins and glycolipids (Damjanov, 1987). They have been used as probes to detect different types of glycoproteins and sugar moieties on cellular membranes and histological structures (Lotan and Nicolson, 1979). The Ulex europeus agglutinin is a lectin with specificity for d-L-fucose residues. It reacts preferentially with the vascular endothelium and is generally accepted as a reliable marker for tumours of vascular origin (Holthofer et al., 1982).

This study shows that peroxidase-labelled Ulex europeus I (UEP) can be used to study the angio-architecture of healing fractures. The method is applicable to routinely fixed and processed specimens but it appears to be species specific. In this study only human materials were positively stained by UEP. Holthofer (1983) who investigated the kidneys of 14 species also found that the lectin would only bind the endothelial cells of man.

In this study, bone trabeculae, osteoblasts and chondrocytes showed no evidence of binding, whereas adjacent blood vessels revealed clearly defined staining of the endothelium. Osteogenic cells lying adjacent to endothelial cells were not stained. If osteogenic cells were derived from endothelial cells, one would have expected some of them to be stained, even if weakly. This lack of staining of bone cells casts some doubt on the proposition by Keith (1927), Trueta (1963) and Brighton and Hunt (1991) that endothelial cells give rise to bone cells.

REFERENCES

Brighton, C.T., R.M. Hunt: Early histological and ultrastructural changes in medullary fracture callus. J. Bone Joint Surg. 73A: 832-847 (1991).

Damjanov, I.: Biology of disease. Lectin cytochemistry and histochemistry. Lab. Investig. 57: 5-20 (1987).

Haller, von A.: Experimentum de ossium formatione. In Opera minora, vol. 2, Lausanne, Francisci Grasset, pp 400 (1763).

Holthofer, H., I. Virtanen, A.L. Kariniemi, H. Hormia, E. Linder, A. Miettinen: Ulex europeus I lectin as a marker for vascular endothelium in human tissues. Lab. Investig. 47: 60-66 (1982).

Holthofer, H.: Lectin binding sites in kidney. A comparative study of 14 animal species. J. Histochem. Cytochem. 31: 531-537 (1983).

Hulth, A., O. Johnell, L. Lindberg, M. Paulsson, D. Heinegard: Demonstration of blood-vessellike structures in cartilaginous callus by antilaminin and antiheparin sulfate proteoglygan antibodies. Clin. Orthop. 254: 289-293 (1990).

Hunter, J.: In works, vol. 1, London, Longman etc., pp 502, 1835.

Keith, Sir A.: Concerning the origin and nature of osteoblasts. Proc. Roy. Soc. Med. 21: 301 (1927).

Lotan, R., G.L. Nicolson: Purification of cell membrane glycoproteins by lectin affinity chromatography. Biochim. Biophys. Acta 559: 329-376 (1979).

Pepper, M.S., J.D. Vassalli, R. Montesano, L. Orci: Urokinase-type plasminogen activator is induced in migrating capillary endothelial cells. J. Cell Biol. 105: 2535-2541 (1988).

Pringle, S., D.P. de Bono: Monoclonal antibodies to damaged and regenerating vascular endothelium. J. Clin. Lab. Immunol. 26: 159-162 (1988).

Stafford, H., O.O.A. Oni, J. Hay, P.J. Gregg: An investigation of the contribution of the extra-osseous tissues to the diaphyseal fracture callus using a rabbit tibial fracture model and in situ immunocytochemical localisation of osteocalcin. J. Orth. Trauma 6: 190-194 (1992).

Trueta, J.: The role of the vessels in osteogenesis. J. Bone Joint Surg. 45B: 402-418 (1963).

Walker, R.A.: Ulex europeus I-peroxidase as a marker of vascular endothelium: its application in routine histopathology. J. Path. 146: 123-127 (1985).

HISTORY OF DISCOVERIES OF BONE MARROW AND BONE VASCULARISATION AND INNERVATION

Z. DABROWSKI, and Z. TABAROWSKI

Laboratory of Experimental Hematology and Toxicology
Jagiellonian University
Kraków, Poland

The tables present the main discoveries in the field of basic knowledge concerning the anatomy and physiology of bone circulation, i.e.: Bone and Bone marrow circulation; Vascular interdependance between bone and bone marrow; Intraosseous pressure; Bone marrow innervation, Bone innervation.

The following section deals with more recent works on bone innervation:
Existing data on the innervation of bone marrow have been recently reviewed by Felten and Felten (1991). However, the most important information on the anatomy of that innervation comes from the work of Yamazaki and Allen (1990). They studied the distribution of nerve endings in mouse bone marrow at the ultramicroscopic level and concluded that the efferent nerve terminals are distributed mainly beside arterial smooth muscle cells or between the layers of periarterial adventitial cells. Less frequently they were also seen in hematopoietic parenchyma and on sinus walls. The nature of neuro-transmitters used by these nerves has been elucidated by Bjurholm (1991). Apart from noradrenergic (TH-positive) sympathetic nerve fibres, he described peptidergic nerves containing substance P (SP), calcitonine generelated peptide (CGRP), vasoactive intestinal peptide (VIP) and neuropeptide Y (NPY).
The majority of nerves were vascular, although several non-vascular endings were observed at the growth plate of bone and amidst marrow cells. Physiological importance of nerves in the marrow is unclear. There is no doubt that they are vasomotor but there is growing evidence that they constitute an anatomical link by which the central nervous system can influence immune function (Weihe et al., 1991). It is postulated that proliferation, differentiation and migration of immune cells can be affected by substances released at nerve endings.
Bone and periosteum are innervated by both sympathetic and sensory nerves (Bjur-holm, 1991; Hill and Elde, 1991; Hukkanen et al., 1992). There are experimental studies on the influence of innervation on bone metabolism (Hill et al., 1991) and bone blood flow (Davis et al., 1987). These experiments include surgical or chemical sympathectomy

Bone Circulation and Vascularization in Normal and Pathological Conditions
Edited by A. Schoutens *et al.*, Plenum Press, New York, 1993

49

Table 1. Bone and bone marrow circulation

YEAR	AUTHOR	SOURCE	SUBJECT OF DISCOVERY
1674	A.van Leeuwenhoek	Microscopical observations from M.Leeuwenhoek concerning blood, milk, bones, the brain, spittle and cuticula etc. Letter to Royal Society Phil.Trans.R.Soc.9,1: 21-128	Description of the small veins on the surface of the shinbone of a cow and little holes passing from without inwards.
1691	C.Havers	Osteologya Nova, or some New Observation of the Bones e.t.c. London, Samuel Smith	Description of large nutrient artery. It pierces the shaft of long bones forming ramifications in bone marrow (BM). First description of minute canals in bone cortex.
1754	B.C.Albinus	De ossibus corporis humani ad auditores suos. Texta exemplar Leidae Batavorum	First application of the injection technique displaying the blood vessels of the human body including the vascularization of the cortex of long bones.
1863	R. Volkmann	Archiv fur Klinische Chirurgie vol. 4 : 437	Discovery of cross connections between osteons.
1868	G.Bizzozero	Gazetta med.Ital.Lomb.No 46 and Zentralbl. Med. Wissensch. 6	Description of marrow veins, shoving that their walls are extremely thin (easy to penetrate by blood cells).
1869	H.Hoyer	Gazeta Lekarska, Warszawa.VII.12.177	Good notes on blood vessels distribution in bones and the weaving of vessels in bone marrow.
1875	K.Langer	Denksch. ksl Akad Wiss. Wien. Math. naturwiss. 3. In Commision bei Karl Gerold's Sohn.Wien.	Vascular organization of bone and bone marrow established in thin microscopic sections obtained from material perfused with Prussian blue and vermilion suspension.
1957	J.Trueta	J. of Bone and Joint Surgery 39 (B) : 358	A concise picture of the modifications of the vasculature of the human femoral head. He divided the vasculature into 5 phases from birth to maturation.
1961	M. Brookes et al	Lancet 1 : 1078	Proved that in physiological conditions the blood flow through bone cortex runs in centrifugal not in centripetal direction.
1965	F.W. Rhinelander	Clin. Orthop. 40 : 12	New data on microcirculation in normal and healing bone.
1979	A.Schoutens et al.	Amer. J. Physiol. 236: H1	Comparison between bone seeker clearance and blood flow using microspheres (45 Ca & 85 Sr).

(guanethidine or 6-hydroxydopamine treatment) as well as destruction of sensory nerves by capsaicin. However it should be stressed that results obtained after surgical sympathectomy must be treated with caution in the light of work recently published by Waris et al. (1991). They stated that the saphenous nerve section or even bilateral lumbar sympathectomy in the rat did not result in complete destruction of peripheral adrenergic nerves around vessels. Frymoyer and Pope (1977), reported enhanced healing of fibula fractures in sciatically denervated rats. On the other hand Hill et al. (1991) concluded that neither chemical sympathectomy nor the selective destruction of fine calibre primary afferent nerves significantly affected radial bone growth or resorption in rat tibia. Nevertheless they stated that sensory and sympathetic nerves affect local bone remodelling. Surgical sympathectomy in dog results in significant but transient increase in bone blood flow (Davis et al., 1987). The peripheral nervous system may play both a sensory and afferent role in the mechanical and inflammatory influences on bone structure. Differences in the number and distribution of SP and CGRP-immunoreactive nerves between healthy and arthritic rats were found by Hukkanen et al., (1992). They suggest that sensory innervation may play a role in pain perception and in the healing and remodelling of bone. An other possible physiological connection arises from recent study of Bjurholm et al. (1992) who studied the effect of several neuropeptides on the cyclic AMP formation in cultured osteoblastic cell lines and primary bone cells. Their results suggest that neurohormones may have a role in the regulation of osteoblastic activity.

Figures 1, 2, 3, 4 derive from our studies on nerves in the bone marrow and their relation to hematopoiesis and healing process of fractured bones.

Table 2. Vascular interdependence of bone and bone marrow

1951	R.Marneffe	Acta Chir.Belgica 7: 469.	Introduced the method of injecting radiopaque media into the main vessels of a limb.
1959	P.I.Branemark et al	Scand.J.Clin. Lab.Invest. Suppl.38.2.	Introduced a new approach, vital microscopy with movie camera, to the investigation of BM circulation in the living rabbit. Description of the existence of direct communications between capillaries and venules called shunting capillaries.
1970	P. De Bruyn et al.	Anat. Rec. 168 : 55	Enlargement of knowledge of bone and BM microcirculation. Terminal capillaries of the nutrient artery enter the bone where they anastomose with the inraosteal vessels which in turn connect with the sinusoidal network at the osteomyeloid junction.
1987	R.Burkhard et al.	Bone 8 : 157	First emphasised changes in trabecular bone, hematopoiesis and bone marrow vessels in aplastic anemia, primary osteoporosis and old age.

Table 3. Pressure in bone and bone marrow

1952	E. Blumenthal et al.	Surg. Gynecol. Obst. 95 : 215.	First examination of bone and BM pressure.
1954	N. Petrakis	J. Clin. Invest. 33: 27.	Conjunction phenomenon between bone marrow pressure and leukemia.
1959	E. Herzig, W. Root	Amer. J. Physiol.196(5) : 1053	Description of relation of sympathetic nervous system to blood pressure of b.m.
1963	N.E. Shaw	Clin. Sci.24 : 311	Finding that the marrow flow can change independently of systemic blood pressure, but that there is a direct relationship between intramarrow pressure, blood pressure and blood flow.
1979	G. Bouteiller et al.	Rev. du Rhumatism,46 : 369	Demonstration that the intramedullar pressure, as an isolated parameter, cannot accurately reflect the rate of bone output.
1979	S. Kumar et al.	Acta orthop. scand. 50 : 507	Noticed that in rapid dynamic loading, a slight rise in intra-medullary pressure was observed. Contraction of the femoral muscles also resulted in a BM pressure increase.
1981	Z. Dabrowski et al.	Acta Physiol. Polonica 32 : 729	This work proved contribution of BM pressure to the egress of mature blood cells from BM to blood stream.

Table 4. Bone innervation

1846	M. Gros	Compte rendu des seances de l'acad. des sciences, 23: 1106	First description of bone innervation.
1941	C.J. Zinn J.Q. Griffith	Proc. Soc. Exp. Biol. Med. 46 : 311	Effect of sympathectomy upon blood supply of bone.
1965	J. Miligram R.A. Robinson	Bull. Johns Hopkins Hosp. 117 : 163	Description of unmyelinated nerves in the Haversian canals of the dog at electron microscopic level.
1968	G. Cooper	Science 160 : 327	Ultrastructural description of mixed (myelinated and unmyelinated nerves).
1986	E. Hohmann et al.	Science 232 : 868	First description of the presence of VIP-containing sympathetic nerves in the periosteum and bone.
1988	A. Bjurholm et al.	J. Auton. Nerv. Syst. 25 : 119 & Peptides 9 : 165	Immunoreactive peptidergic (NPY, VIP, SP, CGRP) nerves in bone.
1991	E.L. Hill et al.	Neuroscience 44 : 747	Studies on physiological importance of bone innervation by means of sympathectomy and capsaicin treatment.

Table 5. Bone marrow innervation

1700	M. Du Verney	Histoire de L'Academie Royale des Sciences	First description of nerves in the marrow cavity.
1880	G. Variot, CH. Remy	J. de l'Anat. et Physiol. 16 : 273	Stated that myelinated and nonmyelinated nerves enter the marrow cavity through the nutrient foramen.
1884	L. W. Tumas	Iejeniedielnai klinicieskaja gazieta 14 : 163	First experimental work on the physiological importance of nerves to marrow.
1901	D. Ottolenghi	Atti della Reale Acad. delle Scienze di Torino 36 : 939	Discovery of nerve fibres that form a plexiform network not only in the wall of arteries and capillaries but also among marrow cells.
1916	C.K. Drinker, K.R. Drinker	Amer. J. Physiol.40: 514	They established the existence of vasoconstrictor nerves in the marrow, as well as the rate of flow of perfusing blood for a variety animals.
1925	F. de Castro	Trav. du lab. de Rech. biol. Madrid. 23 : 26	Application of silver impregnation to the study on marrow nerves and their source in CNS.
1932	F. Rossi	Arch. ital. Anat. Embriol. 29 : 539	Described different types of nerve endings; small brushes, buttons or rings in the marrow parenchyma.
1968	W. Calvo*	Amer. J. Anat. 123: 315	Broad comparative and developmental studies on marrow innervation.
1990	K. Yamazaki T.D. Allen	Amer. J. Anat. 187: 261	First clear description of marrow innervation at the ultramicroscopic level.

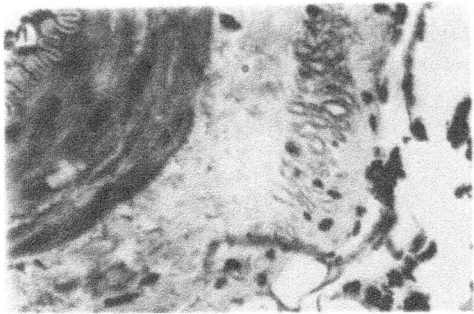 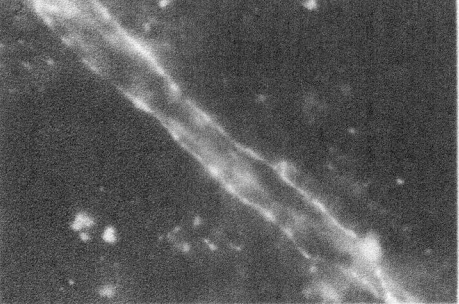

Figure 1. Cross-section on nutrient artery in rat bone marrow. Myelinated and non-myelinated nerves are seen in adventitia. Phase contrast microscopy.

Figure 2. Adrenergic nerve fibers around arteriole in rat bone marrow. Glyxilic acid induced histofluorescence. Epiluminescence.

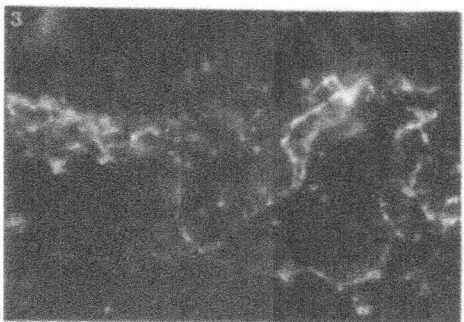 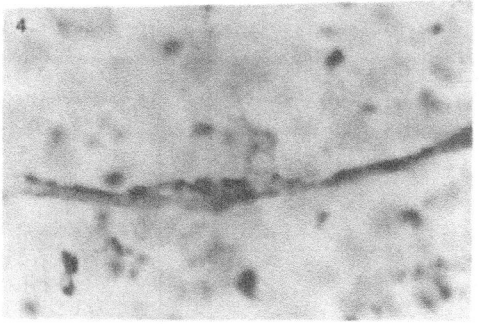

Figure 3. Small adrenergic nervous fibers seen among marrow parenchyma. Epiluminescence.
Figure 4. CGRP-immunoreactive nerve fiber in rat bone marrow associated with small vessel. Nomarski microscopy.

REFERENCES

Bjurholm, A.: Neuroendocrine peptides in bone. Int. Orthop. 15: 325-329 (1991).

Bjurholm, A., A. Kreicbergs, M. Schultzberg, U.H. Lerner: Neuroendocrine regulation of cyclic AMP formation in osteoblastic cell lines (UMR-106-01, ROS 17/2.8, MC3T3-E1, and Saos-2) and primary bone cells. J. Bone Miner. Res. 7: 1011-1019 (1992).

Davis, R.F., L.C. Jones, D.S. Hungerford: The effect of sympathectomy on blood flow in bone. J. Bone Joint Surg. 69A: 1384-1390 (1987).

Felten, S.Y., D.L. Felten: Innervation of lymphoid tissue. In: R. Ader, D.L. Felten, N. Cohen (eds) : Psycho-neuro-immunology II, pp 27-61. Academic Press San Diego, 1991.

Frymayer, J., M.H. Pope: Fructure healing in the sciatically denervated rat. J. Trauma 17: 355-361 (1977).

Hill, E., R. Elde: Distribution of CGRP-, VIP-, DßH-, SP-, and NPY- immunoreactive nerves in periosteum of the rat. Cell and Tissue Res. 264: 469-480 (1991).

Hill, E.L., R. Turner, E. Elde: Effects of neonatal sympathectomy and capsaicin treatment on bone remodelling in rats. Neuroscience 44: 747-755 (1991).

Hukkanen, M., Y.T. Konttinen, R.G. Rees, S. Santavirta, G. Terenghi, J.M. Polak: Distribution of nerve endings and sensory neuropeptides in rat synovium, meniscus and bone. Int. J. Tiss.Reac. 14: 1-10 (1992).

Hukkanen, M., Y.T. Konttinen, R.G. Rees, S.J. Gibson, S. Santavirta, J.M. Polak: Innervation of bone from healthy and arthritic rats by substance P and calcitonin gene related peptide containing sensory fibers. J.Rheumatol. 19: 1252-1259 (1992).

Weihe, E., D. Nohr, S. Michel, S. Muller, H.J. Zentel, T. Fink, J. Krekel: Molecular anatomy of the neuro - immune connection. Int. J. Neurosci. 59: 1- 23 (1991).

Waris, T., O. Kaarela, L. Lasanen, J. Junila, M. Ruuskanen, K. Kyosola: Perivascular sympathectomy does not remove adrenergic nerves from distal vessels. J. Surg. Res. 51: 303-309 (1991).

Yamazaki, K., T.D. Allen: Ultrastructural morphometric study of efferent nerve terminals on murine bone marrow stromal cells, and the recognition of a novel anatomical unit: the " neuroreticular complex ". Amer. J. Anat. 187: 261-276 (1990).

THE DIRECT EFFECTS OF ACIDOSIS AND ALKALOSIS ON LONG BONE VASCULAR RESISTANCE

(Preliminary presentation. Full paper submitted for publication)

T. R. C. DAVIS, F.R.C.S. and M. B. WOOD

Nottingham University, England
Mayo Clinic, Rochester,
Minnesota, USA

INTRODUCTION

Although previous studies have demonstrated that acidosis increases bone blood flow, it is unclear whether this is due to a direct effect of hydrogen ion concentration ($[H^+]$) on bone blood vessels or a central effect on vasoregulatory mechanisms. It is also unclear how sensitive bone blood vessels are to changes in $[H^+]$ and whether the magnitude of pH changes experienced in physiological and pathological situations are sufficient to influence bone blood flow. This study has investigated the direct effects of $[H^+]$ on long bone vascular resistance.

METHOD

Thirty tibiae were obtained from freshly killed adult mongrel dogs and assigned to three groups. After nutrient artery cannulation, each bone was placed in an organ chamber and was perfused at a steady flow rate with aerated (95 % O_2 - 5 % CO_2) Krebs-Ringer solution. For all three groups, resting (unstimulated) perfusion pressure (RPP) was measured and noradrenaline (NE) and sympathetic nerve stimulation (SNS) dose response curves were obtained at normal perfusate pH (7.34 - 7.44). The perfusate of the Acidosis group (n = 10) was then made acidotic (pH 7.2 - 7.33), while that of the Alkalotic group (n = 10) alkalotic (pH 7.47 - 7.58). The pH of the perfusate of the Control group (n = 10), was not altered. After 30 minutes' perfusion at the new pH, the RPP was remeasured and further dose response curves were obtained.

Bone Circulation and Vascularization in Normal and Pathological Conditions
Edited by A. Schoutens *et al.*, Plenum Press, New York, 1993

55

RESULTS

These are summarised in Table 1. Alkalosis increased RPP by 56 % (p < 0.0001). Acidosis attenuated (18 % reduction) and alkalosis enhanced (65 % increase) the vasoconstrictor action of noradrenaline (p < 0.0001). Acidosis also attenuated (11 % reduction) the effect of sympathetic nerve stimulation (p = 0.025).

Table 1. The mean (st dev) percentage changes in Resting Perfusion Pressure and the Noradrenaline and Sympathetic Nerve Stimulation pressure responses

	RPP*		NORADRENALINE°		NERVE STIM§	
ACIDOSIS	+1.5	(12)	-18.0	(20)	-11.0	(32)
ALKALOSIS	+56.3	(27)	+65.5	(34)	+11.5	(22)
CONTROL	+1.2	(4)	+19 8	(18)	+35.2	(35)

*RPP   - sign diff between alkalosis and others.
°NE   - sign diff between all groups.
§SNS   - sign diff between acidosis and control.

CONCLUSIONS

It is concluded that the sensitivity of long bone resistance vessels to circulating noradrenaline and sympathetic nerve stimulation is regulated by local hydrogen ion concentration. As this is a direct effect of $[H^+]$ on bone blood vessels and only small changes in pH are required, hydrogen ion concentration may provide bone with a mechanism to synchronise local blood flow with local metabolic activity.

THE REGULATION OF BLOOD FLOW IN BONE

S.P.F. HUGHES, and I.D. McCARTHY

Orthopaedic Surgery
Royal Postgraduate Medical School
Hammersmith Hospital
London, United Kingdom

INTRODUCTION

The anatomy of the blood supply to bone has been well documented by Brookes (1971) and shown to be centrifugal in direction. Lopez-Curto et al. (1980) investigated the microstructure of the blood supply to bone, and demonstrated that the cortical and medullary vessels in the canine tibia had separate parallel supplies. In our own laboratories we have shown that the end vessel in bone, the capillary, has a continuous endothelium, similar to that seen in cardiac or skeletal muscle.

CAPILLARY PERMEABILITY

We have studied transcapillary exchange of small ions and molecules in the canine tibia, using outflow dilution techniques. The model for these experiments was the perfused nutrient artery of the dog, in which tracers were injected as a bolus into the nutrient artery, and their concentration in the venous outflow collected from the ipsilateral femoral vein measured. If one of the tracers injected is a reference vascular tracer that remains within the vascular system during the time of the experiment, then the extraction of a diffusible test tracer can be studied. Data from such experiments may be analysed mathematically using an equation derived independently by Renkin (1959) and Crone (1963):

$$PS = F_S \log_e (1\text{-}E_{max}) \qquad (1)$$

where PS is the permeability-surface area product, F_S is solute flow, and E_{max} the maximum instantaneous extraction. E_{max} was chosen as being the best indication of extraction before back diffusion occurred (Bassingthwaighte, 1974). Using this analysis, it is possible to estimate the permeability surface area product and to relate this to diffusion coefficient for the molecule.

Bone Circulation and Vascularization in Normal and Pathological Conditions
Edited by A. Schoutens *et al.*, Plenum Press, New York, 1993

By this technique we have demonstrated that transcapillary exchange is a passive event, dependent on molecular size, and unaffected by any active process or electrical charge (Hughes et al., 1977). We have also shown that at normal blood flows capillary permeability rather than blood flow limits transcapillary exchange; only at low flows does exchange become flow limited (McCarthy et al., 1980).

For Equation 1 to be applicable, the extravascular fluid space needs to be sufficiently large for there to be little back diffusion from the extravascular space to the vascular system. Measurements of fluid spaces in bone (Hughes et al., 1978; Morris et al., 1982) have shown that the total fluid space in cortical bone is about 0.28 ml/ml bone, with the extravascular space being 70 % of the total. This is sufficiently large for Equation 1 to be applicable to studies on bone.

With more sophisticated mathematical analysis and appropriate combination of test tracers, it is possible to use the outflow dilution technique to study exchange beyond the capillary endothelium. When applied to bone, it has been shown that transcapillary exchange is not the rate limiting step in the movement of ions from blood to bone (McCarthy and Hughes, 1989).

Studies of macromolecular permeability in bone have produced some surprising results. It appears that cortical bone takes up albumin much more rapidly than skeletal muscle (McCarthy et al., 1990). At the time of writing, it is unclear what proportion of this transport is coupled to volume flow within the interstitial fluid of bone.

Hence the Haversian system the basic transport system in bone consist of a capillary which is a single wall vessel, surrounded by free fluid and separated from bone crystal by bone cells which are linked to together. It seems that solutes leave the capillary at a rate dependent molecular size and then pass into the fluid space. Calcium would leave the capillary quickly because of its low molecular weight and then pass through the fluid space before reaching the bone crystal.

NEUROHUMORAL EFFECTS ON BONE

Early observations on the vascular reactivity of bone were given by Drinker and Drinker (1916). Using a model of the isolated canine tibia perfused via the nutrient artery, they showed that electrical stimulation and injection of epinephrine caused vasoconstriction, which they ascribed to vessels in bone marrow. Using a similar model, Driessens and Vanhoutte have confirmed these results and also shown that acetylcholine and calcitonin have vasoactive effects on bone, and that parathyroid hormone does not effect basal perfusion pressure (Driessens and Vanhoutte, 1979, 1981). Measurement of bone blood flow has also shown that bone vascular resistance increases during exercise and haemorrhage (Gross et al., 1979). Recently, the presence of alphaladrenoceptors, muscarinic receptors, and prostaglandin H_2/thromboxane A_2 receptors in the vasculature of bone were shown (Brinker et al., 1990).

The effects of changes in bone blood flow on mineral exchange have been studied (Cochrane and McCarthy, 1992). They examined the change in blood flow, vascular resistance, and strontium-85 clearance in the rat bone after pharmacological injection of noradrenaline and ATP. These experiments were able to demonstrate that increased doses of noradrenaline increased the vascular resistance in bone and reduced the bone flow as measured by the microsphere technique. On the other hand, using a vasodilator such as ATP the flow was also decreased to bone, as systemic arterial pressure decreased to a

Table 1. The effects of systemic injection of ATP and noradrenaline on bone blood flow, bone vascular resistance, and systemic arterial blood pressure

	ATP(1.25 mg)	Noradrenaline(2.0 μg)
Change in Blood Flow (ml/min/g)	-0.11 ± 0.14	-0.13 ± 0.04
Change in Vascular Resistance	-80 ± 250	442 ± 208
Change in B.P. (mm Hg)	-55 ± 16	39 ± 13

much greater degree than the vascular resistance of bone. This confirms observations in other papers that bone is relatively insensitive to vasodilators, but actively responds to vasoconstrictor mechanisms (Brinker et al., 1990) (Table 1).

FRACTURE

Turning to the events that occur after a fracture into bone, it is known that bone blood flow increases after a fracture (Paradis and Kelly, 1975). We have been able to report these findings in the canine tibia after a controlled osteotomy and using microspheres to measure bone blood flow.

In experiments conducted on the osteotomised canine tibia, it was found that, even if the bone blood flow increased following a fracture, the instantaneous and net extraction of the solute strontium did not change (Hughes et al., 1979). This suggested that an increase in bone blood flow led to a larger surface area available for exchange or that the capillaries became more permeable. From Equation 1, this showed that the product of capillary permeability/surface area (PS) had remained unchanged. It seems likely that the capillary surface area available for exchange had increased by full vasodilatation of the vascular bed and formation of new vessels.

BLOOD FLOW AND BONE REMODELLING

There is evidence of a relationship between bone blood flow and bone remodelling. Studying dogs with different degrees of remodelling (hypoparathyroidism, hypo-thyroidism, hyperparathyroidism, and hyperthyroidism), it was shown that osseous blood flow was closely related with remodelling activity (Sim and Kelly, 1970). Whiteside et al. examined the differences in remodelling and blood flow in cancellous and cortical bone, and concluded that the two were related (Whiteside et al., 1977). Clinical studies of bone blood flow using fluorine-18 have shown a statistically significant correlation between osteoblast work rate and fluorine-18 estimates of skeletal blood flow (Reeve et al., 1988). Recently, McCarthy and Yang (1992) have compared the heterogeneity of blood flow and remodelling within the tibia, and again found a significant correlation (Figure 1).

The most simple explanation for these observations is that cortical bone is capable of autoregulation of blood flow, responding to local metabolic needs. However, Trueta (1963) suggested that vessels have an active role in osteogenesis. There is now increasing evidence of interaction between endothelial cells and bone (MacIntyre et al., 1991; Jones et al. 1992; Streeten and Brandi, 1990). It is possible that the relationship between bone blood flow and bone remodelling is complex, and much interesting work remains to be done in this area.

MINERAL EXCHANGE

If bone blood flow is able to autoregulate, is there any other factor that controls mineral uptake in the bone, such as a membrane with bone itself? McCarthy and Hughes (1986) examined the effect of potassium cyanide injection on mineral extraction in the canine tibia model. It was found that the maximum instantaneous extraction was unaffected for both strontium-85 (a calcium analogue) and rubidium-86 (a potassium analogue), showing capillary permeability to be unchanged. However, net extraction of rubidium-86 was decreased, but the net extraction of strontium-85 was significantly increased. These experiments indicate that the volume of distribution of calcium analogues depends on cellular metabolic activity, and shows that the metabolic activity, is directed towards moving the mineral ions out of bone. These observations agree with recent studies on the effect of PTH and PGE2 on the early uptake of calcium by bone (Shaw et al., 1989).

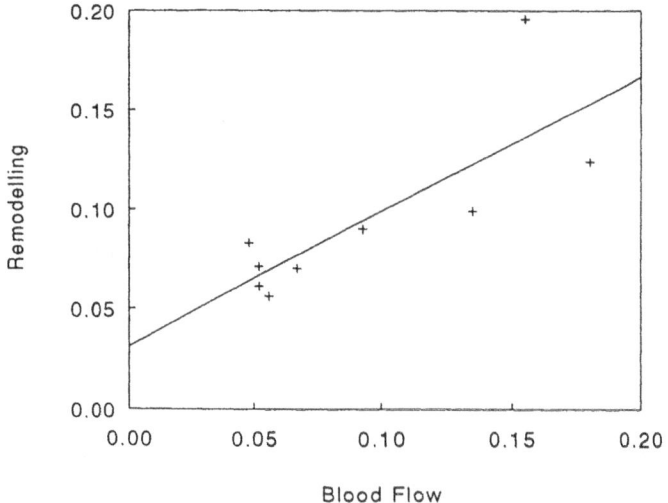

Figure 1. The relationship between bone remodelling (as measured by tetracycline labelling) and bone blood flow (as measured by microspheres) in cortical bone of the canine tibia.

CONCLUSIONS

Bone is a metabolically active structure. Small solutes leave the capillaries in bone by the process of passive free diffusion and pass across the fluid spaces in bone to reach the osseous tissue. There is evidence that extravascular movement of large macromolecules may be assisted by convective flow of the bone fluid.

Blood vessels in bone have been shown to respond to many vasodilators and vasoconstictors, though the magnitude of the responses indicates that bone is relatively insensitive to vasodilators.

Progress has been made in understanding the regulation of blood flow in bone. Exciting directions for future research are in understanding how the vascular system in bone may regulate activity in bone itself.

REFERENCES

Bassingthwaighte, J.B.: A concurrent flow model for extraction during transcapillary exchange. Circ. Res. 35: 483-503 (1974).

Brinker, M.R., H.L. Lippton, S.D. Cook, A.L. Hyman: Pharmacological regulation of the circulation of bone. J Bone Joint Surg. 72-A: 964-975 (1990).

Brookes, M.: The Blood Supply of Bone. Butterworths, London 1971.

Cochrane, E., I.D. McCarthy: Rapid effects of parathyroid hormone (1-34) and prostaglandin E_2 on bone blood flow and strontium clearance in the rat in vivo. J Endocr. 131: 359-365 (1991).

Lopez-Curto, J.A., J.B. Bassingthwaighte, P.J. Kelly: Anatomy of the microvasculature of the tibial diaphysis of the adult dog. J. Bone Joint Surg. 62A: 1362-1369 (1980).

Crone, C.: The permeability of capillaries in various organs as determined by use of the indicator diffusion method. Acta Physiol. Scand. 58: 292-305 (1963).

Driessens, M., P.M. Vanhoutte: Vascular reactivity of the isolated tibia of the dog. Am. J. Physiol. 236: H904-H908 (1979).

Driessens, M., P.M. Vanhoutte: Effect of calcitonin, hydrocortisone, and parathyroid hormone on canine bone blood vessels. Am. J. Physiol. 241: H91-H94 (1981).

Drinker, C.K., K.R. Drinker: A method for maintaining an artificial circulation throught the tibia of the dog, with a demonstration of the vasomotor control of the marrow vessels. Am. J. Physiol. 40: 514-521 (1916).

Gross, P.M., D.D. Heistad, M.L. Marcus: Neurohumoral regulation of blood flow to bones and marrow. Am. J. Physiol.237: H440-H448 (1979).

Hughes S.P.F., D.R. Davies, J.B. Bassingthwaighte, F.G. Knox, P.J. Kelly: Bone extraction and blood clearance of diphosphonate in the dog. Am. J. Physiol. 232: H341-H347 (1977).

Hughes, S.P.F., D.R. Davies, R. Khan, P.J. Kelly: Fluid space in bone. Clin. Orthop. Rel. Res. : 400-409 (1978).

Hughes, S.P.F., G.J. Lemon, D.R. Davies, J.B. Bassingthwaighte, P.J. Kelly: Extraction of minerals after experimental fractures of the tibia in dogs. J. Bone Joint Surg. 61-A: 857-866 (1979).

Jones, A.R., C.C. Clark, C.T. Brighton: The effects of rat microvascular cell products on cultured rat calvarial bone cells. Trans. Orth. Res. Soc. 17: 188 (1992).

McCarthy, I.D., J.S. Orr, S.P.F. Hughes: An experimental model to study the relationship between blood flow and uptake for bone-seeking radionuclides in normal bone. Clin. Phys. Physiol. Meas 1: 135-143 (1980).

McCarthy, I.D., S.P.F. Hughes: Inhibition of bone cell metabolism increases strontium-85 uptake. Calcif. Tiss. Int. 39: 386-389 (1986).

McCarthy, I.D., S.P.F. Hughes: Multiple tracer studies of bone uptake of ^{99m}TcMDP and ^{85}Sr. Am. J. Physiol. 256: H1261-H1265 (1989).

McCarthy, I.D., J.T. Bronk, P.J. Kelly: The measurement of interstitial fluid flow in cortical bone. FASEB Journal 4: A1262 (1990).

McCarthy, I.D., L. Yang: The relationship between bone blood flow, strontium clearance, and bone turnover in the tibial cortex. Trans. Orth. Res. Soc. 17: 11 (1992).

MacIntyre, I., M. Zaidi, A.S.M. Towhidul Alam, H.K. Datta, B.S. Moonga, P.S. Lidbury, M. Hecker, J.R. Vane: Osteoclastic inhibition: an action of nitric oxide not mediated by cyclic GMP. Proc. Natl. Acad. Sci. 88: 2936-2940 (1991).

Morris, M.A., J.A. Lopez-Curto, S.P.F. Hughes, K-N. An, J.B. Bassingthwaighte, P.J. Kelly : Fluid spaces in canine bone and marrow. Microvasc. Res. 23: 188-200 (1982).

Paradis, G., P.J. Kelly: Blood flow and mineral deposition in canine tibial fractures. J. Bone Joint Surg. 57-A: 220-226 (1975).

Reeve, J., M. Arlot, R. Wootton, C. Edouard, M. Tellez, R. Hesp, J.R. Green, P.J. Meunier: Skeletal blood flow, iliac histomorphometry, and strontium kinetics in osteoporosis: a relationship between blood flow and corrected apposition rate. J. Clin. Endocr. Metab. 66: 1124-1131 (1988).

Renkin, E.M.: Transport of potassium-42 from blood to tissue in isolated mammalian skeletal muscles. Am. J. Physiol. 197: 1205-1210 (1959.)

Shaw, A.J., G. Whittaker, C.G. Dacke: Kinetics of rapid ^{45}Ca uptake into chick skeleton in vivo: effects of microwave fixation. Quart. J. Exp. Physiol. 74: 907-915 (1989).

Sim, F.H., P.J. Kelly: Relationship between bone remodelling, oxygen consumption, and blood flow in bone. J. Bone Joint Surg. 52-A: 1377-1389 (1970).

Streeten, E.A., M.L. Brandi: Biology of bone endothelial cells. Bone and Mineral 10: 85-94 (1990).

Trueta, J.: The role of vessels in osteogenesis. J. Bone Joint Surg. 45-B: 402-418 (1963).

Whiteside, L.A., D.J. Simmons, P.A. Lasker: Comparison of regional bone blood flow in areas of differing osteoblastic activity in the rabbit tibia. Clin. Orthop. Rel. Res. 127: 267-270 (1977).

METHODS OF INVESTIGATION

MEASUREMENT OF BONE BLOOD FLOW IN ANIMALS

Peter TOTHILL

Department of Medical Physics
University of Edinburgh
Edinburgh, United Kingdom

INTRODUCTION

It is more difficult to measure blood flow to the skeleton than to any other organ or tissue in the body because of the complex arterial supply and venous drainage of each bone and the heterogeneity of cortical and trabecular bone and bone marrow. Most techniques are invasive and some are lethal, so that they cannot be applied to human studies. On the other hand, animal experiments can give an insight into the metabolism and growth of bone and such topics as fracture healing. The methods to be considered here are (1) indicator fractionation, using radioactive microspheres, (2) the clearance of bone-seeking tracers, (3) the washout of diffusible tracers and (4) laser Doppler measurements. The methods are the same in principle as those used for other organs or tissues, but the skeleton does pose particular problems.

MICROSPHERE TECHNIQUES

The indicator fractionation technique depends on introducing into the circulation particles of such a size that they lodge in capillaries or arterioles. The markers are then distributed in proportion to the blood flow. Radioactive labelling of the particles simplifies their quantitation. Most commonly, plastic microspheres, which are available commercially labelled with a variety of radionuclides, are employed. The particles must be representative of blood flow in respect of behaviour at junctions and arteriovenous shunting, so size is important. It is mostly accepted that 15 μm diameter microspheres are satisfactory for the measurement of bone blood flow, although evidence has been presented that smaller ones may be used. Thorough mixing of the particles with the blood is important and the best results are achieved by injection into the left atrium of the heart. However, this is not easy, and left ventricle injection is usually considered acceptable.

The number of particles injected is influenced by two, possibly conflicting, requirements. The injection should not influence the blood flow itself or any other aspect

Bone Circulation and Vascularization in Normal and Pathological Conditions
Edited by A. Schoutens *et al.*, Plenum Press, New York, 1993

65

of the functioning of the tissue. This requires that the numbers should be such that the proportion of vessels occluded should be small. On the other hand, statistical certainty requires that there should be a reasonable number of microspheres in the specimen examined. A coefficient of variation of 5 % requires about 400 particles. If flow to segments of bone is being investigated, several million particles may need to be injected into the heart. The statistics of radioactivity measurement also need to be considered, but are not usually a limitation if efficient detection is used and there has not been too much decay.

If microspheres are introduced into the heart, the proportion lodging in any tissue is equal to the fraction of the cardiac output flowing to that site. This may be sufficient information for some studies. If cardiac output is determined, flow can be expressed in ml/min. When microspheres are injected into the heart, the reference organ technique can be used, blood being drawn into a syringe at a known rate from a cannulated artery and its radioactivity measured.

$$\text{Then blood flow} = \frac{\text{microsphere activity in bone x pump rate}}{\text{microsphere activity in blood}} \tag{1}$$

It is usual to express results as specific flow, that is the flow divided by the weight of the bone, commonly in the units ml/min per 100 g.

Frequently a carotid artery approach is used for the placement of a catheter in the heart and the reference syringe is connected to a femoral artery, but other combinations are used. It is apparent that quite complicated procedures are required, but results from a large number of bones, or indeed other organs, can be acquired from one injection. Catheter placement has to be under anaesthetic, but the actual measurements may be carried out after recovery, at least in the dog.

A valuable characteristic of the microsphere technique is the ability to study changes in circulation, using sequential injections with different radionuclide labels that can be separated by gamma-ray spectroscopy. It is necessary to correct for crossover between counting channels, but up to four or five tracers could be used in one experiment.

Examples of Microsphere Experiment Results

The microsphere technique is pre-eminent in animal experimentation and there have been many reports of bone blood flow measurements. Taking examples from our own experience, Tothill and MacPherson (1986) found that the total skeletal blood flow in dogs averaged 7.5 % of the cardiac output, in rabbits 4.2 % and in rats 4.5 %. Specific blood flow to parts of the skeleton in rats and dogs is illustrated in Table 1 (Tothill, 1986). This demonstrates that specific flow differs greatly between bones.

The use of multiple injections and the determination of the distribution of blood flow within an individual bone can be illustrated by the results of an experiment to examine the effect of sympathectomy on bone blood flow (Davies et al., 1984). [57]Co labelled microspheres were used to assess the blood flow to the tibiae of dogs before the femoral and sciatic nerves on one side were cut and [113]Sn microspheres after the sympathectomy. The distribution of flow within the bones has subsequently been examined (Tothill et al., 1987). Figure 1 shows the distribution of blood flow to sections of each tibia, the flow being higher at the ends than in the shaft. The overall reduction of blood flow was the same on each side and so not due to sympathectomy. The relative flow distribution between the bone segments did not change. Similar experiments in which one nutrient artery was severed allowed deductions about the proportion of flow contributed by different routes. Although flow per unit mass to the marrow was higher than that to bone, the proportion of cardiac output was much lower.

Table 1. Specific Blood Flow to Parts of Skeleton in Rats With Weights 400-650 g and Adult Dogs.

	Rats (n - 4)	Dogs (n - 2)
Tibiae	16 ± 10	5 ± 3*
Femora	16 ± 7	11 ± 6*
Radii + ulnae	15 ± 8	5 ± 2*
Humeri	29 ± 15	13 ± 6*
Scapulae	31 ± 14	12 ± 5
Skull	25 ± 19	14 ± 0
Lower jaw	40 ± 11	1 ± 0
Spine	28 ± 8	26 ± 10
Pelvis	22 ± 8	14 ± 4
Ribs + sternum	27 ± 11	24 ± 8
Total skeleton	25 ± 9	13 ± 4

Mean values are in mL min^{-1} 100 g^{-1} $\pm$ SD.
* For dog limb bones, n = 9.

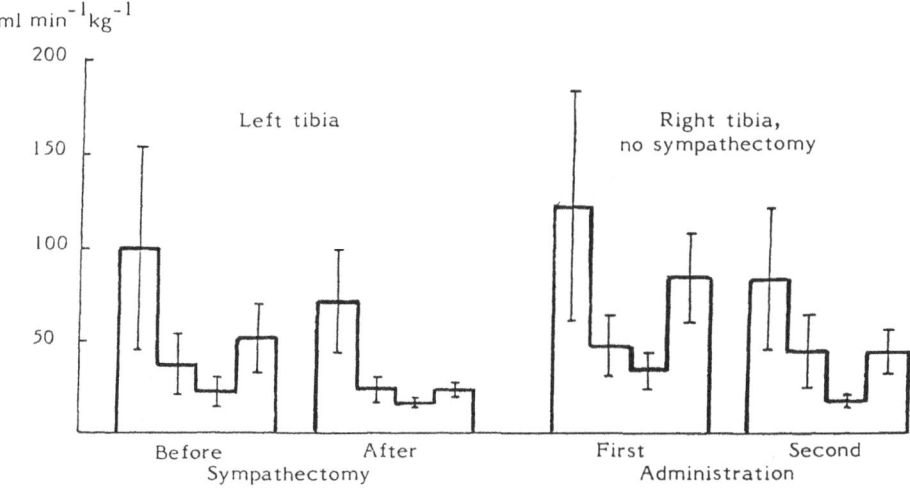

Figure 1. Mean distributions of specific blood flow, with standard errors, in four equal bone sections following systemic injection of microspheres in five dogs. Proximal end of tibia to left.

CLEARANCE OF BONE-SEEKING TRACERS

This technique relies on the Fick principle, namely that the amount of tracer accumulating in a tissue is the difference between input and output. Arterial concentration can be sampled, but the complex drainage from bone precludes the collection of all venous blood, so tracer outflow concentration cannot be measured and one element of the equation is unknown. It has commonly been assumed that the extraction by bone of such tracers as ^{47}Ca, ^{85}Sr or ^{18}F is complete, or at least constant. However, it has become clear that the extraction of diffusible tracers depends on the flow rate. As the flow increases, the transit time along a capillary becomes shorter than the time needed for the tracer to diffuse through the capillary wall.

Single passage extraction measurements suffer from the difficulty that the maximum transit time through bone is greater than the minimum recirculation time. However, the microsphere technique has been used to measure extraction during recirculation, when back diffusion of the tracer can occur. When using the reference organ method, the clearance of ^{85}Sr (for example) can be defined as =

$$\frac{^{85}\text{Sr activity in bone x pump rate}}{^{85}\text{Sr activity in blood}} \qquad (2)$$

The net extraction of ^{85}Sr equals clearance (from equation 2) divided by blood flow (from equation 1).

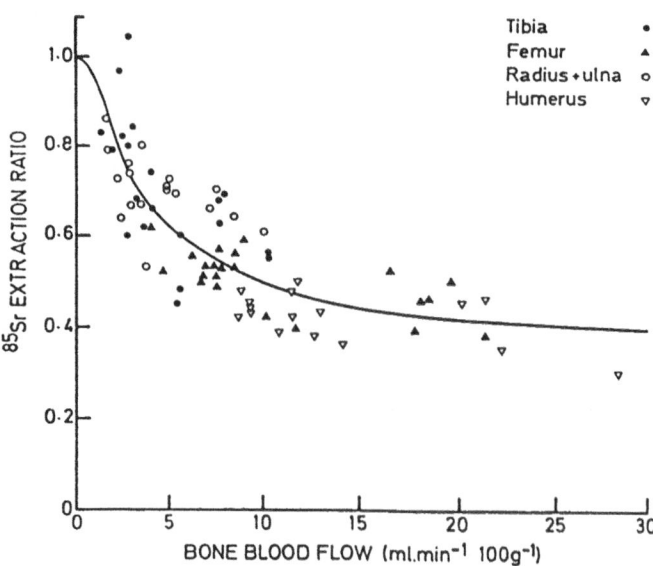

Figure 2. Variation of extraction ratio of ^{85}Sr with blood flow during five-minute recirculation period in dog limb bones.

Schoutens et al. (1979) found evidence of a reduction of ^{45}Ca extraction with increasing flow rate. Tothill et al. (1985) showed that the net extraction of ^{85}Sr in five minutes of recirculation in dog limb bones fell from near unity at low flows to approximately 0.4 at higher flows (Figure 2). Similar results were obtained for ^{18}F and ^{99m}Tc MDP.

Although the variable extraction of bone-seeking tracers throws doubt on quantitative measurement of skeletal blood flow, such tracers can be used non-invasively and can provide qualitative information in humans. Perhaps more importantly, the combined use of microspheres and mineral clearances in animal experiments can enlighten such subjects as the effect of vasoactive agents on the exchange of mineral ions between blood and bone (Cochrane and McCarthy, 1991).

WASHOUT OF DIFFUSIBLE TRACERS

This technique depends on the introduction of a diffusible tracer into the tissue under investigation and the observation of the rate of removal, which is dependent on blood flow. If a radioactive tracer is used, the observation can be performed by a detector outside the body, although interference from other tissues may then be a problem. Best results are obtained using a lipophilic tracer that readily crosses cell membranes. The flow per unit weight of tissue is taken to equal the rate constant of the exponential fall of concentration times the partition coefficient, defined as the solubility of tracer in the tissue relative to that in blood. Ideally, the tracer should be introduced via an artery, but no bone has a single route. Kelly et al. (1971) successfully used ^{125}I-labelled iodoantipyrine introduced through a nutrient artery to measure tibial blood flow in the dog, but the heterogeneous nature of the tissue led to multi-exponential disappearance curves.

Recirculation can cause errors, but can be minimised by the use of a poorly soluble gas, such as ^{133}Xe, which is released during passage through the lungs.

Non-radioactive tracers involve a greater degree of invasiveness in detection. Whiteside et al. (1977) used inhaled hydrogen gas as the tracer and subsequently demonstrated the effects of periosteal stripping and medullary reaming on regional blood flow (Whiteside et al., 1978). The technique still finds application; Takahashi et al. (1990) used it to investigate the acute effect of spinal paralysis on regional bone blood flow in the rat. A small hole is drilled into the bone and a platinum electrode inserted to measure hydrogen concentration. Only a very small volume of bone is examined.

LASER DOPPLER FLOWMETRY

Laser flowmetry uses the Doppler principle that light reflected from a moving object undergoes a frequency shift that is proportional to the difference in velocity between the light source and the object. For blood flow measurements the moving objects are the red blood cells. The processed output signal is proportional to the number of moving scatterers within the illuminated tissue and their velocities and is usually expressed in mV, with no attempt to derive quantitative flow. The technique was first used to assess blood flow in skin (Nillson et al., 1980), but was applied to study blood flow in the cancellous bone of the pig mandible by Hellem et al. (1983) and has since found other applications. The method requires exposure of the bone but has been used in human patients at operation. The volume of blood examined is determined by the penetration of the light into the tissue. Studies of bovine bone by Notzli et al. (1989) showed that the maximum depth at which the probe could evaluate flow was 2.9 mm in cortical bone and 3.5 mm in trabecular bone. The laser probe can be applied to the surface of the bone, when predominantly cortical bone is examined. Readings can be made through articular cartilage, which avoids problems due to bleeding. If trabecular bone is to be assessed, the probe is placed in a drilled hole. In either case, it is necessary to clamp the probe in position.

Comparisons between laser Doppler flowmetry (LDF) and other methods of measuring bone blood flow have not shown agreement, perhaps because something different is being measured. Hellem et al. (1983) found no correlation between LDF and ^{133}Xe clearance in the pig mandible. Swiontkowski et al. (1986) compared LDF and microsphere measurements in rabbit femurs and again found no significant correlation.

Although LDF provides only qualitative results, it is sufficiently non-invasive to find application in human studies. For example, Wannfors and Gazelius (1991) studied circulatory changes in jaw bones affected by chronic osteomyelitis.

CONCLUSIONS

When quantitative results are required and in circumstances when the experimental animal can be killed, there is no doubt that the microsphere technique is the method of choice for measuring bone blood flow. Less invasive methods are being developed, particularly using positron-emission tomography, for human studies and of course these may find application for animal work

REFERENCES

Cochrane, E., I.D. McCarthy: Rapid effects of parathyroid hormone (1-34) and prostaglandin E2 on bone blood flow and strontium clearance in the rat in vivo. J. Endocrinol. 131: 359-365 (1991) .

Davies, R., P. Tothill, G. Hooper, R.H. Fleming, I.D. Mccarthy, S.P.F. Hughes: The early effects of sympathectomy on bone blood flow. Calcif. Tissue Int. 36: 622-625 (1984).

Hellem, S., S. Jacobsson, G.F. Nilsson, D.H. Lewis: Measurement of microvascular blood flow in cancellous bone using Doppler flowmetry and ^{133}Xe-clearance. Int. J. Oral Surg. 12: 165-177 (1983).

Kelly, P.J., T. Lipintsoi, J.B. Bassingthwaighte: Blood flow in canine tibial diaphysis estimated by iodoantipyrine ^{125}I washout. J. Appl.Physiol. 31: 38-47 (1971).

Nilsson, G.E., T. Tenland, P.A. Oeberg: Evaluation of a laser Doppler flowmeter for measurement of tissue blood flow. IEEE Trans. Biomed. Eng. 27: 597-604 (1980).

Notzli, P., M.F. Swiontkowski, S.T. Thaxter, G.K Carpenter III, R. Wyatt: Laser Doppler flowmetry for bone blood flow measurements: helium-neon laser light attenuation and depth of perfusion assessment. J. Orthop. Res. 7: 413-424 (1989).

Schoutens, A., P. Bergmann, M. Verhas. Bone blood flow measured by [85]Sr microspheres and bone seeker clearances in the rat. Am. J. Physiol. 236: Hl-H6 (1979).

Swiontkowski, M.F., S. Tepic, S.M. Perren, R. Moor, R .Ganz, B.A. Rahn. Laser Doppler flowmetry for bone blood flow measurement: correlation with microsphere estimates and evaluation of the effect of intracapsular pressure on femoral head blood flow. J. Orthop. Res. 4: 362-371 (1986).

Takahashi, H., T. Yamamuro, H. Okumura, R. Kasai, K. Tada. Bone blood flow after spinal paralysis in the rat. J. Orthop. Res. 8: 393-400 (1990).

Tothill, P. Methods of assessing blood flow in bone. Seminars in Orthopaedics 1: 138-146 (1986).

Tothill, P., G. Hooper, I.D. McCarthy, S.P.F. Hughes. The variation with flow rate of the extraction of bone-seeking tracers in recirculation experiments. Calcif.Tissue Int. 37: 312-317 (1985).

Tothill, P., G. Hooper, I.D. McCarthy, S.P.F. Hughes. The pattern of distribution of blood flow in dog limb bones measured using microspheres. Clin. Phys. Physiol. Meas. 8: 239-247 (1987).

Tothill, P., J.N. McPherson. The distribution of blood flow to the whole skeleton in dogs, rabbits and rats measured with microspheres. Clin. Phys. Physiol. Meas. 7: 117-123 (1986).

Wannfors, K., B. Gazelius. Blood flow in jaw bones affected by chronic osteomyelitis. Br. J. Oral Maxillofac. Surg. 29: 147-153 (1991).

Whiteside, L.A., P.A. Lesker, D.J. Simmons. The measurement of regional bone and bone marrow blood flow in the rabbit using the hydrogen washout technique. Clin. Orthop. 122: 340-346 (1977).

Whiteside, L.A., K. Ogata, P. Lesker. The acute effects of periosteal stripping and medullary reaming on regional bone blood flow. Clin. Orthop. 131: 266-272 (1978).

ARTERIOLAR BLOCKADE REVISITED: COMPARISONS BETWEEN THE USE OF RESIN PARTICLES AND MICROSPHERES FOR BONE HAEMODYNAMIC STUDIES

W.J. REVELL and M. BROOKES

Academic Department of Orthopaedics
Rayne Institute, St.Thomas' Hospital
London, United Kingdom

INTRODUCTION

The term "arteriolar blockade" was first used by Brookes (1970) to denote a method of blood flow measurement whereby arterioles are blockaded by labelled particles introduced into the arterial system. The principle of the method is that if a large number of particles of appropriate size are injected into the left ventricle, they will lodge in the arteriolar beds in proportion to the fraction of the cardiac output perfusing a given tissue or organ. If these particles are labelled isotopically, then the proportion of cardiac output delivered to the tissue is given by the ratio of tissue counts to the total counts injected.

Modern microspheres have a narrow size distribution and are resistant to isotope leeching in physiological fluids. Their specific gravity however is still more than that of red blood cells. They are expensive, and this factor may deter smaller laboratories from undertaking haemodynamic studies. On the other hand, cationic exchange resin particles are cheap and easily labelled. They have been used in many studies (Brookes 1970; Brookes and Gallennaugh 1975; Tothill and McCormick 1976; Brueton, Revell and Brookes 1991), often in situations where isotope activities in a limb, following a surgical procedure, are compared with the contralateral unoperated limb. Particles were introduced via the right carotid artery, but apart from Tothill and McCormick, injections were not made into the left ventricle; absolute flow rates have not been determined using resin particles.

This study aims to compare blood flow rates in whole bones and bone segments measured with cationic exchange resin labelled with ^{59}Fe, and microspheres labelled with ^{85}Sr. The exchange resin used was from the same batch, and prepared in the same way, as the particles used by Brookes and his associates. The current investigation also aims to characterise the properties of the resulting tracer, in terms of size distribution, mass and number of particles injected. The overall aim was to develop and validate a cheap and routine method for measuring skeletal haemodynamics in the rat.

Bone Circulation and Vascularization in Normal and Pathological Conditions
Edited by A. Schoutens *et al.*, Plenum Press, New York, 1993

73

MATERIALS AND METHODS

^{59}Fe Resin particles: A quantity (0.47 g) of cation exchange resin particles (Amberlite CG120; BDH Ltd, UK) were shaken with 100 ml distilled water in a measuring cylinder, and allowed to stand for 10 minutes. The liquid suspension was carefully decanted from the sediment, which contained the largest particles. This liquid was allowed to stand for a further 1 hour. After this time the supernatant, containing only fine resin "dust" was discarded. The precipitate was re-suspended in 10 ml distilled water and shaken mechanically with 120 μCi ^{59}Fe ferric chloride solution (Amersham International) for 30 minutes. The particles were repeatedly washed with distilled water and centrifuged at 3000 rpm until only background counts were recorded in the supernatant. The ^{59}Fe labelled resin particles were finally suspended in 5 ml of distilled water with 0.01 % of Tween 80.

Samples of the suspension were spread on a glass slide and the long axes of the spindle shaped particles measured using a Digithurst Microscale image analysis system. A particle size distribution was plotted from this data. Further batches were dried to constant weight at 45°C to obtain the weight of particles injected per unit volume. The suspension was also examined in a Coulter size analyser to determine the number of particles per unit volume.

^{85}Sr microspheres: These were NENTRAC (Dupont) microspheres, 15.5 μm ±0.1 μm in size, resin coated after ^{85}Sr labelling, to minimize isotope leeching. They were suspended in 0.9 % mammalian saline to which 0.01 % Tween 80 was added to minimise aggregation. Two batches were used to determine blood flow; 1 mg per rat, and 3 mg per rat, equivalent to 350,000 and 1,050,000 microspheres per animal respectively. The spheres were suspended in 0.5 ml saline-Tween 80 for each injection. Samples of microspheres were streaked across graph paper and counted microscopically. ^{85}Sr activity was determined in a gamma counter. This enabled a plot to be made of the specific activity per sphere, in order to estimate the number of microspheres residing in tissue samples.

Eleven week old male Wistar rats (weight 270 ± 18 g) were used, 10 in each group. Anaesthesia was by intramuscular Hypnorm (0.1 ml), followed by an intraperitoneal injection of Diazepam (0.3 ml). The appropriate Home Office Regulations were observed throughout. The left and right carotid arteries were exposed by sectioning the overlying sternomastoid and omohyoid muscles. A flexible nylon catheter (0.63 mm o.d.) was introduced into the right carotid. The catheter was connected via a three way tap to a Camino fibre optic pressure transducer, linked via a differential amplifier to a digital storage oscilloscope. This compact arrangement enabled intra-arterial pressure changes to be monitored. The catheter was advanced until the tip was positioned in the left ventricle; the location was unequivocally identified by observing the characteristic change in the pressure waveform on entering the heart.

Another catheter (1.02 mm o.d.) was placed in the left carotid artery. This was connected to a Harvard pump, calibrated to withdraw a blood reference flow sample into a 1 ml syringe at a rate of 1ml.min^{-1}. Particle suspensions were ultrasonicated at 40° C for 30 minutes and then agitated on a vortex mixer just before injection. The withdrawal pump was started and allowed to run for 10 seconds, at which point blood was seen to enter the syringe. Injection of particles was then made over a 20 second period, followed by a 0.5 ml saline flush, again taking 20 seconds. Withdrawal continued for a further 10 seconds. The total withdrawal period was therefore 1 minute, during which 1 ml of blood was taken, and a total of 1 ml of particles/saline injected. The blood sample was ejected into 2 ml of 1 % ammonium hydroxide in distilled water, to ensure haemolysis. The animal was then terminated by an injection of barbiturate through the intracardiac catheter.

The tibiae and femora were removed from both limbs, weighed, stripped of soft tissues, and placed separately in scintillation counting vials. In the case of animals receiving 1 mg microspheres per rat, the humeri were also removed. Reference blood samples and bones were placed in racks, and the activity of the contained ^{59}Fe or ^{85}Sr measured on an LKB 1582 Compugamma automatic scintillation counter. After subtracting the background count, and correcting for height in the tube (Heyman et al., 1977) the measured activities were used to calculate the absolute flow rate through the bone specimens :

$$\text{Flow} = \frac{\text{activity in bone x reference pump rate}}{\text{reference artery count}} \tag{1}$$

In this case the reference pump rate was 1 ml.min^{-1}
Therefore, standardising for bone weight;

$$\text{Flow rate per 100 g bone} = \frac{\text{bone counts}}{\text{(reference artery count x bone weight)}} \times 100 \tag{2}$$

All calculations were performed using a Unistat statistical spreadsheet.

The distal femoral epiphysis and metaphysis and the proximal tibial epiphysis and metaphysis were separated from each of the paired bones. A segment of diaphysis was also isolated; the marrow was extruded using a blunt probe and the cortical bone cleaned with a pressurised water jet. All samples were weighed, placed in 2 ml buffered formalin, and the activity counted as before. The reference blood samples were re-counted with the bone segments; therefore it was not necessary to allow for isotope decay. Flow rates in the bone segments were calculated as before.

One of the assumptions of the method is that tracer particles in a sample drawn from the reference artery is representative of tracer delivery to the region of interest. In 10 rats, simultaneous withdrawals were made with two Harvard pumps calibrated at 1ml.min-1, from the left carotid and femoral arteries. Identical 0.63 mm diameter catheters were used in each case, and injections of ^{85}Sr microspheres into the left ventricle were made using the same technique and time schedule as before. The activity in the two arterial samples were compared.

Samples of blood were taken from the femoral vein after injection, to determine the extent of any re-circulation of particles or spheres, or leeching of isotope. The microspheres are coated by the manufacturers with resin to prevent leeching. It has been suggested however that plasma transferrin has a higher affinity for iron than exchange resin (Tothill and McCormick, 1976), and that the addition of unlabelled iron to the resin particles would overcome this problem, by competitive binding. Therefore, a batch of resin particles was prepared as above, but with the addition of 2 mg unlabelled ferric chloride, during the labelling procedure. Three rats were injected with these particles, and the femoral vein activity determined. The resin used in the current blood flow measurements was not treated with unlabelled iron, as it was intended to compare and evaluate published results of the use of cation exchange resin, where this treatment was not performed.

RESULTS

Resin Particle Characteristics

The particles are irregular in shape. The long axis of 1004 particles was measured. The mean length was 32.28 μm ± 15.5 μm. The size distribution is shown in Figure 1. The dry weight of particles in 5 ml injectate (5 ml) was 14 mg ± 2.4 mg (n=4). As 0.5 ml of this suspension was injected into each rat, it follows that each animal receives approximately 1.4 mg of particles. The number of particles in the prepared suspension, measured on a Coulter analyser, was 800,000 ± 73,000 (n=4) particles per ml, in the size range 10-70 microns. This suggests that each animal received in the order of 400,000 particles per injection.

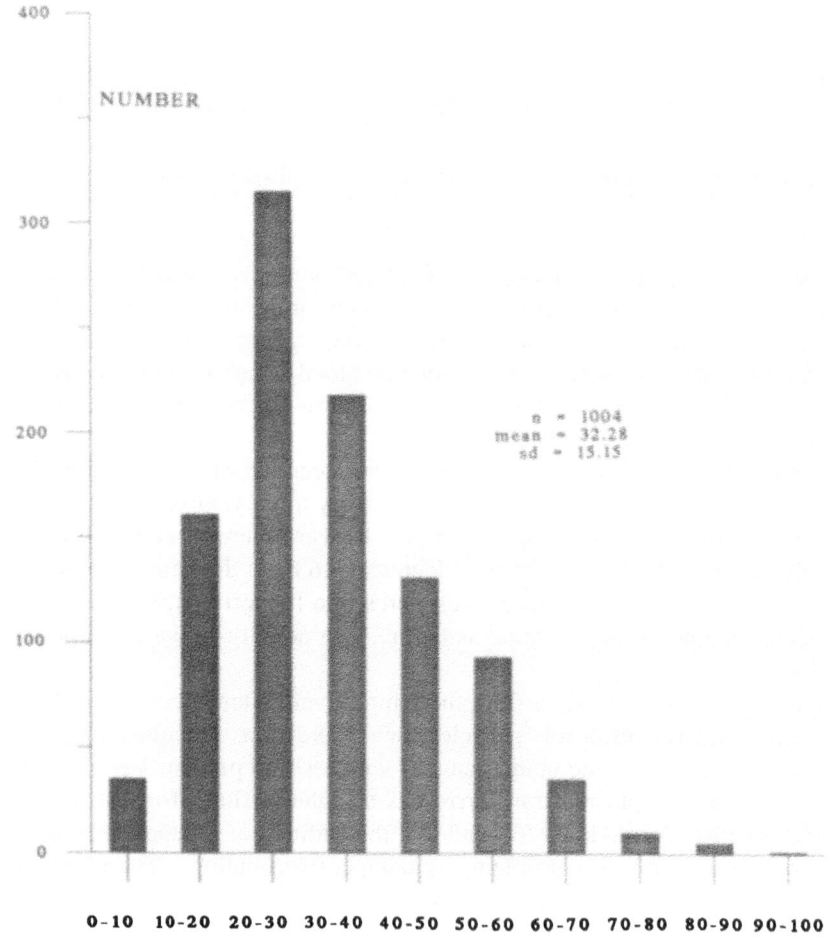

Figure 1. Size distribution of resin particles.

Blood samples (0.3 ml) from the femoral vein after injection of ^{59}Fe labelled resin particles revealed a large number of counts above background; $2,467 \pm 688$ per 0.3 ml, or 8,223 counts per ml. This suggests a high level of leeching to plasma. As a comparison, the number of counts recorded for the reference blood sample was $26,550 \pm 4,875$ per ml. This suggests an upper limit of 30 % of the bound iron isotope being extracted by plasma. Femoral vein samples from animals injected with resin particles labelled with ^{59}Fe in the presence of unlabelled iron, showed no significant activity above background; leeching was effectively eliminated.

Blood Flow Determinations

The blood flow values for 3 categories of tracer particles have been calculated for the following groups:

- ^{59}Fe labelled cationic exchange resin (1.4 mg/400,000 particles per rat);
- ^{85}Sr labelled microspheres (1 mg/350,000 spheres per rat);
- ^{85}Sr labelled microspheres (3 mg/1,050,000 spheres per rat).

The blood flows in $ml.min^{-1}100\ g^{-1}$ for left and right whole femora and tibiae are given in Table 1, and for bone segments in Table 2. Flow rates for left and right humeri, using 1mg microspheres, are also shown in Table 1. Left and right values were compared, and no significant difference was found between any pair, using a matched paired t test.

Table 1. Whole bone blood flow (ml/min/100g). Left vs right

Fe59/resin

L.FEMUR	S.D.	< p >	R.FEMUR	S.D.
34.6	12.3	0.65	36.7	19
L.TIBIA	S.D.	< p >	R.TIBIA	S.D.
32.7	10.2	0.3	30.9	10.6

Sr85/microspheres (1mg)

L.FEMUR	S.D.	< p >	R.FEMUR	S.D.
39.3	18.4	0.85	40	11
L.TIBIA	S.D.	< p >	R.TIBIA	S.D.
35.3	13.6	0.09	36.8	14.3
L.HUMERUS	S.D.	< p >	R.HUMERUS	S.D.
36	11.6	0.06	40.1	12.8

Sr85/microspheres (3mg)

L.FEMUR	S.D.	< p >	R.FEMUR	S.D.
38.1	6.1	0.93	37.8	7.9
L.TIBIA	S.D.	< p >	R.TIBIA	S.D.
36.5	8.3	0.74	37.4	8.6

Table 2. Femoral segment blood flow (ml/min/100g). Left vs right

Fe59/resin	LEFT	S.D.	< p >	RIGHT	S.D.
EPIPHYSIS	41	16	0.84	41.5	17.1
METAPHYSIS	57.3	26	0.21	67.1	25.6
MARROW	26.8	15	0.8	28	16.6
CORTEX	16.5	7.2	0.06	17.5	7

Sr85/microspheres (1mg)	LEFT	S.D.	< p >	RIGHT	S.D.
EPIPHYSIS	50.4	19.8	0.65	51.7	18
METAPHYSIS	52	11.7	0.06	58.9	9.9
MARROW	47.2	11	0.93	46.5	5.7
CORTEX	4.9	1.5	0.83	4.5	1.2

Sr85/microspheres (3mg)	LEFT	S.D.	< p >	RIGHT	S.D.
EPIPHYSIS	45.4	10.1	0.11	48.4	13.7
METAPHYSIS	67.8	18.1	0.96	67.6	10.2
MARROW	61.2	16.2	0.25	55.2	14.3
CORTEX	6.1	2	0.72	6.2	2.1

TIBIAL SEGMENT BLOOD FLOW (ml/min/100g): LEFT vs RIGHT

Fe59/resin	LEFT	S.D.	< p >	RIGHT	S.D.
EPIPHYSIS	49.9	17.5	0.79	51.2	21.6
METAPHYSIS	50.3	22.2	0.66	51.2	21.6
MARROW	23	13.2	0.12	27.5	13.5
CORTEX	15	8.6	0.86	15.2	11

Sr85/microspheres (1mg)	LEFT	S.D.	< p >	RIGHT	S.D.
EPIPHYSIS	56.6	27.5	0.33	63.6	18.7
METAPHYSIS	62.6	27.7	0.23	67.4	28.5
MARROW	45.6	26.8	0.12	34.6	21.8
CORTEX	7.4	4.1	0.26	5.8	2.1

Sr85/microspheres (3mg)	LEFT	S.D.	< p >	RIGHT	S.D.
EPIPHYSIS	50	11.9	0.13	54	10
METAPHYSIS	71.2	11.9	0.1	74.3	10.7
MARROW	60.2	17.2	0.73	61.1	17.1
CORTEX	10.2	2.4	0.2	9.2	2.4

As there were no significant differences between left and right samples, these values were combined to compare whole bone and segment flow rates, resulting from the use of the three tracer groups. The results of these comparisons are shown graphically for the femur in Figure 2 and for the tibia in Figure 3. Note that the error bars show standard deviations; this is to emphasise the considerable range of values found within each group. There was no significant difference between blood flows of whole femora and tibiae measured with resin particles and microspheres (1 mg/rat), or between blood flows measured using 3 mg and 1 mg microspheres per rat. When segment results obtained from the use of resin particles were compared with those from the 1 mg dose of microspheres, no significant difference was found between the metaphyses and epiphyses of femur or tibia. In both femoral and tibial marrow, however, the flow measured with resin particles was reduced compared with 1 mg microspheres, whilst in the diaphyseal cortices the flow was significantly elevated. Segment flow comparisons between the 1 mg and 3 mg microsphere doses showed no significant difference between epiphyseal flows in femur or tibia, and flows for the tibial metaphysis were also not significantly different. Femoral metaphyseal flow was significantly elevated using 3 mg microspheres per rat. The femoral marrow and cortex were not significantly different. In the tibia, both diaphyseal marrow and cortex had elevated flows in the 3 mg dose animals, compared with those receiving only 1 mg.

In order to validate the use of the left carotid artery as a reference sampling site for calculating blood flow in the lower limb, simultaneous and identical blood reference samples were taken from the femoral and carotid arteries of 10 rats. After injection of [85]Sr labelled microspheres, counts obtained from the two sites were compared. A scatter diagram is shown in Figure 4, with a simple regression line fitted. Pearson's correlation coefficient for the two data sets was 0.91.

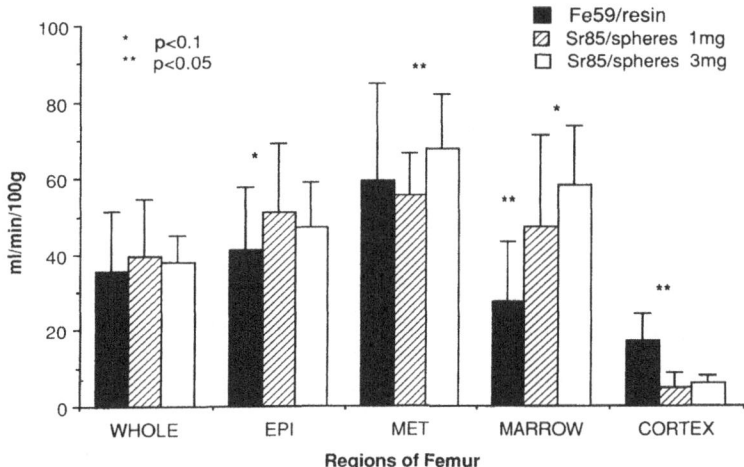

Figure 2. Blood flows to whole femora and femoral segments; summary of data from all groups.

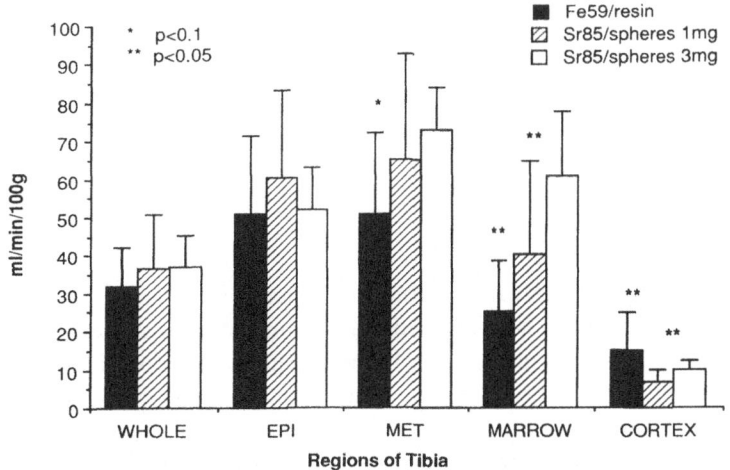

Figure 3. Blood flows to whole tibiae and tibial segments; summary of data from all groups.

Table 3. Comparison between microsphere numbers found using 1 mg and 3 mg doses

Numbers of microspheres - whole bones and segments				
	1 mg	s.d.	3 mg	s.d.
whole femur	254	98	1382	429
femoral epiphyses	86	41	332	118
femoral metaphyses	95	64	413	140
femoral marrow	20	10	157	72
femoral cortex	7	6	50	18
whole tibia	223	143	1200	337
tibial epiphyses	51	24	216	60
tibial metaphyses	102	71	514	195
tibial marrow	12	8	129	53
tibial cortex	8	4	73	24

Estimates of the number of microspheres in the various bone samples were made, and the results are shown in Table 3. A substantial elevation of sphere numbers was obtained using 3 mg batches (>1,000,000 spheres) injected into each animal.

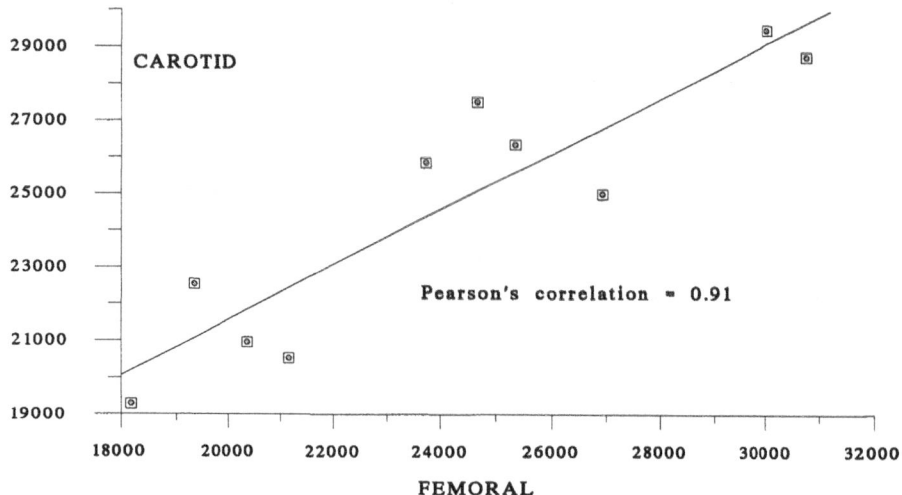

Figure 4. Radioactivity, counts per minute, recorded from simultaneous carotid and femoral reference samples.

DISCUSSION

The rat is widely used in experimental biological research; its cheapness and small size enable large numbers to be used, thus facilitating statistical validity. Furthermore, the radioactivity of whole organs or component parts can conveniently be counted, to obtain estimates of regional blood flow. Arteriolar blockade, combined with reference artery sampling in order to determine absolute organ perfusion rates, has been extensively utilised for soft tissue measurement in the rat (Malik et al., 1976). Surprisingly, very few haemodynamic investigations of rat bones have been performed. The only report which is directly comparable with the present investigation was by Kirkeby and Berg-Larsen (1991). These authors injected 15.5 μm microspheres into the ascending aorta of the rat and calculated regional bone blood flows to femora and tibiae. Normal bone values were derived only from the right limb in each animal; left/right comparisons were not made, although microsphere counts from contralateral iliac bones were found to be not significantly different. Blood flow rates to whole bones were similar to those found here, as were those in the femoral metaphyses. Tibial metaphyseal flows were considerably higher in our study compared with theirs, as were the tibial and femoral epiphyses. It is difficult to explain these discrepancies found in absolute values, as the methods used were so similar. Interestingly, however, it has been reported that roughly equivalent numbers of microspheres were found in epiphyses and metaphyses of the rat after intraventricular injection (Schoutens et al., 1979), supporting the findings here of high flows in cancellous bone. The spheres were counted by direct microscopical observation in bones made transparent by chemical treatment; the sample, however, was very small.

It should be noted that the present study, like that of Kirkeby and Berg-Larsen (1991) and Brookes (1971), has produced high values for regional blood flow. Schoutens et al. (1979) determined plasma flows in rat femora and tibia. These values, adjusted to whole blood flow rates, are in the order of 20 ml.min^{-1}100g^{-1} for the femur, and 19 ml.min^{-1}100g^{-1} for the tibia. Tothill and McPherson (1986) determined a value of 4.46 % of cardiac output going to the whole skeleton in the rat, with 0.57 % of cardiac output perfusing the combined contralateral femora and tibiae. This suggests a much lower specific flow rate than measured here. These authors, however, boiled the carcass to aid cleaning of the bones. Bone flows were probably underestimated, as approximately half the microsphere radioactivity may be lost as a result of boiling (Wootton, 1988). Brookes (1970), estimated that 27 % of the cardiac output perfused the total rat skeleton, using arteriolar blockade with the same labelled resin particles as utilised here. The distribution to individual bones was not measured, so it is difficult to extend precisely the effects of this high cardiac output proportion to individual bone flow rates. However, it is possible that an overestimate may have resulted from the injection technique. If resin particles are injected into the right carotid artery, distal to the origin of the subclavian artery, then a large proportion of the injected suspension is carried by counterflow into the tissues of the right forelimb, thereby being lost to the circulation (unpublished observation, WJR). Tothill and McCormick (1976) introduced the same resin particles directly into the aorta. They reported 3.1 % of the cardiac output went to the whole skeleton. This figure was similar to that produced by administration of ^{99m}Tc labelled microspheres, in the same investigation. These figures are almost certainly underestimates, however, as the carcasses were again boiled in a pressure cooker. Interestingly, the same experiment (Brookes, 1970) which produced the value of 27 % cardiac output to the skeleton also produced values for the liver and spleen which were very similar to those quoted by Tothill and McCormick (1976).

Another factor influencing absolute flow values using microspheres, is the assumption of homogeneity of sphere concentration throughout the arterial system. In the present study the left carotid artery was used as a reference. The catheter was only inserted about 4-5 mm, and no attempt was made to enter the aorta. The reference sample was therefore collected in its normal flow direction. Kirkeby and Berg-Larsen (1991) collected blood from the distal aorta, sampling against the arterial flow. They also injected microspheres into the ascending aorta, a procedure which does not achieve complete mixing as obtained from a left ventricular injection, and no attempt was made to verify that the sample was representative. In the present investigation all tracer particles were injected into the left ventricle. To determine that the reference activity was representative of blood perfusing the lower limb bones, reference flow from the left carotid artery was compared with an identical and simultaneous sample withdrawn from the femoral artery. The results (Figure 4) show a good correlation (r = 0.91) and a matched paired t test showed the two data sets were not significantly different. Malik et al. (1976) also found good agreement between left carotid and femoral artery reference samples. The technique described here therefore appears to be adequate, thus removing the need to utilise more awkward reference sources such as the brachial or renal arteries. Schoutens et al. (1979) used the tail artery of the rat to collect a reference flow sample. In our experience, this is by no means a simple procedure, and no validation was given for its use.

The above discussion emphasises the variability of skeletal blood flow measurement obtained from different laboratories, even when using similar methods and materials. Just as striking is the dispersal of flow values obtained within any particular experiment. The coefficient of variance for whole femoral and tibial blood flows using 3 mg microspheres per rat was 18 % and 23 % respectively; and for the 1 mg microsphere dose the figures were 37 % and 38 % respectively. The coefficient of variance for whole bone flow, using resin particles were 44 % for the femur and 32 % for the tibia. If the statistical distribution of microspheres in the bone follows Poisson's Law, then the expected coefficient of variance will equal the reciprocal of the square root of the number found in any particular bone sample. If there are sufficient numbers of spheres in the reference sample, then the variance in flow measurement should be similar to variance in sphere numbers. The mean number of spheres found in the reference arterial sample was 4,600 in the 3 mg dose animals, and 1,200 in the rats receiving 1 mg spheres. However, the disparity between these two variances does not vary linearly with the number of spheres found in the reference sample. Dole (1982) has calculated that for a tissue sample containing 400 spheres, increasing reference sample sphere numbers from 400 to 2,000 will decrease relative flow error from 13.9 % to 10.7 %. An increase in reference sphere number to 10,000 only reduces the relative flow error to 10 %. From Table 3 it can be calculated that the expected variance in whole bones, from microsphere numbers, are 6.2 % and 6.7 % for femur and tibia respectively in the 1 mg dose category, and 3 % for femur and tibia when 3 mg of spheres were injected. These figures suggest that measured variance in blood flow rate far exceeds the expected variance from the numbers of resident spheres, and that flow variation may be a reflection of a real dissimilarity, either in a given instance in time in an individual animal, or in different animals in the population.

Another consideration gained upon inspection of Table 3, is the very low numbers of microspheres found, particularly in the low dose animals. It has been calculated that 400 is the minimum number of spheres required in a tissue sample to give 10 % precision at the 95 % confidence level (Buckberg et al., 1971), although 100-200 have given acceptable results in low flow rate situations in the dog (Riggi et al., 1990). Ensuring a minimum 400 particles is not difficult in larger animals. In the rat one is limited by tissue size, particularly of bone segments. This study uses a maximum dose rate of 3.9×10^6 microspheres per kilogram rat weight (based on a mean rat weight of 270 g), and $1.3 \times 10^6 Kg^{-1}$, for animals receiving 1 mg spheres. The number of ^{59}Fe labelled resin particles injected per Kg rat weight (1.48×10^6) was similar to the 1mg microsphere dose. Only whole bones and the metaphyses of rats receiving 3 mg microspheres exceeded the statistical minimum requirement of 400 spheres. For animals receiving 1 mg of microspheres, and similarly for the resin particles, the numbers recorded were woefully small, especially in the marrow and cortical samples. The marrow sample it should be noted has a very low weight (0.05 g is typical), while the cortical sample has a greater mass but a comparatively low flow rate. Kirkeby and Berg-Larsen (1991) used a microsphere dose rate of 5.26×10^6 Kg^{-1} rat weight. This ensured a minimum tissue number of 200, found in the proximal tibial epiphyis. The diaphysis was not separated into cortical and marrow components and therefore it is not possible to estimate numbers found in these segments, but by extrapolation it is likely that the numbers in the cortex remained small. Large numbers of particles need to be infused to achieve acceptable confidence levels.

The rat, fortunately, is known to tolerate intraventricular catheterisation well, and large injections of microspheres are permissable without haemodynamic perturbation. Flaim (1978), for instance, has reported the injection of 850,000 microspheres in one bolus, without noting systemic changes. Cumulative doses of up to 1,440,000 spheres were injected into the left atrium of the rat by Stanek et al. (1983), again without haemodynamic change. The mean rat weight used in their study was 278 g, giving a dose rate of 5.18×10^6 Kg^{-1}. This dose was exceeded by Kirkeby and Berg-Larsen, who also reported no significant haemodynamic change. In the present study, detailed physiological parameters were not monitored, but intracardiac pulse amplitude and frequency did not change as a result of injections containing up to 1,050,000 microspheres.

The great variation between individual flow rates makes it desirable to use large numbers of animals to determine the mean flow rate within a population, and with a reasonable confidence interval. This favours the rat as an experimental animal. It is also apparent that investigation of regional bone blood flow requires the injection of large numbers (at least 5×10^6 Kg^{-1} rat weight) of particulate iontophores, resin or spheres, to obtain statistical confidence. The results of the comparison made here between flow values obtained using labelled resin particles and microspheres are of great practical importance, in spreading bone blood flow measurement to non-specialised laboratories. In the whole femur and tibia, and the epiphyses and metaphyses, no statistical difference was found in flow rates measured by the injection of ^{59}Fe labelled resin particles or an equivalent dose of microspheres. Only in the cortex and bone marrow were flow values obtained from the two materials significantly different (Figures 2,3). Cortical flow was elevated when measured by resin particles, while marrow flows were reduced; both in comparison with microsphere results. Plasma leeching of ^{59}Fe from the resin would have the effect of reducing the counts in areas of moderate flow, but giving more counts in regions of low flow, such as the cortex. Because of leeching, an element of bone blood volume is included in the calculated flow value. It seems statistically insignificant in most cases, given the very large intrinsic standard deviations. We have shown that labelling the resin with ^{59}Fe in the presence of additional unlabelled ferric chloride, eliminated in-vivo leeching from resin particles. It is probable therefore that such treatment (attributed to N.Veall, in Tothill and McCormick, 1976), would eliminate these discrepancies. A systematic pre-injection of ferrous gluconate might possibly produce a similar effect by saturating iron receptors in plasma transferrin.

Comparing 3 mg to 1 mg doses of microspheres, a dose dependant increase in measured flow rate is suggested by the tibial marrow and cortex ($p<0.05$) and the femoral metaphysis ($p<0.05$). The femoral marrow flow is also raised but not significantly. This may reflect an insufficient number of particles in the low dose animal to give adequate blockade of the arteriolar beds, and emphasises the importance of injecting as large a number of particles as possible.

Because of the large coefficient of variance found within any single population, comparisons of absolute blood flow between different populations demand large numbers to obtain an acceptable confidence interval for the population means. It must be emphasised, however, that irrespective of the absolute flow rates measured, comparisons between left and right limbs in all animals examined in this investigation, were never significantly different. This is remarkable when one examines the very low numbers of microspheres found in bones injected with 350,000 (1 mg) microsphere batches, and indicates an extremely homogeneous dispersal of the available particles in the perfusing arteries. From the findings presented here, therefore, we suggest that the best strategy for determining haemodynamic effects of orthopaedic procedures is always to compare the operated limb with its contralateral control. If left-right compari-sons are taken as an index of haemodynamic change over time, it obviously becomes unnecessary to determine absolute flow rates; counts per unit weight of bone are suffi-cient for comparison. The use of resin particles generates similar data to that produced by the much more expensive microspheres, with an equivalent variance of results. If care is taken to produce an appropriate particle size distribution; injection is made into the left ventricle to ensure adequate mixing; a sufficiently large dose is infused; and procedures adopted to prevent leeching of the isotope label to plasma; then the results of using cationic exchange resin particles as a tracer for bone blood flow measurement by arteriolar blockade are indistinguishable from those produced by the use of microspheres.

REFERENCES

Brookes, M.: Arteriolar blockade: a method of measuring blood flow rates in the skeleton. J. Anat. 106 (3): 557-563 (1970).

Brookes, M.: Blood supply of Bone. London, Butterworths (1971).

Brookes, M., S.C. Gallennaugh: Circulatory depression in bone after acrylic implantation. Clin.Orthop. 107: 274-276 (1975).

Brueton, R.N., W.J. Revell, M. Brookes: Haemodynamic change during bone healing in a model fracture. Proceedings of the Anatomical Society. J. Anat. 179: 219-220 (1991).

Buckberg, G.D., J.C. Luck, D.B. Payne, L.I.E. Hoffman, J.P. Archie, D. Fixler: Some sources of error in measuring regional blood flow with radiactive microspheres. J. Appl. Physiol. 31: 598-604 (1971).

Dole, W.P., D.L. Jackson, J.I. Rosenblatt, W.L. Thompson: Relative error and variability in blood flow measurements with radiolabelled microspheres. Am. J. Physiol. 243: H371-H378 (1982).

Flaim, S.F., Z.Q. Morris, T.J. Kennedy: Dextran as a radioactive microsphere suspending agent: severe hypotensive effect in the rat. Am. J. of Physiol. 235: H587-591 (1978).

Kirkeby, O.J., T. Berg-Larsen: Regional blood flow and strontium-85 incorporation rate in the rat hindlimb skeleton. J. Orthop. Res. 9: 862-868 (1991).

Malik, A.B., J.E. Kaplan, T.M. Saba: Reference sample method for cardiac output and regional blood flow determinations in the rat. J. Appl. Physiol. 40: 472-475 (1976).

Riggi, K., M.B. Wood, D.M. Ilstrup: Dose-dependant variations in blood flow evaluation of canine nerve, nerve graft, tendon, and ligament tissue by the radiolabelled-microsphere technique. J. Orthop. Res. 8: 909-916 (1990).

Schoutens, A., P. Bergmann, M. Verhas: Bone blood flow measured by [85]Sr microspheres and bone seeker clearances in the rat. Am. J. Physiol. 236: H1-H6 (1979).

Stanek, K.A., T.L. Smith, W.R. Murphy, T.G. Coleman: Haemodynamic disturbances in the rat as a function of the number of microspheres injected. Am. J. Physiol. 245: H290-H293 (1983).

Tothill, P., J.StC. McCormick: Bone blood flow in the rat determined by the uptake of radioactive particles. Clin. Sci. Mol. Med. 51: 403-406 (1976).

Tothill, P., J.N. MacPherson: The distribution of blood flow to the whole skeleton in dogs, rabbits and rats measured with microspheres. Clin. Physics and Phys. Meas. 7: 117-123 (1986).

Wootton, R.: Errors in bone blood flow measured with microspheres due to sample preparation technique. Clin. Physics and Phys. Meas. 9: 273-276 (1988).

MEASUREMENT OF BONE BLOOD FLOW IN HUMANS

R. WOOTTON

Department of Medical Physics
Hammersmith Hospital
London, United Kingdom

INTRODUCTION

From the perspective of the measurement scientist, the ideal organ in which to measure blood flow is one comprising a single homogeneous tissue supplied by a single artery and drained by a single vein. The kidney comes close to this ideal, and estimates of renal blood flow have been widely accepted for decades. Bone on the other hand is a much more difficult proposition. It is heterogeneous, being composed of various kinds of calcified tissue and marrow, which have a complex, interlinked circulation, and there are multiple arterial and venous connections to any given bone of the skeleton. The wide variety of methods which have been used to measure bone blood flow in animals is testimony to the technical difficulties. It is hardly surprising therefore that the estimates for bone blood flow in humans are spread over a very wide range and that there is no generally accepted value.

This paper reviews the published work on the measurement of bone blood flow in man and attempts to arrive at a consensus position.

METHOD

Relevant publications were identified by performing both manual and computerised literature searches. The Embase and Medline databases were used for the latter purpose. In addition, citation searches of some of the earlier methodological papers were carried out using the Science Citation Index.

For reasons of the methods by which they have been obtained, published results for bone blood flow have been reported in one of several different units e.g. % of cardiac output or ml/min/100g of bone. Before a comparison could be made it was necessary to convert them to a common system of units. This was done by assuming that in an adult male the skeletal mass is 8 kg (see Table 1), that blood volume is 5.2 litres (ICRP, 1975) and that resting cardiac output is 6.7 l/min (Williams and Leggett, 1989). That is, a result for bone

Bone Circulation and Vascularization in Normal and Pathological Conditions
Edited by A. Schoutens *et al.*, Plenum Press, New York, 1993

85

Table 1. Composition of the skeleton in reference man (ICRP, 1975)

Component	Weight (kg)
Bone	
cortical	4
trabecular	1
Marrow	
red	1.5
yellow	1.5
Skeletal cartilage	1.1
Periarticular cartilage	0.9
Total	10

blood flow reported as, say F ml/min, is equivalent to (F/C)*100 % of cardiac output, where C is cardiac output in ml/min, or to (F/V)*100 % of blood volume/min, where V is blood volume in ml, or to (F/W)*100 ml/min/100g of whole bone, where W is the skeletal weight.

RESULTS

Papers relating to seven different techniques were identified. With a single exception, plethysmography, all studies were based on the measurement of tracer uptake by bone or tracer washout from bone. Three different tracers were employed. Regional bone blood flow was measured by the uptake of the bone-seeking isotopes ^{18}F or ^{85}Sr, or by the washout of the freely-diffusible tracer ^{133}Xe. Whole body measurements were based on the skeletal clearance of ^{18}F.

Table 2 shows the six published values for bone blood flow obtained by tracer methods interconverted to a common system of units. Figure 1 shows these mean values expressed as % of cardiac output, and their confidence intervals calculated from the SEM by standard methods.

DISCUSSION

Terminology

Blood flow can be defined as the volume of blood flowing in a blood vessel past the point of measurement in unit time. Organ blood flow in general is therefore the volume of blood flowing to an organ in unit time, which under steady state conditions is the same as that leaving it. If the organ in question is the complete skeleton, then the blood flow of interest is the total flow rate to all the bones comprising the skeleton; if the organ is an individual bone then the blood flow rate is that to the individual bone. In either case the definition is not complicated by the diverse pathways for arterial supply and venous drainage: bone blood flow is simply defined as the total volume of blood flowing into the organ as a whole in unit time.

Table 2. Bone blood flow in normal human subjects

Values underlined are authors' measurements; other figures are estimated as described in the text.

Reference	Method	Bone studied	Number of subjects	Skeletal blood flow (% cardiac output)	(% blood vol/min)	(ml/min)	Perfusion (ml/min/100g whole bone)
Shim et al. (1971)	^{85}Sr clearance	Distal leg	1	3.5	4.4	230	2.9
Van Dyke et al. (1971)	^{18}F clearance (3 compartments)	Whole skeleton	4	3.3 (SD 1.1)	4.3	220	2.8
Wootton et al. (1976)	^{18}F clearance (impulse analysis)	Whole skeleton	8	4.3	5.1 (SD 0.5)	290 (SD 40)	3.6
Charkes et al. (1978)	^{18}F clearance (5 compartments)	Whole skeleton	60*	16.8 (SD 1.0)	24.6 (SD 1.1)	1280	16.0
Lahtinen et al. (1982)	^{133}Xe washout	Head of femur	7	8.8	11.4	590	7.4 (SD 1.2**)
Hawkins et al. (1992)	^{18}F clearance (PET)	Lumbar vertebrae	11	9.1	11.7	610	7.6 (SD 3.9)

* average data from 9 published studies

** SD estimated from published range by the method of Kendall & Stuart (1969)

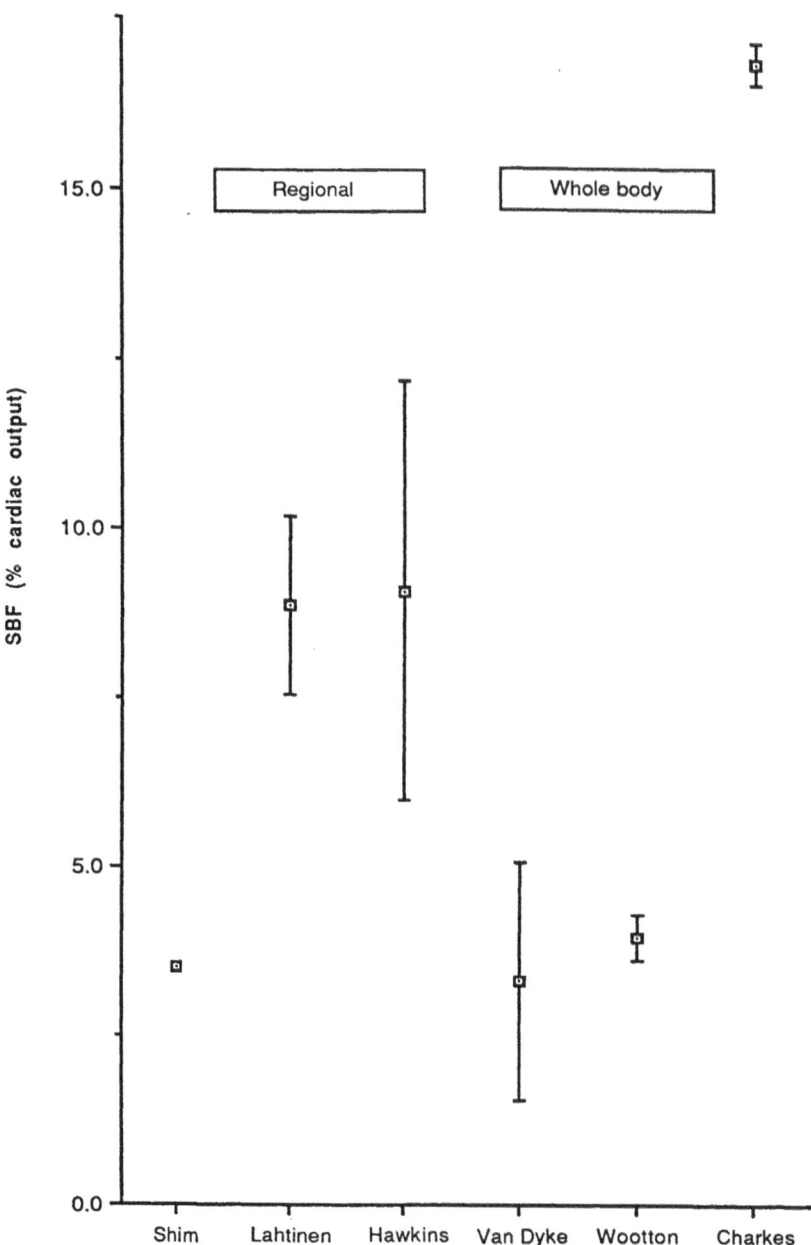

Figure 1. Mean values for skeletal blood flow and 95% confidence limits.

Perfusion is defined as the organ blood flow per unit mass of organ. This is a straightforward definition when the organ is reasonably homogeneous, but in the case of bone - which is not - a certain amount of confusion has been perpetrated in the literature by authors who have not been careful to distinguish between blood flow rates per unit mass of total bone, of mineralised bone alone, or of marrow.

If bone is deemed to be the sum of, say, mineralised bone and marrow, then organ blood flow is the total volume of blood flowing to the combined tissue in unit time. Note that if a substantial part of the venous drainage from the marrow takes place via the surrounding cortex, cortical perfusion will comprise two components: the direct arterial supply and that draining via the marrow. This is analogous to the blood supply of the liver. In a similar way, it may be necessary to take this into account when interpreting the results obtained by any particular method of measurement.

Plethysmography

The earliest attempt to measure bone blood flow in man was by Edholm et al. using plethysmography (Edholm et al., 1945). They reported perfusion values of 0.5 - 1 ml/min/100g of whole bone in the distal humerus and in the tibia and fibula of normal subjects. Plethysmographic methods depend on measurement of the initial rate of increase in limb volume following the release of a tourniquet. In principle therefore, such a method can be used to measure peripheral bone blood flow. However, a number of unresolved questions of technique arise: it is not easy to define accurately the mass of bone tissue being perfused; the reactive hyperaemia following the release of an arterial tourniquet may make the results difficult to interpret; the possibility exists of intramedullary venous drainage past the occluding cuff. For these reasons Edholm et al. subsequently came to regard plethysmography as a qualitative technique only (Edholm and Howarth, 1953).

Regional Measurements

Shim et al. (1971) measured the clearance of ^{85}Sr in the bones of the lower limb of a single subject prior to amputation. This method depends on there being little reflux of tracer from bone during the measurement period, so that net clearance is approximately the same as unidirectional clearance from blood to bone. During the 3 minute measurement period which was used by Shim et al., it is likely that significant reflux of tracer would have occurred (Copp and Shim, 1965). In addition, if clearance is to be equated with blood flow, it is necessary to assume 100 % extraction of tracer by bone. There is evidence that the extraction efficiency of radiostrontium is high, although probably not 100 %. It is not as high as ^{18}F, for example (Schumichen et al., 1979). Shim et al.'s value of 2.9 ml/min/100g of whole bone is therefore - as the authors themselves recognised - likely to underestimate true perfusion in the lower limb. If their value is extrapolated to the entire skeleton it should properly be seen as a lower bound to true skeletal perfusion.

Lahtinen et al. (1982) measured bone perfusion in the head of the femur using the ^{133}Xe washout technique. This method depends on saturating the tissue of interest with ^{133}Xe following an intravenous injection, and then monitoring its disappearance with an external detector. It is necessary to know the partition coefficient between bone and blood, and between marrow and blood, in order to convert the observed exponential washout curves into values for perfusion. In comparison with animal data from similar anatomical regions, Lahtinen et al.'s value for perfusion of 7.4 ml/min/100g of whole bone appears wholly plausible. It is not easy however to use this regional value to obtain an estimate for total skeletal blood flow in man. Perfusion of trabecular bone in animals is 5 - 10 times higher than cortical bone (Morris and Kelly, 1980) and the head of the femur contains relatively more trabecular bone than the skeleton as a whole (ICRP, 1975). Lahtinen et al.'s value is therefore likely to overestimate total skeletal perfusion.

Hawkins et al. (1992) measured the clearance of ^{18}F in the lumbar vertebrae using positron emission tomography. Their value for perfusion of 7.6 ml/min/100g of whole bone is low in comparison with that measured by Nahmias et al. (1986) in dogs using a similar technique, though not unreasonable. The vertebrae, like the head of the femur, contain a relatively high proportion of trabecular bone. Extrapolating Hawkins et al.'s value to the whole skeleton is likely to produce an upper estimate for skeletal perfusion.

Whole Body Measurements

Van Dyke et al. (1965) were the first group to use ^{18}F to estimate skeletal blood flow in a number of species including man. They measured the net clearance by the skeleton using a 3 compartment model but recognised that the net clearance could only provide a minimum value for the flow because of the reflux of tracer from bone to blood following uptake. They estimated minimum skeletal blood flow as 3.3 % of cardiac output in a group of normal adults (Van Dyke et al., 1971).

Charkes et al. (1978) used a 5 compartment model to estimate the skeletal clearance of ^{18}F. For the purpose of their analysis they constructed a single average blood curve based on 9 published studies relating to 60 different subjects. This composite blood curve was then analysed using the compartment model. Unfortunately this will not produce the same result as analysing 60 individual data-sets and then averaging the answers. A further complication is that their chosen model, while perfectly reasonable on physiological grounds, is not strictly identifiable in the mathematical sense: there are in fact an infinite number of possible compartmental models which would fit the averaged data as well as (or better than) that chosen. These are technical reasons why one would tend to attach a low weight to their result. It must also be pointed out that if 16.8 % of the cardiac output were really perfusing the skeleton in resting man, that the accepted values for organ blood flow (Williams and Leggett, 1989; ICRP, 1975) must be seriously in error. That is, if the contributions from the other major organs are summed, it is simply impossible for skeletal blood flow to be more than about 5 or 6 % of the cardiac output.

The ^{18}F methods described above are based on the use of compartment models. Such models assume that ^{18}F behaves in the body as though it were distributed in a series of well-mixed pools, from which first order transport takes place. While these assumptions may be tenable in the case of pools with a rapid turnover, such as blood, there is no evidence to suggest that this is the case for ^{18}F in bone. Indeed, the behaviour of other longer-lived bone-seeking nuclides in the skeleton, such as calcium or the rare earths, is known to be quite different (ICRP, 1972; Jung et al., 1978).

An alternative, and more general approach, is the use of impulse analysis to establish the unidirectional non-renal clearance of ^{18}F from the blood. This does not require the assumption of compartmental behaviour. The measured clearance can be corrected for the non-bone component by use of a second, extra cellular fluid tracer such as ^{51}Cr-EDTA. Full details of the method are given in Wootton et al. (1976). Use of this technique in 8 normal subjects gave values of 5.1 % of blood volume/min or about 4.3 % of cardiac output.

Extraction Ratio

All of the methods using ^{18}F attempt to estimate, to a greater or lesser accuracy, the unidirectional clearance of the tracer by bone from blood. To convert this clearance to a blood flow rate it is necessary to know the single passage extraction ratio of ^{18}F, i.e. the fraction of a bolus of ^{18}F which is removed from blood on first passage through bone. The original experimental data relating to the extraction ratio of ^{18}F (Wootton, 1974) were subsequently criticised, mainly on the grounds that transit times through bone were longer than the period of measurement (Tothill & Hooper, 1984). Although this does not invalidate the results, it might explain some of the observed variation. A further and more accurate series of experiments was therefore carried out using a different and much more accurate technique. The results confirmed that, in rabbit bone at least, the single passage extraction ratio of ^{18}F is approximately 100 % (Wootton and Doré, 1986). There is also evidence for 100 % extraction in the rat (Schumichen et al., 1979).

Other measurements of the extraction ratio of ^{18}F have also been made using Crone's technique. These suggest that the extraction ratio is less than 100 % (Kelly and Bassingthwaighte, 1977). There are however some difficulties of interpretation associated with the use of Crone's method. First it depends on there being no significant backflux of tracer during the initial period following injection, when the slope of the graph of tracer extraction is being established. Crone himself said "A lot of misunderstanding could have been obviated if it were remembered that the method can only deal satisfactorily with cases where one has some confidence in believing that there is no back-diffusion ... the sign that back-diffusion is negligible is a horizontal part of the extraction curve, E(t)." (Crone, 1970). ^{18}F, in common with many other bone-seeking tracers, is well known to be released quickly following uptake and the observed E(t) curves for these tracers are not in fact horizontal.

Second, in the case of bone, all such measurements are made by injecting the tracer into the nutrient artery of a long bone and collecting the venous outflow. It has yet to be established whether such a technique can gain information which is representative of the bone in general, or only of the capillary bed of the nutrient artery, and whether the long bone in question is representative of the skeleton as a whole.

Other Methods

At the time of writing the use of ^{15}O-labelled water and positron emission tomography to measure bone blood flow appears promising (Ashcroft et al., 1992). In addition, the possibility exists of using magnetic resonance scanning techniques to derive quantitative values for bone perfusion (Tsukamoto et al., 1992). Although neither technique is currently capable of yielding quantitative results, it appears likely that they will be able to do so in future.

Finally, it is worth noting that O'Flaherty has recently reviewed the literature on bone blood flow in the rat (O'Flaherty, 1991a). Based on her best estimate for bone blood flow in that species, she extrapolated to the human using an allometric model and obtained a value for skeletal blood flow of 5.3 % of cardiac output, or 4.1 ml/min/100g of whole bone (O'Flaherty, 1991b).

Variation in Health and disease

Published standard deviations for bone blood flow measured by a given technique in normal adults (Table 2) represent a combination of variation due to inter-individual differences and to experimental error. Without further information it is not possible to separate them. However, in a study based on the impulse analysis method, it was found that the experimental error was relatively small in comparison with differences between individuals (Wootton et al., 1976). Analysis of the results of paired measurements showed that the precision of a single determination corresponded to a SE of 16 %.

Studies using both ^{18}F and ^{133}Xe show that large changes in skeletal blood flow occur in metabolic bone disorders (Green et al., 1987), especially Paget's disease (Wootton et al., 1978), and in haematological disorders affecting the marrow (Lahtinen et al., 1982; Van Dyke et al., 1971).

CONCLUSION

In summary, the published data lead one to conclude that skeletal blood flow in adult man at rest:

- has an estimated lower limit of 3.5 % of the cardiac output;
- has an estimated upper limit of 9 % of the cardiac output;
- has an estimated mean value of 4 % of the cardiac output. This corresponds to a blood flow of 270 ml/min, or 5.2 % of the blood volume/min, or a mean skeletal perfusion of 3.4 ml/min/100g of whole bone.

The latter values are largely based on ^{18}F blood clearance studies and would need revision upwards were it to be shown conclusively that the single passage extraction ratio of ^{18}F in the skeleton is significantly lower than unity.

REFERENCES

Ashcroft G.P., N.T.S. Evans, D. Roeda, M. Dodd, J. Mallard, R.W. Porter, F. Smith: In vivo bone blood flow measurement with positron emission tomography (a study of patients with tibial fracture). J. Bone Jt. Surg (B), (in press) (1992).

Charkes N.D., P. Todd Makler, C. Philips: Studies of skeletal tracer kinetics. 1. Digital-computer solution of a five-compartment model of [^{18}F] fluoride kinetics in humans. J. Nucl. Med. 19: 1301-1309 (1978).

Copp D.H., S.S. Shim: Extraction ratio and bone clearance of Sr85 as a measure of effective bone blood flow. Circ. Res. 16: 461-467 (1965).

Crone C.: Capillary permeability - techniques and problems. In: C CRONE, NA LASSEN (eds): Capillary Permeability, pp 15-31. Academic Press (1970).

Edholm O.G., S. Howarth, J. McMichael: Heart failure and bone blood flow in osteitis deformans. Clin. Sci. 5: 249-260 (1945).

Edholm O.G., S. Howarth: Studies on the peripheral circulation in osteitis deformans. Clin. Sci. 12: 277-285 (1953).

Green J.R., J. Reeve, M. Tellez, N. Veall, R. Wootton: Skeletal blood flow in metabolic disorders of the skeleton. Bone 8: 293-297 (1987).

Hawkins R.A., Y. Choi, S.C. Huang, C.K. Hoh, M. Dahlbom, C. Schiepers, N. Satyamurthy, J.R. Barrio, M.E. Phelps: Evaluation of the skeletal kinetics of fluorine-18-fluoride ion with PET. J. Nucl. Med. 33: 633-642 (1992).

ICRP: Report of the task group on reference man. ICRP Publication 23. Pergamon Press, Oxford (1975).

ICRP: Alkaline earth metabolism in adult man. ICRP Publication 20. Pergamon Press, Oxford (1972).

Jung A., P. Bartholdi, B. Mermillod, J. Reeve, R. Neer: Critical analysis of methods for analysing human calcium kinetics. J. Theor. Biol. 73: 131-157 (1978).

Kelly P.J., J.B. Bassingthwaighte: Studies on bone ion exchanges using multiple-tracer indicator-dilution techniques. Fed. Proc. 36: 2634-2639 (1977).

Kendall M.G., A. Stuart: The advanced theory of statistics. Vol. 1. Charles Griffin, London (1969).

Lahtinen R., T. Lahtinen, T. Romppanen: Bone and bone-marrow blood flow in chronic granulocytic leukemia and primary myelofibrosis. J. Nucl. Med. 23: 218-224 (1982).

Morris M.A., P.J. Kelly: Use of tracer microspheres to measure bone blood flow in conscious dogs. Calcif. Tissue Int. 32: 69-76 (1980).

Nahmias C., W.P. Cockshott, L.W. Belbeck, E.S. Garnett: Measurement of absolute bone blood flow by positron emission tomography. Skel. Radiol. 15: 198- 200 (1986).

O'Flaherty E.J.: Physiologically based models for bone-seeking elements. I. Rat skeletal and bone growth. Toxicol. App. Pharmacol. 111: 299-312 (1991a).

O'Flaherty E.J.: Physiologically based models for bone-seeking elements. III. Human skeletal and bone growth. Toxicol. App. Pharmacol. 111: 332-341 (1991b).

Schumichen C., H. Rempfle, M. Wagner, G. Hoffmann: The short-term fixation of radiopharmaceuticals in bone. Eur. J. Nucl. Med. 4: 423-428 (1979).

Shim S.S., S. Mokkhavesa, G.D. McPherson, J.F. Schweigel: Bone and skeletal blood flow in man measured by a radioisotopic method. Can. J. Surg. 14: 38-41 (1971).

Tothill P., G. Hooper: Invalidity of single-passage measurements of the extraction of bone-seeking tracers in rats and rabbits. J. Orthop. Res. 2: 75-79 (1984).

Tsukamoto H., Y.S. Kang, L.C. Jones, M. Cova, C.J. Herold, E. McVeigh, D.S. Hungerford., E.A. Zerhouni: Evaluation of marrow perfusion in the femoral head by dynamic resonance imaging. Effect of venous occlusion in a dog model. Invest. Radiol. 27: 275-281 (1992).

Van Dyke D., H.O. Anger, Y. Yano, C. Bozzini: Bone blood flow shown with F18 and the positron camera. Am. J. Physiol. 209: 65-70 (1965).

Van Dyke D., H.O. Anger, H. Parker, J. McRae, E.L. Dobson, Y. Yano, J.P. Naets, J. Linfoot: Markedly increased bone blood flow in myelofibrosis. J. Nucl. Med. 12: 506-512 (1971).

Williams L.R., R.W. Leggett: Reference values for resting blood flow to organs of man. Clin. Phys. Physiol. Meas. 10: 187-217 (1989).

Wootton R.: The single-passage extraction of ^{18}F in rabbit bone. Clin. Sci. Mol. Med. 47: 73-77 (1974).

Wootton R., C. Doré: The single-passage extraction of ^{18}F in rabbit bone. Clin. Phys. Physiol. Meas. 7: 333-343 (1986).

Wootton R, J. Reeve, E. Spellacy, M. Tellez-Yudilevich: Skeletal blood flow in Paget's disease of bone and its response to calcitonin therapy. Clin. Sci. Mol. Med. 54: 69-74 (1978).

Wootton R., J. Reeve, N. Veall: The clinical measurement of skeletal blood flow. Clin. Sci. Mol. Med. 50: 261-268 (1976).

SKELETAL FLUORIDE KINETICS OF ^{18}F$^-$ AND POSITRON EMISSION TOMOGRAPHY (PET): In-vivo Estimation of Regional Bone Blood Flow and Influx Rate in Humans

C. SCHIEPERS

Nuclear Medicine
University Hospitals
KU Leuven, Belgium

In this chapter the skeletal kinetics of fluoride will be described from a more clinical point of view. Our main objective is to estimate regional bone blood flow in vivo. In the preceding section, Wootton has presented an overview of the methods and techniques presently available to determine skeletal blood flow. Here, the technique with the positron emitter ^{18}F$^-$ and tomography both in the imaging and modeling domain will be dealt with. The application of Positron Emission Tomography (PET) to human studies will be discussed, the method to evaluate fluoride kinetics and the mathematical modeling for parameter estimation, i.e. regional flow, influx rate, and blood volume.

Blau et al (1962) demonstrated the clinical usefulness of ^{18}F$^-$ for bone scanning. In the classic study of Van Dyke et al. (1965) on ^{18}F$^-$ bone uptake, the initial radionuclide distribution was regarded as an indicator of blood flow to bone. Charkes (1980) described the implications to routine bone scanning, and emphasized the relationship between tracer uptake and skeletal flow. Moreover, uptake could be used to classify disease states. Wootton and his group (1974, 1976, 1986) showed the unidirectional extraction fraction of bone to be 1 in animals with a negligible marrow uptake. The initial fluoride uptake, therefore, appears to reflect skeletal flow. Morris and Kelly (1980) found that fluoride is only partially extracted, leading to an underestimation of bone blood flow. This issue of extraction fraction has also been touched upon in Wootton's contribution.

Compartmental modeling performed by Charkes et al. (1978, 1979, 1980) yielded rate constants indicating loss of fluoride from the bone compartment at late times. Nahmias et al. (1986) who investigated ^{18}F$^-$ kinetics in dogs with PET, however, concluded that there is no net loss of ^{18}F$^-$ from the bone, and provided evidence to support the general assumptions of Van Dyke's (1965) work on ^{18}F$^-$: a) no loss from bone and b) no saturation effects. Thus, regional parameters estimated with PET do not suggest fluoride loss from the bone. On the other hand, methods based on clearance by the total skeleton suggest a depletion of fluoride, especially over long periods of time (Wootton et al., 1976; Charkes et al., 1979). However, deconvolution techniques as Wootton's and

Bone Circulation and Vascularization in Normal and Pathological Conditions
Edited by A. Schoutens *et al.*, Plenum Press, New York, 1993

95

modeling with explicit diffusion rates for uptake and release as proposed here, yield estimates of instantaneous clearance and bone blood flow.

In a recent article we described the regional skeletal kinetics of fluoride in normal male volunteers measured with PET and $^{18}F^-$ (Hawkins, 1992). We found a three compartment model the preferred structure to describe the measured data. The parameter that describes fluoride loss from the bone compartment appeared to be small.

Given the relatively short half life of $^{18}F^-$ and PET acquisition times, reliable estimates of regional skeletal blood flow should be possible. Considering the controversies about extraction fraction and depletion of $^{18}F^-$, the technique will still be able to estimate in vivo the distribution of skeletal flow, i.e. in a regional sense, with a much higher degree of accuracy than conventional whole body clearance techniques.

METHODS

Imaging is performed with a PET scanner, a device that generates tomographic images of the body. These images represent the activity distribution of the radiopharmaceutical that was administered, e.g. functional images of fluoride uptake proportional to flow. In order to quantify the biochemical processes involved, the following requirements need to be met: 1) knowledge of the tracer clearance curve, 2) dynamic imaging to measure in vivo uptake in human bone, 3) correction for attenuation, and 4) calibration of count rates from PET scanner to blood sampling device. In addition, a reasonable patient "immobility" compliance is required to perform a study.

The standard physiological assumptions about vascular, perivascular and cellular spaces are made. With mathematical modeling the rate constants of transport between the different compartments are estimated. Hereto, the disappearance of tracer from the plasma is measured by sampling arterial blood. This so-called input function may be obtained by determining the freely diffusable fluoride concentration in plasma. Since fluoride is not evenly distributed over the red blood cells and plasma, a correction is necessary in case whole blood counting is used (plasma concentration is twice that of red blood cells). Since we measure parameters per volume element (voxel), small (or moving) structures may only partially fill a voxel necessitating a correction. Therefore, in case of a measured input function over a vascular structure (e.g. heart, aorta), both blood partition and partial volume effects need to be accounted for, and appropriate scaling is needed to generate the plasma input function. A graphic representation of the modeling approach:

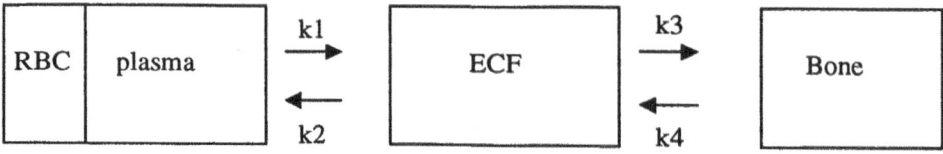

The micro-parameters k1 and k2 represent forward and reverse transport from plasma, k3 and k4 the uptake and release from bone. k1 is equal to the regional skeletal flow if the extraction fraction of tracer equals 1. Macro K, defined as K=k1.k3/(k2+k3) is the net forward transport of fluoride to the bone compartment, i.e. influx rate. In this model the second compartment comprises both the ECF and the un-bound bone compartment as defined by Charkes (1979, 1980). Hawkins et al (1992) showed the above scheme the preferred model structure. Charkes (1979) also demonstrated that this model described the data adequately, but they opted for an additional bone-ECF compartment on empirical grounds. With the PET scanner the concentration of tracer in all 3 compartments combined is measured. Curves of regional fluoride activity can be generated by drawing regions over the areas of interest, e.g. output curves of bone or other tissues. With these input and output functions the differential equations of the model that describe the transport of $^{18}F^-$ between compartments can be solved and the regional parameters are obtained. This is done mathematically with non-linear regression.

Because the uptake of fluoride from plasma to bone is primarily uni-directional, the Patlak Graphical Analysis method may also be used to calculate macro K. For short, macro K is estimated as the slope of the graph that relates the cumulative activity in plasma from start to time T, to the activity in the second and third compartment combined at the same time T. This appears to be a much simpler computational procedure. Excellent correlation has been reported between K's obtained with non-linear regression and Patlak Graphical Analysis [Schiepers et al. (1990), Hawkins et al. (1992)].

The parameters k1 and K are expressed per unit volume as ml/min of plasma concentration per ml. If the bone density and the plasma fluoride concentration are known, the net mass flux of fluoride to the bone compartment in nmol/min/g may be calculated (Equation 1).

$$\text{Flux} = [F^-] \cdot K \qquad\qquad\qquad\qquad\qquad\qquad 1$$

The parameter k1 is proportional to bone blood flow, as has been shown in a qualitative way by Van Dyke et al. (1965) and Charkes (1980). Quantitative estimates of Hawkins et al. (1992) correspond favorably to animal data. Kelly's group showed in dogs a flow of 5 ml/min/100g in cortical and 16 ml/min/100g in trabecular bone. We found 0.11 ml/min/ml, which may be converted to 8 ml/min/100g by using a density for vertebrae of 1.4 g/ml (determined with QCT). Additionally, Kelly's group has shown a vertebral blood volume of 0.05 ml/ml in dogs with microspheres, while our PET method gave 0.08 ml/ml in humans.

Limitations

Three kinds of problems may be encountered: a) insufficient spatial resolution to identify structures; b) extraction fraction of fluoride less than 1; c) input function determination.

The spatial resolution of the present generation of PET scanners is about 7 mm, which is usually not sufficient to resolve cortical from trabecular bone. Therefore, determination of the specific mass in the volume may pose difficulties.

A discussion of the extraction fraction issue can be found in Wootton's chapter.

In the here described technique we used a centrally sampled input function, because it appeared difficult to derive a local one. As local tracer clearance may differ, distortion of the input function may occur. This distortion can be "modeled" by incorporating dispersion.

A direct comparison between $^{18}F^-$ - PET and microsphere flow data in a well controlled animal model is not available yet.

GENERAL ACQUISITION PROTOCOL

The patient's skeletal area of interest is positioned in the PET scanner, e.g. femoral head. A transmission scan is acquired to allow for correction of attenuation of tissues between origin of photons and the detector. A dose of 250-350 MBq of ^{18}F as sodium fluoride is injected in an ante-cubital vein. The input function is generated from the plasma clearance curve. This curve may be calculated from: 1) the plasma ^{18}F counts obtained from arterial or arterialized venous blood, or 2) measuring the clearance curve with the PET scanner over vascular areas (heart, aorta) and correction of the obtained count rates for partition and partial volume effects.

A dynamic acquisition is started simultaneously with the tracer administration. In case of a measured input function, high temporal sampling is necessary for measuring this input function, e.g. 12x15, 4x30, 4x60, 2x120 second frames followed by 5 min frames up to 1 hour. Otherwise a temporal sampling of 5x1, 5x2, 9x5 minute frames is adequate for the first hour. Between 1 and 2 hours the bone uptake is high and the plasma concentration sufficiently decreased to image different areas. Ten minute frames will supply high quality images. A total investigation takes 2 to 3 hours.

By drawing region of interests, appropriate time activity curves may be obtained that serve as output functions. Estimation of the various diffusion rates between the vascular, extra-vascular and bone compartments, is performed with mathematical modeling as described above.

CLI N ICAL EXAMPLES

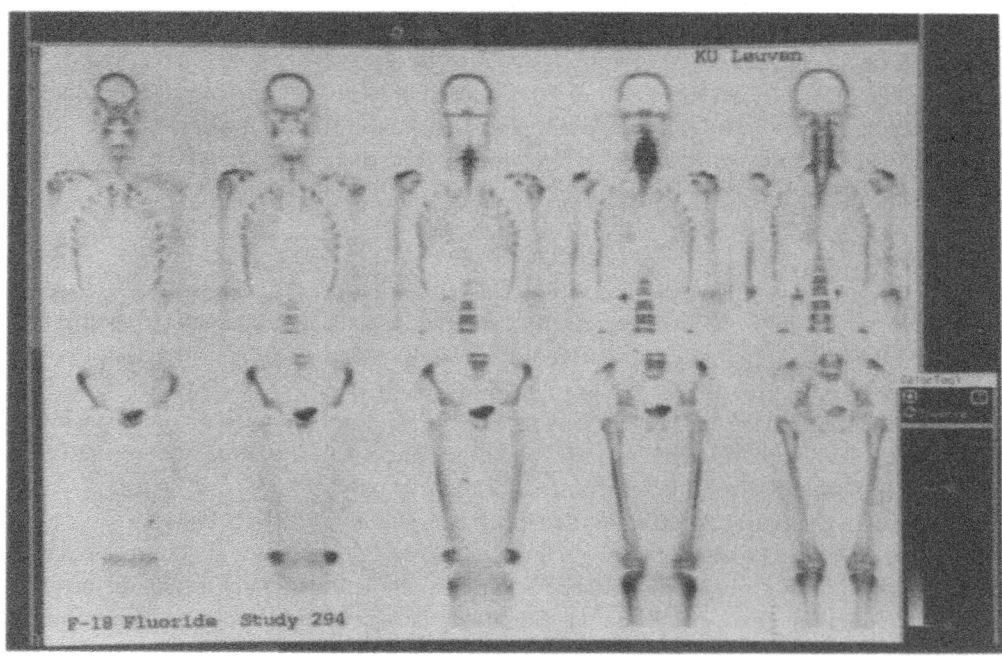

Figure 1. Normal Volunteer. The coronal images of a total body PET scan of a 30 year old white male is given. The compartmental modeling is applied to regions drawn on transaxial images. Estimated parameters for the thoracic vertebrae were: k1 = 0.09 ml/min/ml and K = 0.04 ml/min/ml. Note the detail in the spine.

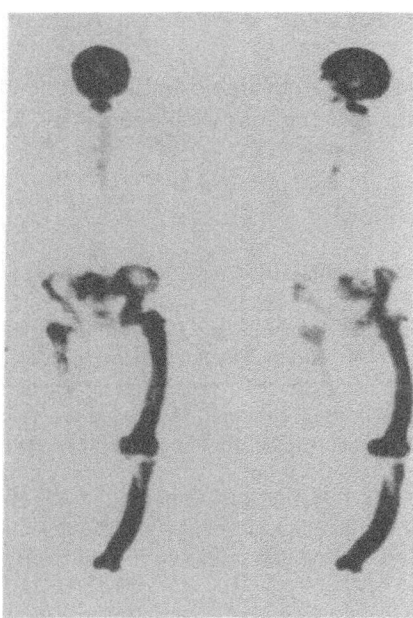

Figure 2. Paget's Disease. Projection images in anterior and left anterior oblique views of a 66 year old white female with longstanding Paget's disease. The flow estimates in the pathologic hemi-pelvis ranged from 0.11 to 0.18 ml/min/ml, which is significantly increased relative to the normal contra-lateral side with 0.06 ml/min/ml.

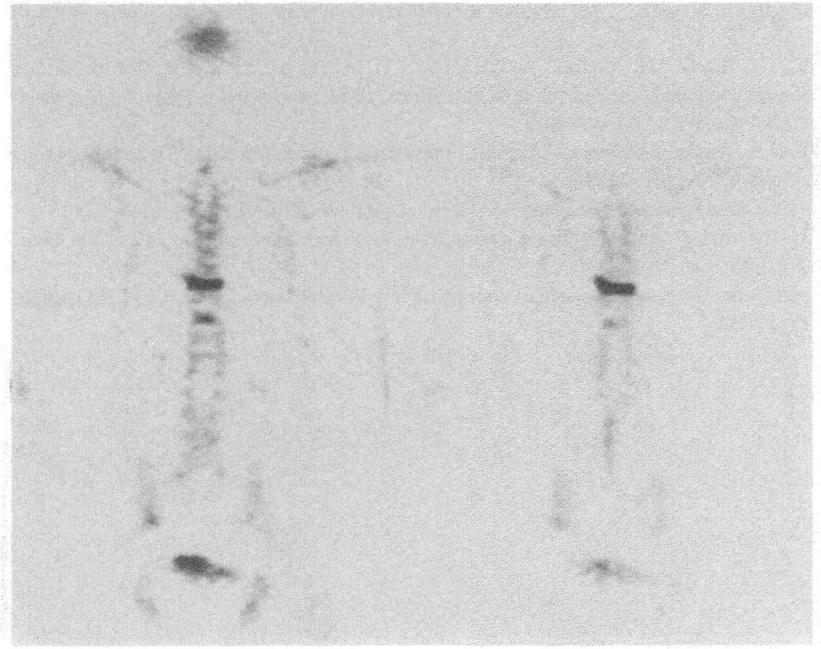

Figure 3. Osteoporosis. In this 63 year old female, a compression fracture was diagnosed in the thoracic spine. The coronal planes reveal the compression fracture. An additional unsuspected lesion, not evident on the routine bone scan, was discovered in a vertebra 2 levels lower.

In Figures 1-3 PET studies with ^{18}F are shown. Areas of abnormality can easily be identified.

Macro-K may be used to separate high from low "bone turnover" states and might be useful to monitor therapeutic interventions (Schiepers et al., 1991).

REFERENCES

Blau, M., W. Nagler, M.A. Bender: Fluorine-18: A new isotope for bone scanning. J. Nucl. Med. 3: 332 (1962).

Charkes, N.D., P.T. Makler, C. Philips: Studies of skeletal tracer kinetics: I Digital-computer solution of a five-compartment model of (^{18}F) fluoride kinetics in humans. J. Nucl. Med. 19: 1301-1309 (1978).

Charkes, N.D., M. Brookes, P.T. Makler: Studies of skeletal tracer kinetics: II Evaluation of a five-compartment model of (^{18}F) fluoride kinetics in rats. J. Nucl. Med. 20: 1150-1157 (1979).

Charkes, N.D.: Skeletal blood flow: implications for bone scan interpretation. J. Nucl. Med. 21: 91-98 (1980).

Hawkins, R.A., Y. Choi, S.C. Huang, C.H. Hoh, M. Dahlbom, C. Schiepers, N. Satyamurthy, J.R.Phelps M.E.Barrio: Evaluation of the Skeletal Kinetics of ^{18}F-fluoride ion with PET. 33: 633-642 (1992).

Li, G., J.T. Bronk, P.J. Kelly: Canine blood flow estimated with microspheres. J. Ortho. Res. 7: 61-67 (1989).

Morris, M.A., P.J. Kelly: Use of tracer microspheres to measure bone blood flow in conscious dogs. Calcif. Tissue Int. 32: 69 (1980).

Nahmias, C., W.P. Cockshott, L.W. Belbeck, E.S. Garnett: Measurement of absolute bone blood flow by positron emission tomography. Skel. Radiol. 15: 198-200 (1986).

Schiepers, C.W.J., R.A. Hawkins, Y. Choi, C. Koh, S.C. Huang, M. Dahlbom, E.J. Hoffman, J.R. Barrio, M.E. Phelps: Kinetics of Bone Metabolism assessed with ^{18}F and PET. Eur. J. Nucl. Med. 16: 450 (1990).

Schiepers, C., P. Geusens, S. Vleugels, L. Mortelmans, M. De Roo, J. Dequecker: Positron Emission Tomography (PET) with ^{18}F to evaluate metabolic rate in bone disorders. J. Min. Bone Res. 6: S243 (1991).

Simonet, W.T., J.T. Bronk, M.T. Pinto, E.A. Williams, T.H. Meadows, P.J. Kelly: Cortical and cancellous bone: age related changes in morphological features, fluid spaces and calcium homeostasis in dogs. Mayo Clin. Proc. 63: 154-160 (1988).

Van Dyke, D., H.O. Anger, Y. Yano, C. Bozzini: Bone blood flow shown with ^{18}F and the positron camera. Am. J. Physiol. 209: 65-70 (1965).

Wootton, R.: The single-passage extraction of ^{18}F in rabbit bone. Clin. Sci. Mol. Med. 47: 73-77 (1974).

Wootton, R., J. Reeve, N. Veall: The clinical measurement of skeletal blood flow. Clin. Sci. Mol. Med. 50: 261-268 (1976).

Wootton, R., C. Dore: The single-passage extraction of ^{18}F in rabbit bone. Clin. Phys. Physiol. Meas. 7: 333-343 (1986).

INTRAOSSEOUS PRESSURE, GAS TENSION AND BONE BLOOD FLOW; IN NORMAL AND PATHOLOGICAL SITUATIONS: A SURVEY OF METHODS AND RESULTS

Hakon KOFOED

Department of Orthopaedics
Frederiksberj Hospital
Copenhagen, Denmark

INTRODUCTION

In situ monitoring of intraosseous parameters may give information about haemodynamic features, diffusion gradients and nutrition in the bone under normal and pathological conditions. Measurement of the intraosseous pressure (IOP) was first performed about 80 years ago and has ever since been a classical parameter in the evaluation of different bone and joint disorders.

In situ measurement of bone blood flow (BBF) is another parameter of distinct interest and several methods have been applied. Microembolisation with microspheres, cannot be used in man for ethical reasons. Different forms of wash-out techniques have therefore been suggested (Whiteside et al., 1977, Kiær and Grønlund, 1989). Recently, laser doppler flowmetry has been introduced as an in situ method for BBF in man (Swionkowski et al., 1987). In spite of the results obtained from IOP and blood flow measurements these alone can only give clues to their effect on bone nutrition. Therefore knowledge of the environment of bone cells is of vital interest in order to change pathological conditions either by pharmacological, electrochemical, surgical or other means. Measurement of respiratory gases and pH and their behavior under different circumstances would seem essential. Mass spectrometry is a rather new method for the in situ study of respiratory gases in bone (Kofoed et al., 1983). It has the potential for analysing several gases simultaneously in different regions. This enables analysis of the interrelationsship of such parameters ignoring the classical physiological set-up where all parameters must be kept constant except for the one under study. Thereby a much more differentiated view on the intraosseous environment can be obtained.

The purpose of the present paper has been to give an introduction to the use of in situ measurements of IOP and intraosseous respiratory gases and bone blood flow. Also it will high-light some physiological and pathological conditions of interest for further studies of the intraosseous environment.

Bone Circulation and Vascularization in Normal and Pathological Conditions
Edited by A. Schoutens *et al.*, Plenum Press, New York, 1993

THE METHOD OF CONTINUOUS INTRAOSSEOUS PRESSURE MEASUREMENT

The IOP is measured in a pool of fluid with sources from mainly arterial and venous blood. Measurements are performed via a trochart which may be a hollow screw, a conical cannula or any other device which in its penetration of the cortical bone can be proven not to leak intraosseous fluids. This device (Figure 1) is connected to a fluid filled tube which again is connected to an electronic pressure transducer (Figure 2) and a writer. The demands for the system is that it does not influence the pressure positively or negatively.

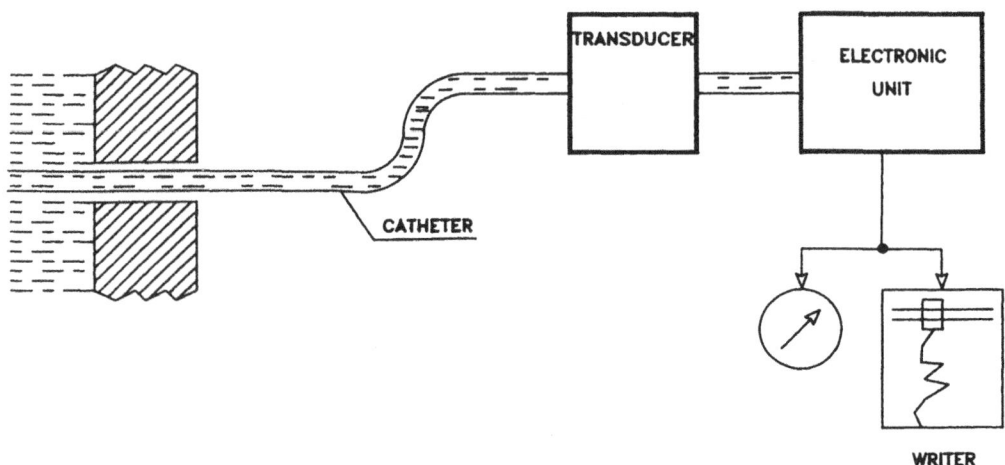

Figure 1. IOP monitoring. A schematic drawing.

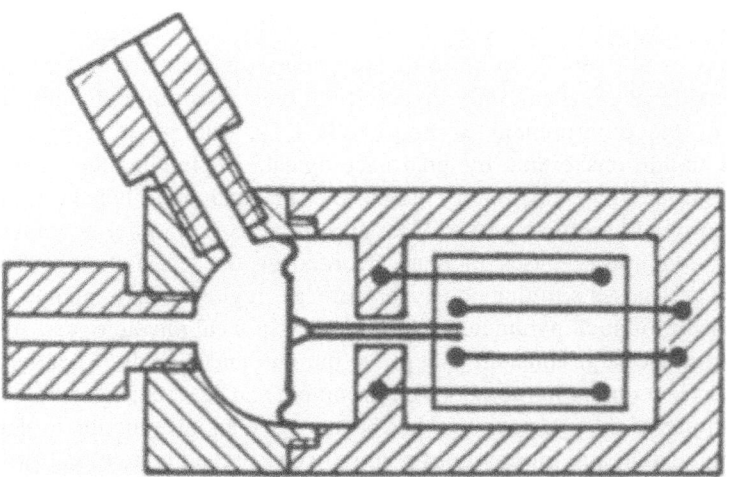

Figure 2. Illustration of an extravascular pressure transducer using strain gauges.

The Calibration Procedure

A reference point for the pressure recording is essential. Usually the left margin of the middle of the sternum is recommended as the level for the pressure transducer. As the target for measurements is a mixed pool of arterial and venous blood also the regional venous pressure should be recorded in order to be able to corret for the intraosseous venous pressure. The regional venous pressure is assumed to be equal to the intraosseous venous pressure.

In cases where the regional venous pressure for reasons of size or influence on the effluent flow cannot be used, the central venous pressure may be used as a reference. The calibration of the equipment can - if the device is not a self-calibrating electronic unit - be performed aginst different known levels from the reference point of the pressure of a bottle containing the same liquid as the tubes. In practice this is done by closing and opening a three-way stop-cock against the atmosphere and against the connection to the bottle.

Pitfalls

Air bubbles in the tubes and in the transducer may seriously disturb the compliance in the system and thereby pressure recordings. Such bubbles should let out via the three-way stopcock to the atmosphere. Another hazard is clotting of the cannula in the bone. This will record non-pulsative signals. Mostly this may be repaired by a slight turn of the cannula or by a short flushing of the cannula in situ. Some use a side connection to their cannula with constant flushing. Another possibility to avoid the phenomenon is to heparinisize both the cannula and the liquid (in man) or in cases of animal experiments, the blood of the animal. Under no circumstances should the cannula be withdrawn. If the system after all does not work with the specific cannula, it is better to plug the cannula and insert a new one from another direction. Under all circumstances: pulsative curves is a must. The position of the bone may be of importance and must be defined (Figure 3).

Measurements

Usually measurements are performed in several places simultaneously i.e. both femoral heads, the regional veins and possibly the corresponding joints. Also the arterial pressure should be measured simultaneously. The recorder should therefore have several channels. A digitalised computerised system would on the other hand be able to work with a single channel recorder.

MASS SPECTROMETRY. A METHOD OF CONTINOUS IN SITU MEASUREMENT OF PARTIAL GAS PRESSURES

A mass spectrometer is used to analyse vapours and gases, and separate them according to their molecular weights. Figure 4 shows a schematic drawing of the principle. The method is based upon gas diffusion from a medium at ambient pressure through a membrane at rates proportional to their partial pressures and into the vacuum of tubing that leads to the mass spectrometer. In the vacuum of the mass spectrometer the gases are ionized and separated and analysed according to their mass. According to Woldring (1970) the concentration distribution in a medium surrounding the blood gas

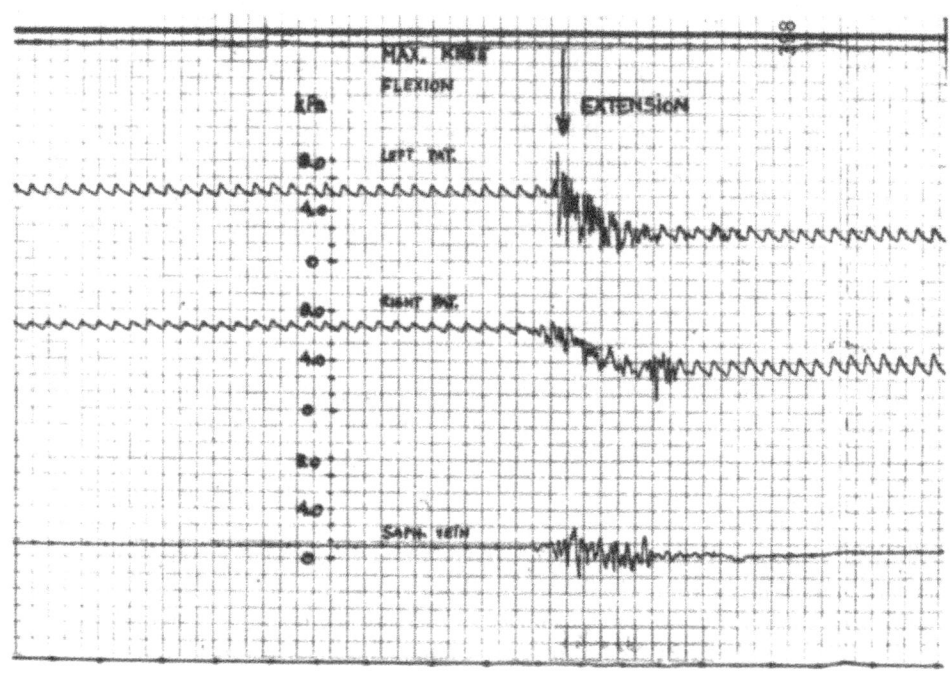

Figure 3. Illustration of the effect on IOP caused by changing the position of a knee joint.

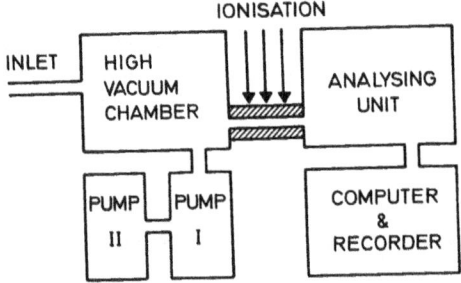

Figure 4. A schematic drawing of the principle in a mass spectrometer.

catheter membrane is a multiple function of 1) membrane conductance, 2) diffusion coefficient in the medium, 3) the stirring rate, 4) time, and 5) the distance from the medium. The reader is referred to such work for further details. The mass spectrometer enables simultaneous and continuous analysis of several gases from several measuring points.

Calibration Procedure

The calibration procedure has been dealt with in details elsewhere (Kofoed et al., 1983). Basically, it uses the signal of an inert gas, argon, as the reference gas towards oxygen because these two gases seem to share the same diffusional properties. The reason why the oxygen signal in itself must be corrected is the presence of a socalled stirring layer of infinite thickness adjacent to the membrane. This layer is characterised by delivering gases through the membrane, and at the same time receiving gases from the medium. The rate and the speed of these processes are unknown and therefore the correction of the signal is necessary by use of an inert gas with a constant and well-known partial pressure. Recently, a modification of the membrane covered catheter tip has made it possible to avoid the use of a reference gas (Kiær et al., 1988). Calibration of the apparatus can be performed in well-stirred venous or arterial blood samples with known values of the gases of interest. The practical measurements are performed by introducing the tip of the blood gas catheter through a bone cannula. Thus the target of measuring is the same intraosseous pool of venous and arterial blood as in IOP measurements.

Pitfalls

As measuring IOP, clotting on the membrane may be a problem. It is easily recognised on a scope recording the partial pressures because the argon signal will fall from its steady level. The clotting effect may be corrected by turning the tip of the blood gas catheter or move it to and for a few times. On the other hand, the membrane is extremely thin and by changing its position it may be scratched by the bone. A leak of the membrane will make the catheter unusable because the diffusion gradient between the medium and the mass spectrometer is lost. Like in IOP measurements flushing of the catheter membrane have been recommended by some. While it may prevent clotting of the membrane surface, it certainly also changes the composition of the stirring layer and therefore such flushing should not be used.

Measurements

Following calibration the measurements can be performed simultaneously at several places depending on how many catheters the apparatus is equipped with.

MASS SPECTROMETRICAL BONE BLOOD FLOW (BBF) MEASUREMENTS

The mass spectrometer can be used for spongious BBF estimations either as qualitative measurements (Grønlund et al., 1984) or as quantitative measurements (Kiær and Grønlund, 1989). In both cases the methods are based on registration of the wash-out curve of an inert gas.

Combined Measurements of IOP and Mass Spectrometry

So far simultaneous measurements in the same place is not possible. The reason is that IOP measurements should be performed without leaks in the system, whereas mass spectrometrical measurements in a closed system have a high tendency for clotting. Therefore one must either perform serial measurement with the IOP measured firstly, in order not to disturb the IOP, or measure IOP in the given region and simultaneously partial gas pressures in the corresponding region of the opposite limb.

DISCUSSION OF INTRAOSSEOUS PHYSIOLOGY

Correlation between IOP and Arterial Pressure

Under normal circumstances no correlation seems to exist between IOP and arterial pressure (T¢ndevold, 1983). This would suggest autoregulatory mechanisms at work.

Correlation between IOP and Bone blood Flow

Within normal physiological limits it has never been proven that positive or negative correlations exist between IOP and BBF. Often the assumption is forwarded that IOP is proportional to the blood flow. However, if the bone marrow or spongious bone is considered under the view that it is a multi-inlemulti-outlet system todays methods for measuring IOP and BBF are not expected to show any relationship because the assumption is based on Poiseuille's law which cannot be used for a multi-inlet/multi-outlet system.

Intraosseous Partial Pressures of Oxygen (pO_2) and Carbon Dioxyde (pCO_2) and Correlation with Arterial Gases.

Eriksen et al., (1979) using extraction of bone blood in dogs was able to demonstrate a positive semilogarithmic correlation between intramedullary pO_2 and arterial pO_2. Intramedullary pCO_2 and arterial pCO_2 showed a positive linear correlation. Using mass spectrometry in rabbits Kofoed et al. (1983) found a positive linear correlation between both pO_2 and pCO_2 in bone and the corresponding arterial values.

Correlation between IOP and Arterial pO_2 and PCO_2

T¢ndevold (1983) found no correlation between IOP and apO_2 and $apCO_2$ when apO_2 was > 10 Barr. When paO_2 was < 10 Barr and holding MAP constant > 11 Barr, IOP fell significantly. IOP was not affected by $paCO_2$. This was interpreted as a shut-off of the intraosseous circulation, but no measurements were performed to support this conclusion. In fact the phenomenon might reflect an autoregulation giving rise to vasodilation intramedullary.

Arterial-Interosseous Oxygen Gradients

In rabbits Kofoed (1986 c) demonstrated a positive linear correlation between the arterial/bone pO_2 gradient as a function of the change in arterial pO_2 (Figure 5)

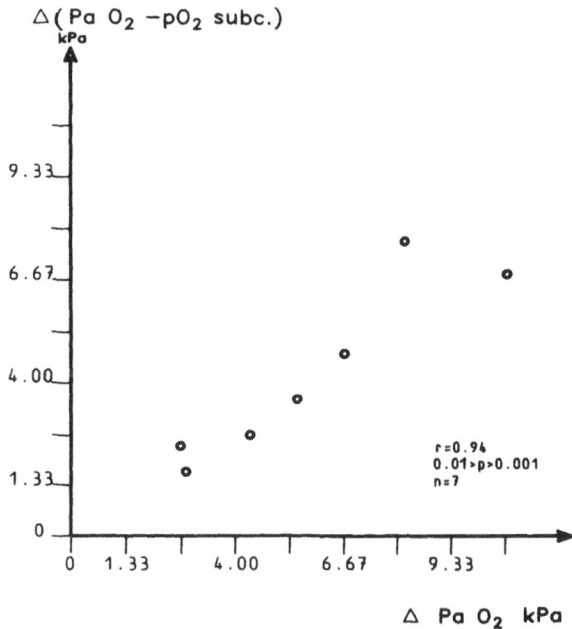

Figure 5. Differences in arterial-bone (subchondral) oxygen gradients as a function of the arterial oxygen gradient.

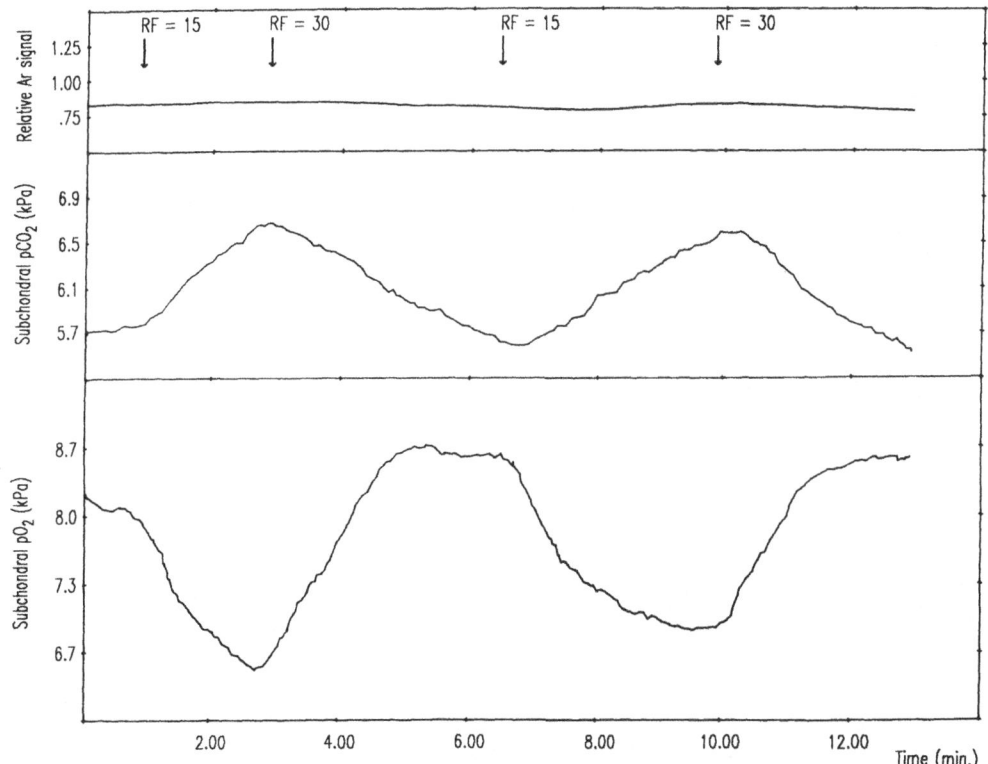

Figure 6. In situ recordings of intraosseous pArgon, pCO_2 and pO_2 during changes of the respiratory frequence (RF).

accomplished by changing the respiratory frequency (Figure 6). No correlation existed between oxygen gradients and actual apO_2. This points at an active autoregulation which tends to preserve the level of oxygen in the bone marrow even in the presence of arterial hypoxia i.e. < 10 Barr.

DISCUSSION OF INTRAOSSEOUS PATHOPHYSIOLOGY

Influence of Increased Outflow Resistance on IOP and Respiratory Gases

Arnoldi et al., (1979) were the first to demonstrate that in rabbits increased intraarticular pressure (IAP) immediately was reflected in an increased IOP. In extensive studies in puppies the same mechanism has been shown by Bünger (1987). He also measured BBF by microspheres and found unchanged BBF in the epiphysis of arthritic knees. Contrary to this, Grønlund et al. (1984) demonstrated by in situ mass spectrometry in adult juxtaarticular rabbits bones that increased IAP also was accompanied by subchondral hypoxia and a decrease in BBF. In a 3 weeks old experimental knee synovitis of adult rabbits Kofoed (1986 a) found hypoxia and acidity in the juxtaarticular bone. These findings suggest that neither outflow resistance, nor inflow resistence, is immediately corrected and this has recently been confirmed by Hansen et al. (1991) in microsphere measured puppies knees and by Kofoed and Levander (1991) using mass spectrometry and intraosseous phlebography in the spine of adult goats.

Correlation between Increased IOP and Bone Necrosis/Arthrosis

One of the few pathological conditions in which IOP measurements is a useful diagnostic tool, is femoral head necrosis. (Arlet et al., 1968). The result is not absolutely specific but in combination with core biopsy excellent. The value of this test seems more specific than scintigraphy and radiography but does exclude arthrosis which is a differential diagnosis. In arthrosis IOP is increased (Arnoldi and Reimann, 1979) and merely always in connection with a venous engorgement syndrome (Brookes and Helal, 1968; Lemperg and Arnoldi, 1978). Pain is a predominant symptom in both syndromes but the pain is a mere function of the increased IOP (Lemperg and Arnoldi 1978, Kofoed 1986 b).

Correlation between Bone Blood Flow, IOP and Intraosseous pO_2 and pCO_2

Recently Kiær et al. (1990) demonstrated a negative linear correlation between mass spectrometrically measured intraosseous pO_2 and IOP in human femoral heads with necrosis. They found no changes in intraosseous pCO_2 and no correlation towards IOP. After core biopsy intraosseous pO_2 increased, pCO_2 was unaffected and IOP decreased. Based on computer simulation, where one of the assumptions were that IOP is proportional to BBF, they estimated that with MAP fixed at 80 mm Hg an increase in IOP from 26 mm Hg to 45 mm Hg would imply a reduction in intraosseous BBF by a factor 1.6. In femoral shaft osteotomies in dogs where the bone successively was deprived of its main nutritional sources - thereby becoming hypoxic - no correlation could be established between intraosseous pO_2 and IOP. On the other hand, a negative linear correlation existed between pCO_2 and BBF (Kofoed et al., 1985) (Figure 7). This would seem to point at one possible regulatory mechanism in as much as hypercapnia has been shown to be an active

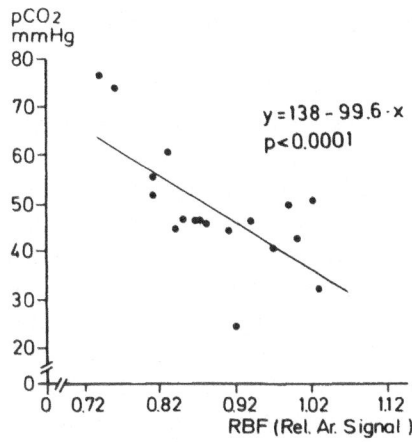

Figure 7. Correlation between qualitative bone blood flow (the relative mass spectrometrical argon signal) and pCO_2 in bone during progressive hypoxaemia.

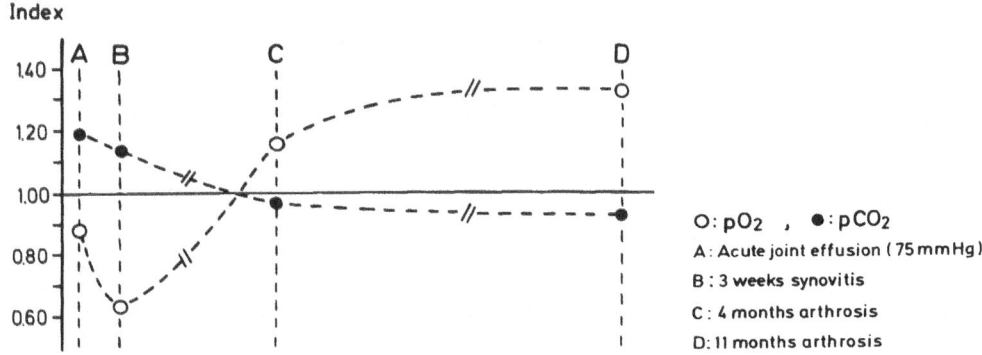

Figure 8. Arthrosis/normal-index of the intraosseous pO_2 and pCO_2 in normal bone and experimental arthrosis.

vasodilator. Regulatory mechanism must exist because initially dead bone do come to live (Catto, 1965; Gardeniers, 1988).

Respiratory Gases, IOP and Bone Blood Flow in Arthrosis

The pathomechanism of arthrosis is well-documented from radiographic and histological studies. But from a pathophysiological point of view it is still unclear which mechanisms that initiate the arthrosis process. In classical works the association with venous engorgement and increased IOP has been established (Brookes and Helal, 1968; Arnoldi and Reimann, 1979). Both phenomenons have been ascribed to intravascular disorders, either thrombo-embolism or fat embolism (Jones, 1979) leading to an arthrosis/necrosis cascade. No matter the initial injury the vascular response seems of utmost importance; but probably not so much the status of IOP and BBF as that of the possibilities for the local respiratory system. In that respect knowledge of pO_2 and pCO_2 seems essential. In an 11-month experimental rabbit model Kofoed(1986 d) found venous congestion, increased IOP, and increased intraosseous pO_2. pCO_2 and pH were unaffected. However, multiple regression analysis showed a positive linear correlation between IOP and pCO_2. The increased pO_2 was ascribed to hyperemia. Kiær et al. (1988) on the other hand, found intraosseous hypoxia and increased IOP in human coxarthrosis. pCO_2 and pH were unaltered, while lactate was increased. They ascribed their findings to either an increase in oxygen consumption or anaerobic metabolism. A direct comparison of the two series is not possible. Knowing that the arthrosis process is rather slow the differences might merely reflect different stages in the process. At least in the experimental arthrosis process such differences are found at intervals (Figure 8).

CONCLUSIONS

Progress and new knowledge are often connected with new methods. This is definitely the case with bone physiology. In the search for physiological events and exposure of pathomechanisms we should not forget that we at all times are prisoners of the available methods. With that in mind all results should be looked upon as true for what ever the method is worth. The interpretation of the results is the jumping point - and until all agree our patients cannot benefit from the clinical exploratation of the results. This survey has shown that there are still many problems yet to be solved.

REFERENCES

Arlet, J., P. Ficat, D. Sebbag.: Intérêt de la mesure de la pression intramédullaire dans le massif trochanté-rien chez l'homme, en particulier pour le diagnostic de l'ostéonécrose fémoro-capitale. Rev. Rhum. 35: 250-256 (1968).

Arnoldi, A., I. Reimann, S.B. Christensen, S. Mortensen: The effect of increased intraarticular pressure on juxtaarticular bone marrow pressure. IRCS Med. Sci. 7: 471 (1979).

Arnoldi, A., I. Reimann: The pathomechanism of human coxarthrosis. A synthesis. Acta Orthop. Scand. (suppl. 181) (1979).

Brooke,s M., B. Helal: Primary osteoarthritis, venous engorgement and osteogenesis. J Bone Joint Surg[Br] 50-B: 493-504 (1968).

Bünger, C.: Hemodynamics of the juvenile knee. Acta Orthop Scand. (Suppl. 222) (1987).

Catto, M.: A histological study of avascular necrosis of the femoral head after transcervical fracture. J. Bone Joint Surg. [Br] 47-B: 749-76 (1965).

Eriksen, J., E. Tøndevold, E. Jansen, J.E. Petersen: Relationships between oxygen and carbon dioxide tensions and acid-base balance in arterial blood and medullary blood from long bones in dogs. Acta Orthop. Scand. 50: 519-525 (1979).

Gardeniers, J.W.M.: Behavior of normal, avascular and revascularizing cancellous bone. Thesis. Nijmegen 1988.

Grønlund, J., H. Kofoed, E. Svalastoga: Effect of increased knee joint pressure on oxygen tension and blood flow in subchondral bone. Acta Physiol. Scand. 121: 127-131 (1984).

Hansen, ES., V.E. Hjortdal, D. Kjølseth, S.Z. He, K. Høy, K. Søballe, C. Bünger: Arteriovenous shunting is not associated with venous congestion in bone. Acta Orthop. Scand. 62(3): 268-275 (1991).

Jones, J.P. Jr.: Osteonecrosis. In: J. McCarty (Eds) Arthritis and allied conditions, pp. 1121-1131. Lea & Febiger, Philadelphia 1979.

Kiær,T., J. Grønlund, K.H. Sørensen: Subchondral pO_2, pCO_2, pressure, pH and Lactate in human osteoarthritis of the hip. Clin. Orthop. 229: 149-155 (1988).

Kiær,T., J. Grønlund: A new method of measuring blood flow distribution in bone. In: J. Arlet , B. Mazieres (eds.): Bone circulation and bone necrosis, pp. 109-113. Springer-Verlag 1989.

Kiær,T., N.W. Pedersen, K.D. Kristensen, H. Starklint.: Intraosseous pressure and oxygen tension in avascular necrosis and osteoarthritis of the hip. J. Bone Joint Surg. [Br] 72-B: 1023-1030 (1990).

Kofoed, H., E. Svalastoga, J. Grønlund, B. Jensen, J. Kofod, P. Svendsen: Continuous measurement of subchondral pO_2 and pCO_2 by mass spectrometry. IRCS Med. Sci. 11: 583-584 (1983).

Kofoed, H., E. Sjøntoft, S.O. Siemssen, H.P. Olesen: Bone marrow circulation after osteotomy. Acta Orthop. Scand. 56: 400-403 (1985).

Kofoed, H.: Synovitis causes hypoxia and acidity in synovial fluid and subchondral bone. Injury 17: 391-394 (1986 a).

Kofoed, H.: Positive correlation between osteoarthritic ankle pain and bone marrow pressure. J. Rheumatol. 13: 801-803 (1986 b).

Kofoed, H.: Oxygen gradients between arterial blood and normal synovial fluid and subchondral bone. IRCS Med. Sci. 14: 250-251 (1986 c).

Kofoed, H.: Hemodynamics and metabolism in arthrosis. Acta Orthop. Scand. 57: 119-122 (1986 d).

Kofoed, H., B. Levander: Epidural venous stasis of the lumbar spine. ARCO 3 (1): 34-43 (1991).

Lemperg, R.K., C.C. Arnoldi: The significance of intraosseous pressure in normal and diseased states with special reference to the intraosseous engorgement-pain syndrome. Clin. Orthop. 136: 143-156 (1978).

Swiontkowski, M.F., R. Ganz, U. Schlegel, S.M. Perren : Laser doppler flowmetry for clinical evaluation of femoral head osteonecrosis. Clin. Orthop. 218: 181-185 (1987).

Tøndevold, E.: Haemodynamics of long bones. Acta Orthop. Scand. (suppl. 205) (1983).

Whiteside, L.A., B.S. Lesker, D.J. Simmons: Measurement of regional bone and bone marrow blood flow in the rabbit using hydrogen washout technique. Clin. Orthop. 122: 340-346 (1977).

Woldring, S.: Tutorial: Biomedical application of mass spectrometry for monitoring partial pressures. J. Assn. Advan. Med. Instrum. 4: 43-56 (1970).

FRACTURE HEALING AND
BONE GRAFTS

THE ROLE OF THE VASCULATURE IN FRACTURE HEALING

S.P.F. HUGHES, I.D. McCARTHY, and M.F. BROWN

Orthopaedic Surgery
Royal Postgraduate Medical School
Hammersmith Hospital
London, United Kingdom

INTRODUCTION

A fracture is the bony manifestation of an injury to part of the body, usually a limb, and is necessarily associated with a soft tissue injury. The severity of the overall injury may bear little relation to, for example, the radiographic appearance of the fracture itself. The degree of disruption of the soft tissues affects the degree of interference with the blood supply of the injured area, and this is influenced by such factors as the velocity of injury.

Ham (1955) laid down the descriptive process for fracture repair in which he defined several phases of repair. Initially, a haematoma develops; this is followed by granulation tissue, followed by the formation of callus and finally bone, which then remodels to the original contours of the bone.

In the section Anatomy and Physiology, on page 57, we outlined the physiological structure of bone, and the relationship between the blood supply, the fluid spaces and the bone itself. The effect of flow was also examined, and it was possible to show that in normal bone solutes necessary for bone formation leave the capillary by the process of free passive diffusion and that this movement is flow limited at low flows and diffusion limited at higher flows (McCarthy et al., 1980).

Blood flow changes after an experimental osteotomy have been measured (Paradis and Kelly, 1975; McCarthy and Hughes, 1984). It was shown that in the animal fracture model there is an increase in flow reaching a maximum at two weeks after osteotomy, and that this flow gradually returns to near normal by 12 weeks.

More recently, Heppenstall (1980) has examined the process of fracture repair and has suggested that following a fracture there are a series of events which take place: impact, induction, inflammation, soft callus, hard callus, remodelling. This process implies the concept of impact having an affect unique to the injury, resulting then in the cascade of events which promote the process of fracture healing.

Recent studies have also demonstrated the importance of the process of induction on both

Bone Circulation and Vascularization in Normal and Pathological Conditions
Edited by A. Schoutens *et al.*, Plenum Press, New York, 1993

115

the endosteal and periosteal cells which are present in bone (Chapman, 1987). Osteoinduction is a process whereby cells of mesenchymal origin, which ordinarily do not form bone, acquire that property through influence of intrinsic or extrinsic factors. These influences are indeed various and include a host of known factors which promote osteoinduction. Some of these factors are: altered oxygen tension, biophysical-electricity, bone morphogenic protein, growth factors, micromotion.

There is now an increasing research effort directed towards understanding the role of growth factors in the control of fracture repair. Platelet derived growth factor (PDGF) and transforming growth factor beta (TGF-b) have been shown to be released by platelets into the fracture haematoma, and it has been shown that TGF-b can induce differentiation in mesenchymal cells to form osteoblasts and chondrocytes (Joyce et al., 1991).
Hence in the process of fracture healing it would appear that osteoprogenitor cells which are present on the endosteal and periosteal surface transform into mature osteoblasts under the influence of these inductive factors. This results in the formation of bone and the fracture repair process.

THE VASCULATURE

The paper of Paradis and Kelly (1975) was of importance in confirming clinical experience that there was an important time in the fracture healing process when the blood supply was reaching its maximum and when the products necessary for the fracture to heal were being transported to the fracture site. Smith from Birmingham had laid down the principal of the optimal biological environment for the fracture, in his original observations on the healing of forearm fractures (Smith, 1959).

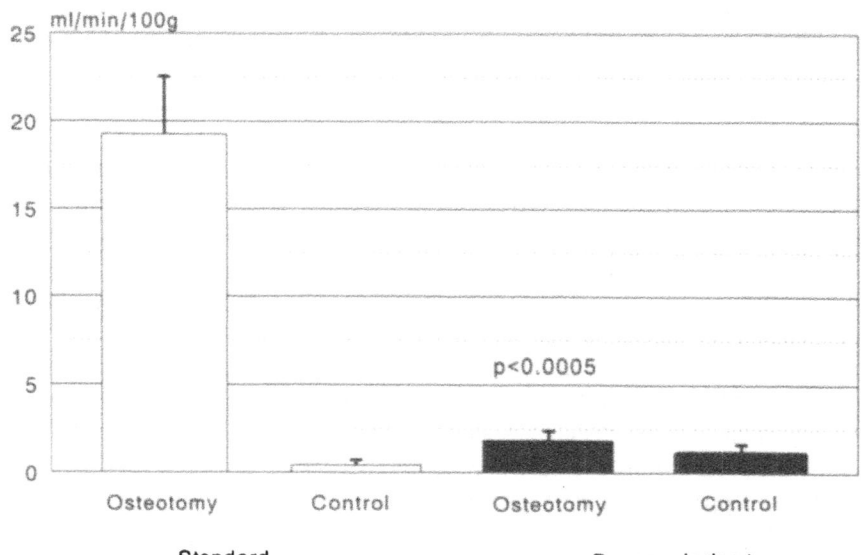

Figure 1. Cortical blood flow as measured by microspheres in the vascularised and devascularised model of ovine tibial osteotomy at two weeks after the osteotomy.

Strachan et al. (1990) examined in detail the role of the periosteal blood supply in the osteotomised canine tibia and emphasised the importance of this source of blood supply in the callus and cortex. This work complemented the observations of Brookes, in which he showed that both the periosteum and indeed the muscle were important factors in supplying the outer third of bone (Brookes, 1971).

Recently Wallace et al. (1991) examined the effect of the blood flow on an experimental osteotomy in the ovine tibia. They studied the blood flow in two groups of animals, one in which the periosteum was removed from 2 cm either side of the osteotomy and a silicone sleeve then placed round the exposed cortex to prevent revascularisation, and the other in which the periosteal blood supply remained intact. Using the microsphere technique, the authors were able to show that the blood flow to the tibial diaphysis as a whole was increased in both cases, but that adjacent to the fracture site blood flow in the devascularised model was significantly reduced compared with the standard osteotomy, particularly in the first two weeks after the fracture. Medullary flow was increased at 2 weeks in both groups, not only in the bones as a whole but also at the osteotomy site. In recent studies Wallace et al. (1992) have been able to confirm this trend at 6 weeks (Figures 1 and 2).

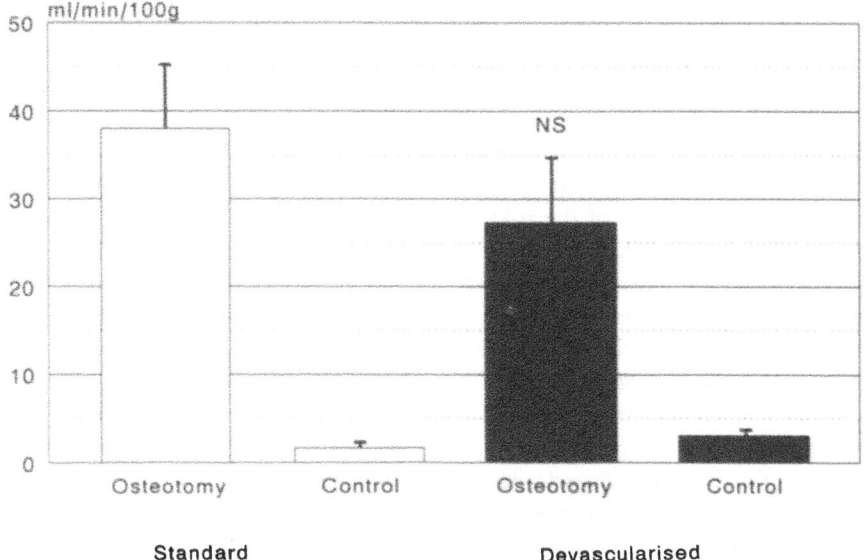

Figure 2. Marrow blood flow as measured by microspheres in the vascularised and devascularised model of the ovine tibial osteotomy at two weeks after the osteotomy.

Hence it can be concluded that the periosteum and adjacent soft tissues are of real importance to the fracture healing processs, and that if damaged, stripped or removed in any way, the revascularisation that is required for successful fracture repair is very considerably diminished. Indeed, Nather et al. (1990) went even further, to show that the surrounding muscle bed seemed to be more important than the periosteum.

This concept of the importance of the blood supply on fracture healing has been recognised since the pioneering work of Trueta (1963), who postulated that blood vessels played a fundamental role in osteogenesis.

CLINICAL APPLICATION

The experimental evidence can be and is applied to clinical practice. Firstly in the classification of fractures, Gustilo (1982) produced an important classification of open fractures, whereby he related the outcome of the fracture to the degree of soft tissue damage. These observations have been developed further by Oestern and Tscherne (1983) in the outcome of closed fractures. They have shown that success in treating these fractures is also related to the degree of soft tissue damage.

McQueen and Court-Brown (1987) conducted a retrospective study of the rate of union of lower limb fractures in patients who were known to have raised compartmental pressure, and showed that those who had compartment syndrome had a longer period of fracture healing then a similar aged group who did not. This again is further evidence that impairment of the blood flow through loss of the muscle and perhaps the periosteum retards the process of fracture healing. Clearly, it is important to cover exposed bone in an open fracture and use has been made of this concept in the clinical use of myocutaneous flaps.

Further work is now being carried out in our own laboratories and in other centres on the role of intramedullary reaming on the fracture healing process. It may well be that intramedullary reaming has a significantly detrimental effect on the endosteal blood supply, which if combined with the effects on the soft tissue of a high velocity injury, could produce significant long term effects on bone fracture healing.

An exciting new development in the study of fracture healing is the role of molecular biology. With the awareness of the role of growth factors and cytokines in transforming the multipotential cell into the osteoprogenitor cell large areas of research are opening up. These factors appear to be released in response to the injury. Several factors have been examined in relationship to the vasculature in this area, including endothelial cell stimulating and activation factor (ESAF), which appears to be specific for endothelial cells of the microcirculation, and which has been identified at elevated levels in the circulation of patients with fractures (Wallace et al. 1991). The role of substances such as endothelin and nitric oxide, which are produced by endothelial cells, but have been shown to have direct effects on bone, needs also to be investigated.

The area of research therefore is vast, and the methods available for studying the fracture repair process are increasing and improving. The contribution of Trueta on the role of the periosteum, and Brookes on the value of the muscle attachment and the centrifugal nature of flow are of real importance. Kelly and his co-workers have defined the structure and the function of the blood supply to bone, whilst Smith applied clinical observation to this important area.

It would seem that within the fracture healing process there really is an optimal biological environment which needs to be maintained in order that the fracture heals.

REFERENCES

Brookes, M. : The Blood Supply of Bone, Butterworths, London, 1971.
Chapman, M.W.: Induction of fracture repair: osteoinduction, osteoconduction, and adjunctive care. In: J. Lane (ed): Fracture Healing, Churchill Livingstone. New York 1987.

Gustilo, R.B.: Management of open fractures and complications. American Academy of Orthopaedic Surgeons, Instructional Course Lectures 31: 64-75 (1982).

Ham, A.W.: Histology. Piman, London, 1955.

Heppenstall, R.B: Fracture healing. In: R.B. Heppenstall (ed): Fracture Treatment and Healing. WB Saunders, Philadelphia 1980.

Joyce, M.E, S. Jingushi, S.P. Scully, M.E. Bolander: Role of growth factors in fracture healing. In: Clinical and Experimental Approaches to Dermal and Epidermal Repair: Normal and Chronic Wounds. pp 391-416, Wiley-Liss 1991.

McCarthy, I.D., J.S. Orr, S.P.F. Hughes: An experimental model to study the relationship between blood flow and uptake for bone-seeking radionuclides in normal bone. Clin. Phys. Physio. Meas. 1: 135-143 (1980).

McCarthy, I.D., S.P.F. Hughes: Extraction of ^{99m}Tc-methylene diphosphonate as a function of bone blood flow. In: J. Arlet, R.P. Ficat, D.S. Hungerford (eds): Bone Circulation. Williams and Wilkins, Baltimore 1984.

McQueen, M.M., C. Court-Brown: Compartment syndrome delays tibial union. Acta Orthop. Scand. 58: 249 (1987).

Nather, A., P. Balasubramanian, K. Bose: Healing of non-vascularised diaphyseal bone transplants. J. Bone Joint Surg 72-B: 830-834 (1990).

Oestern, H-J., H. Tscherne: Pathophysiology and classification of soft tissue injuries associated with fractures. In: H. Tscherne and L. Gorgen (eds): Fractures with Soft Tissue Injuries. Springer Verlag, Berlin 1983.

Paradis, G.R., P.J. Kelly: Blood flow and mineral deposition in canine tibial fractures. J. Bone Joint Surg. 57-A: 220-226 (1975).

Smith, J.E.M.: Internal fixation in the treatment of fractures of the shafts of the radius and ulna in adults: the value of delayed operation in the prevention of non-union. J. Bone Joint Surg. 41-B: 122 (1959).

Strachan, R.K., I.D. McCarthy, R.H. Fleming, S.P.F. Hughes: The role of the tibial nutrient artery: microsphere estimation of blood flow in the osteotomised canine tibia. J. Bone Joint Surg. 72-B: 391-394 (1990).

Trueta, J.: The role of vessels in osteogenesis. J. Bone Joint Surg. 45-B: 402-418 (1963).

Wallace, A.L., E.R.C. Draper, R.K. Strachan, I.D. McCarthy, S.P.F. Hughes: The effect of devascularisation upon early bone healing in dynamic external fixation. J. Bone Joint Surg. 73-B: 819-825 (1991).

Wallace, A.L., E.R.C. Draper, R.K. Strachan, I.D. McCarthy, S.P.F. Hughes: The vascular response to fracture micromovement. (submitted).

Wallace, A.L., B. McLaughlin, J.B. Weiss, S.P.F. Hughes: Increased endothelial cell stimulating angiogenesis factor in patients with tibial fractures. Injury 22: 375-376 (1991).

HAEMODYNAMICS OF BONE HEALING IN A MODEL STABLE FRACTURE

R.N. BRUETON, W.J. REVELL and M. BROOKES

Academic Orthopaedic Research Laboratories
Rayne Institute, St.Thomas' Hospital
London, United Kingdom.

INTRODUCTION

It is generally held that the normal vascular response to the occurence of a bone fracture is a prominent elevation of the blood supply to the bone as a whole, and especially the fracture site (Lexer et al., 1904; Axhausen, 1926).
Utilizing the distribution of radioisotope labelled particles for bone blood flow measurement in experimental osteotomies of the humerus, Brookes et al. (1970) observed a 4-fold increase in blood flow rate in the reparative site, during the brief inflammatory phase from 1-3 weeks after osteotomy. Thereafter the blood flow rate in the reparative site of this unstable fracture was some 10 % below the contralateral quasi-normal flow. Simultaneously, the flow rate in the humerus as a whole was elevated for 6 months, long after the bone had healed 8 weeks postoperatively.
Brookes (1977) made a bore-hole in the rat tibia, and monitored its healing progress histologically. Serial flow studies using arteriolar blockade showed that a profound fall (-50 %) in flow rate occurred at the reparative site of this stable fracture during the 3 days following construction of the tibial defect, and before any new bone trabeculae could be detected. Flow elevation coincided with soft callus in the hole. Flow rate fell with consolidation; and returned to near normal flow rate with bone remodelling.

Because flow measurement suggests that ischaemia is also a feature of bone consolidation and remodelling (Brookes, 1977), it was decided to investigate experimentally the repair of a larger but stable fracture, which was readily reproducible and did not require fixation apparatus. The aim of the investigation was to correlate angiographic, histological and haemodynamic data at various stages in the repair process in order to amplify the role of the blood supply in fracture repair. It was further intended that simultaneous blood flow rate and blood volume measurements might point to changes in the pH of the fracture site, which are basic to its changing morphology during the healing process.

Bone Circulation and Vascularization in Normal and Pathological Conditions
Edited by A. Schoutens *et al.*, Plenum Press, New York, 1993

MATERIALS AND METHODS

A hundred and thirty Wistar rats, weighing 200 g ± 25, and 10 weeks old were used. Under combined Thalamonal and Diazepam anaesthesia, a longitudinal incision was made in the skin of the medial aspect of the leg, exposing the anterior tibial crest and medial surface. A longitudinal 1 cm slot was cut into the rat tibia just above the tibio-fibular synostosis, using a dental drill with a circular saw blade attachment. The wound was brought together in layers using Mersilk 3.0 sutures.

Haemodynamic Investigations

Blood Flow Rate and Volume. These measurements were made at intervals from 1 day to 8 weeks after fracture. Initially groups of 4 animals were studied at 1, 2, 3, 5 days, and 1, 2, 3, 4, 6 and 8 weeks, and a further 12 animals were studied at 2 days and 1, 3, 4 and 8 weeks. Flow rates were measured by arteriolar blockade (particle distribution method utilising radioactive ^{59}Fe-iron chloride adsorbed on to cationic resin particles. Bone blood volumes were measured by haemodilution, utilizing red blood corpuscles labelled with radioactive ^{51}Cr chromate.

In practice, both blood flow rates and volumes were measured simultaneously. A quantity of 0.3 ml radioactive ^{51}Cr-chromate-labelled packed red blood corpuscles was injected into the jugular vein of the anaesthetised rat. After a 10 minute mixing time, a known radioactive dose of cationic resin particles, whose maximum length varies from 10-60 μm (median 35 μm) and which were labelled with radioactive ^{59}Fe-iron chloride, was injected into the carotid artery in a single bolus towards the heart. Three blood samples of mixed arterio-venous tail blood per rat were then gathered in microhaematocrit tubes. The circulation was then stopped instantly by placing the anaesthetized animal in a Dewar flask of freezing acetone fluid at - 50°C. These procedures enable one to calculate the blood volume in any defined organ below the diaphragm, by analysis of the ^{51}Cr-chromate counts; and also the blood flow rate in the organ, by analysis of the ^{59}Fe-iron chloride counts (Brookes, 1970, 1971). Preparations were made of whole rat tibiae and femora, and of transverse sections cut through the slotted zone of the tibia. All specimens were weighed and their radioactivities calculated simultaneously on 2 channels of a LKB 1582 Compagumma automatic scintillation counter.

The pH of Bone. The pH of the healing slot was estimated from the flow data by applying the formula:

$$pH_t = pH_c + k.\log_e \frac{Flow \ \%}{Vol \ \%} \tag{1}$$

developed by one of us (Prof.M.Brookes), where t is for tissue, c is for control bone, and k is a constant taken as unity. pH_c was arbitrarily fixed at 7.4. Flow % change and Vol % change were measured directly.

Angiographic Investigations. Three animals were set aside at each time period for angiographic and histological examination. These were perfused under general anaesthesia intra-arterially with barium sulphate suspension so that the bone vasculature might be studied, after due processing, by microfocal angiography. In practice, the rat hind limbs were perfused with 45 % barium sulphate suspension after cannulation of the aorta. The tibiae were removed, decalcified in 5 % nitric acid in 10 % formol saline solution, and transverse sections through the slotted zone were radiographed in a Hilger & Watts microfocal X-ray unit, using Ilford maximum resolution film.

Histological Investigations. After angiography the transverse tibial sections through the slotted zone were embedded in paraffin. Sections were taken at 5 μm and routinely stained with either haematoxylin and eosin, or by a one-step trichrome stain, allowing the histological features of the healing slot to be recorded.

RESULTS

Angiography and Histology

From 1 - 3 days. Angiographically, the slot was easily visualized. The cortex and marrow in relation to the slot appeared bloodless, in contrast to the well vascularized marrow opposite to the slot in the tibial section. Histologically, blood clot, fibrin clot, initial invasion of large and small cellular elements successively made their appearance (Fracture haematoma and cell immigration).

At one week angiography showed revascularization of the marrow and the presence of blood vessels in the tibial cortical slot (Figure 1). Histologically, the tissue in the slot was an osteogenic blastema (Figure 2) (Early inflammatory phase).

At 2 weeks blood vessels in the slot were prominent, and the cortex was hyperaemic. Angiography also showed external callus deposition on the periosteal surface of the section, the callus being supplied by periosteal blood vessels.

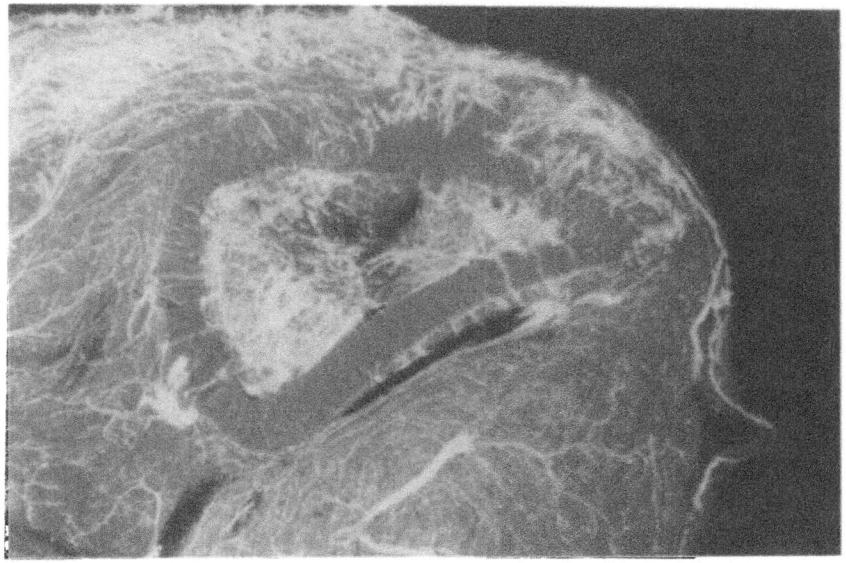

Figure 1. Angiograph of a transverse section taken through the tibial slot at 1 week postoperatively, showing revascularization of the marrow cavity and cortical hyperaemia.

Histologically, enlargement of vascular canals in the cortex had occurred and fine bone trabeculae were present on the periosteal bone surface. Cartilage was always absent (Stage of soft callus formation).

At 4 weeks the fracture site, i.e. the profile of the slot in transverse section, was readily observable in angiographs on account of several small arterial vessels crossing the entire cortical thickness. External callus was also present, associated with an intensely vascular periosteum (Figure 3). Histologically, most of the medial cortex of the tibia was riddled with large spaces filled with haemopoietic marrow. The site of the slot, however, was filled in with bone trabeculae orientated radially, and not circumferentially in congruence with the medial tibial cortex as a whole (Figure 4). A thin layer of hard callus lies on the external surface of the section (Stage of consolidation).

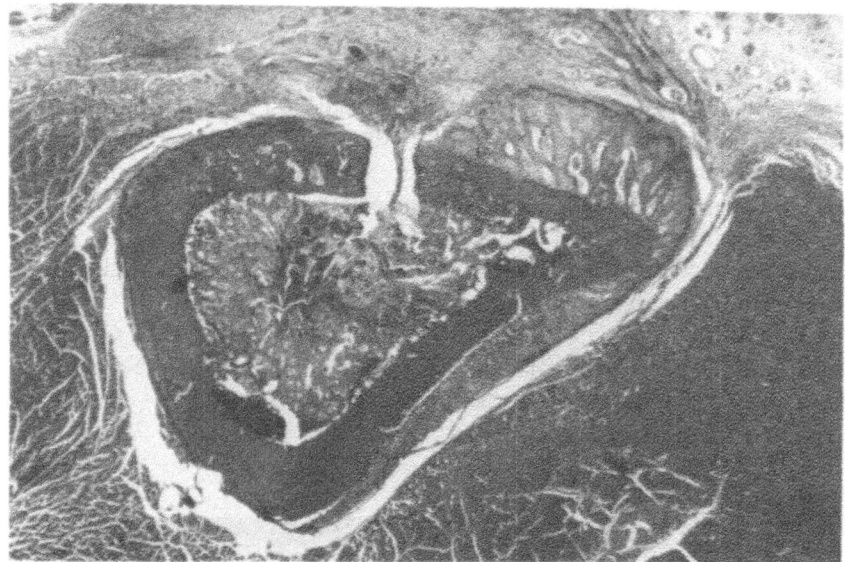

Figure 2. Photomicrograph of transverse section taken from specimen in Figure 1. The slot is occupied by a vascularized osteogenic blastema. Note vessels filled with barium perfusate, and new bone deposits on the periosteal surface.

At 6 weeks the slot is angiographically hypervascular, but both marrow and cortex seem to have defervesced from the vascular point of view, and histologically cortex and marrow are near normal.

At 8 weeks the site of the slot is discerned with difficulty. Some hypervascular spots are still in evidence in the cortex. The bone section as a whole is undistinguished, both angiographically and histologically.

Bone Blood Flow Measurement

The Slotted Tibial Zone. Table 1 shows the percent change in blood flow rates and volumes observed during the repair of a tibial slot, produced in the left tibia of the experimental rats. Because different numbers of rats were used at 2 days and 3 and 4

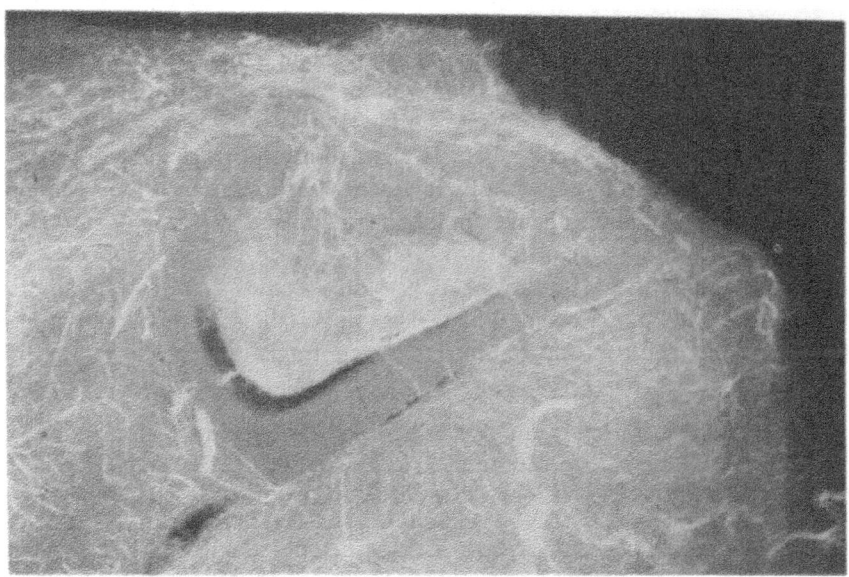

Figure 3. Angiograph through the tibial slot at 4 weeks postoperatively. Arteries cross the full thickness of the cortex at the site of the slot in the medial cortex (uppermost). The cortex as a whole is hyperaemic, and the medial cortex is covered with hard callus supplied by periosteal arteries.

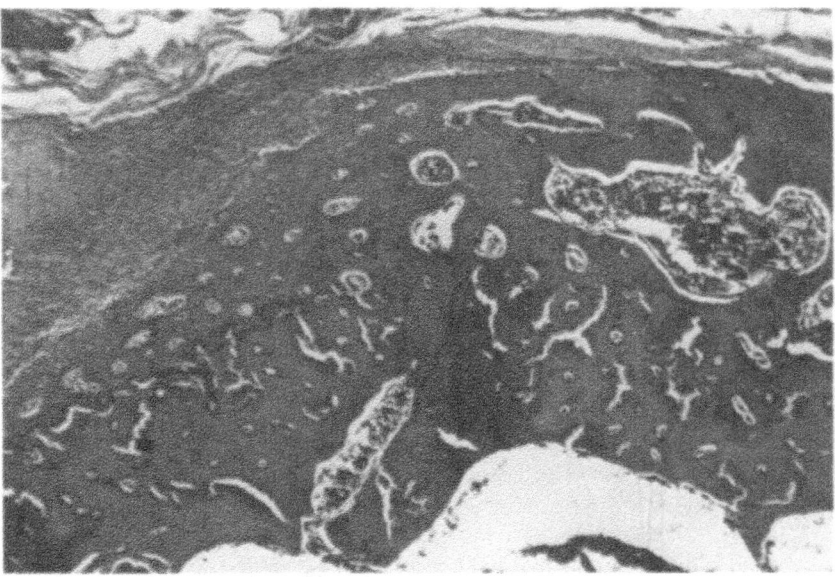

Figure 4. Photomicrograph of a transverse section taken from Figure 3, showing medullization cavities in the remodelling cortex. The external surface is coated with hard callus. Note the radially oriented bone trabecula marking the site of the slot.

weeks, weighted averages were calculated for these time intervals. Figure 5 is a graph showing the changing blood flow rate and blood volume measured in the zone of the tibial slot in the course of its repair. The graph shows that from 1-3 days there is a profound reduction in both flow rate and blood volume in the zone of the slot. At one week, flow rate has increased by 30 % over normal but declines to 20 % below normality at 4 weeks. From 4-8 weeks there is a gradual return of blood flow rate to normal values. A similar pattern is shown in the per cent change in bone blood volume in the tibial slot, which lags behind the changes in blood flow rate.

Whole Femora and Tibiae. The flows L/R % and volumes L/R % were similarly measured in the rat whole femora and tibiae. The flow tables however, showed no significant deviation from the flows and volumes in the contralateral bones.

Table 1. Left to right haemodynamic tibial slot comparisons.

Period post-op	No.of Rats	Flows L/R %	Vols L/R %
1 d	4	92.3	90.9
2 d	12	46.0 ± 4.4	58.3 ± 4.7
	4	44.2	77.4
	weighted average	45.6	63.1
3 d	4	107.6	99.7
5 d	4	105.6	100.3
1 w	12	130.5 ± 5.8	105.5 ± 3.5
11 d	4	–	115.0
2 w	4	109.8	119.0
3 w	12	82.7 ± 5.0	100.8 ± 4.1
	4	78.0	98.0
	weighted average	81.5	100.1
4 w	12	84.7 ± 4.0	83.5 ± 3.2
	4	94.4	82.8
	weighted average	87.1	83.3
6 w	4	86.6	82.4
8 w	4	94.4	91.6

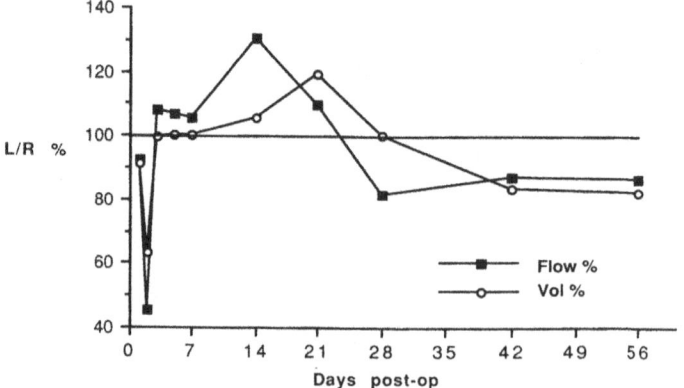

Figure 5. Graph of the haemodynamic data shown in Table 1.

126

The pH of the Healing Tibial Slot. The pH graph (Figure 6) suggests that in the stage of cell immigration an unusually low pH is present. This is followed by a short-lived interval of elevated pH accompanying soft callus formation. Hard callus and bone consolidation are histologically observable when the pH is abnormally low. Remodelling of the fracture site (slotted zone) takes place with the pH mildly elevated above the normal.

Whole Bone pH. The pHs derived from the flow rate and volume data for whole femora and tibiae showed no significant deviation on the left operated side from their equivalents on the right.

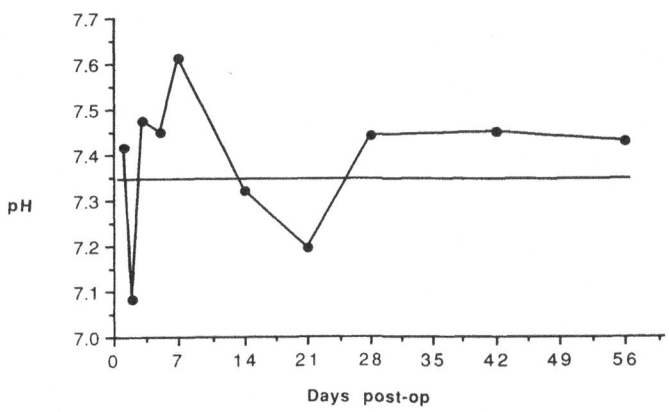

Figure 6. Graph of the pH changes calculated from data shown in Table 1.

Discussion

The results show that in the repair of a model stable fracture, represented by a 1 cm long slot in the medial tibial cortex, angiographic, histological and haemodynamic changes coherently advance and coincide with major structural changes taking place on the fracture site. These are: - Stage 1. Blood clot, cell immigration, formation of an osteogenic blastema: Fall in flow rate and volume; low pH (acid drift); - Stage 2. Soft callus formation: Raised flow rate and volume; elevated pH (alkaline drift); - Stage 3. Hard callus with bone consolidation: Reduced flow rate; low pH (acid drift); - Stage 4. Remodelling: Subnormal flow rate and volume; raised pH (alkaline drift). In general all these stages overlap in time to some extent, and different areas may be picked out histologically in the same field, showing for example, fibrin clot, blastemal tissue and fine bone trabeculae, cheek by jowl.

The immediate ischaemic Stage 1 of fracture repair is associated in time with the production of numerous cytokines and growth factors which stimulate macrophage activity. But it is also a period of cell differentiation, resulting in the emergence of osteoblasts from primitive precursor cells, and the evolution and adhesion of endothelial elements to form blood islands in the blastemal tissue. These coalesce to form capillaries, containing plasma and blood elements, which fuse with surviving peripheral capillaries in the fracture site (Hasan and Brookes, 1990). The pH is unusually low in Stage 1 of slot repair at day 2, and yet antimony chloride electrodes have registered a pH of 6.8 in the dorsal lip of the blastopore and the neural tube of Triton taeniatus (Buytendijk and Woerdemann, 1927), areas where cell differentiation is prominent in the amphibian embryo. Newman et al. (1985) find a pH 7.2 for fracture haematoma in unstable rat tibial fractures, using ^{31}P MRI spectroscopy.

Stage 2 lasts about 3 weeks in our stable fracture during which time fine bone trabeculae are formed, at first on the external and internal cortical surfaces and lastly in the fracture site (slot) itself. Clearly, the raised blood supply during this period supplies the energy and building blocks for soft callus formation. The enzyme lysozyme has a maximum activity at pH 7.6 and is implicated in soft callus formation (Kuettner et al., 1975). Newman et al. (1985) using ^{31}P MRI spectroscopy for non-invasive pH estimations in unstable rat tibial fractures, find soft callus peaks in the 3rd week at pH 7.5. The elevated pH 7.6 calculated from flow data in this investigation, appears to make possible the intense osteoblast proliferation and the reticular character of the trabeculae that are formed (Brookes and Helal, 1968).

Stage 3 sees a slow decline from hyper- to hypovascularity, accompanied by thickening and coalescence of bone trabeculae into hard callus, and the obliteration of the slot (fracture gap) by solid bone. Bone consolidation undoubtedly is the cause of reduced blood volume; there is less space for capillaries. The reduced flow rate is a sure indication of reduced bone metabolism in the slot. Acid drift of the pH is the possible cause of both these 2 chief characteristics of fracture consolidation.

Stage 4 is the prolonged remodelling period which has few investigators. Prima facie the mildly reduced flow rates and volumes recorded here would seem to offer an explanation for the slow bone removal and deposition that persists at a healed fracture site. A diminished blood supply alone might lead to bone sclerosis, which does not occur. Significantly the computed pH in this investigation of 7.4 suggests that bone turnover is not depressed, and therefore sclerosis occurring in conditions of severely reduced vascularity cannot develop.

REFERENCES

Axhausen, G.: Die aseptische Knochennecrose und ihre Bedeutung für die Knochen und Gelenkchirurgie. Acta Chir. Scand. 60: 369-396 (1926).

Brookes, M.: Arteriolar blockade: a method of measuring blood flow rates in the skeleton. J. Anat. 106: 557-563 (1970).

Brookes, M.: The Blood Supply of Bone: An Approach to Bone Biology. pp-327. London: Butterworths (1971).

Brookes, M.: Correlates of experimental bone ischaemia. In: Arlet, P. Ficat, B. Mazières (eds): Circulation Osseuse. Symposium International sur la Circulation Osseuse, II. pp 369-377. Université Paul Sabatier, Toulouse 1977.

Brookes, M., B. Helal: Primary osteoarthritis, venous engorgement and osteogenesis. J. Bone Jt. Surg. 50B: 336-345 (1968).

Brookes, M., D. Richards, M. Singh: Vascular sequelae of experimental osteotomies. Angiology. 21: 355-367 (1970).

Buytendijk, F.J.J., M.W. Woerdemann: Arch. f. Entwicklungsmech. 112: 387-395 (1927).

Hasàn, A., M. Brookes: Angiogenesis in early fracture repair. In: J. Arlet, B. Mazières (eds): Bone Circulation and Bone Necrosis. pp 69-175. Springer-Verlag, Berlin 1990.

Kuettner, K.E., R. Eisenstein, N. Sorgente: Lysozyme in calcifying tissues. Clin. Orthop. 112: 316 (1975).

Lexer, E., P. Kuliga, W. Türk: Untersuchungen über Knochenarterien. Berlin: Hirschwald (1904).

Newman, R.J., R.B. Duthie, M.J.O. Francis: Nuclear magnetic resonance studies of fracture repair. Clin. Orthop. Rel. Res. 198: 297-303 (1985).

PERTURBATIONS OF VASCULARIZATION AND CIRCULATION DUE TO OSTEOSYNTHETIC METHODS

Patrick J. KELLY and James T. BRONK

Department of Orthopedic Research
Mayo Clinic and Mayo Foundation
Rochester, Minnesota, USA

INTRODUCTION

It is axiomatic to say that an adequate blood supply is necessary for initiation of the healing response (Trueta, 1963). The fracture in itself will alter blood flow and flow will be changed if treated by osteosynthetic methods. The surgeon treating fractures finds tibial fractures the most worrisome for two reasons: the tibia is so commonly a site of infection following trauma (Kelly et al., 1990) and it is a frequent site of compartment syndrome (Blick, 1986). Fixation of fractures leads to changes in bone structure. This has been principly emphasized in the tibia and in response to plate fixation. It has been postulated that this change in bone mass is secondary to alterations of fracture stability by various forms of fixation. Stated more clearly, changes in stability lead to decreased formation of bone and to increased resorption of existing bone, perhaps by decreasing the feedback mechanism of strain related potentials that may control bone remodeling, or alternatively, it may decrease transcapillary fluid flow to osteoblasts which contain necessary substrates for cell nutrition (Kelly and Bronk, 1990). Studies have shown that plates interfere with cortical remodeling (Akeson et al., 1976; Jacobs et al., 1981; Uhtoff et al., 1983). Effects of fixation on fracture healing have been recently reviewed (O'Sullivan et al., 1989). While there have been studies on the effects of intramedullary fixation on diaphyseal bone circulation (Dankwardt, 1969), no studies have systematically examined the effect of the three common forms of fracture fixation, external fixation (EF), intramedullary rod (IMR), and plate (PL), on bone remodeling and blood flow in a single study.

Each form of fracture fixation, be it an intramedullary nail, a plate, or an external fixator, has its advantages and disadvantages. The plate provides rigidity of fixation and early function, but infection is a complication that occurs especially in the tibia. Plate removal can be difficult and refracture occurs not infrequently. Nonetheless, plates are used in forearm fractures and often in humeral fractures. The intramedullary form of fixation clearly is the choice for femoral diaphyseal fractures. A disadvantage is the less

Bone Circulation and Vascularization in Normal and Pathological Conditions
Edited by A. Schoutens *et al.*, Plenum Press, New York, 1993

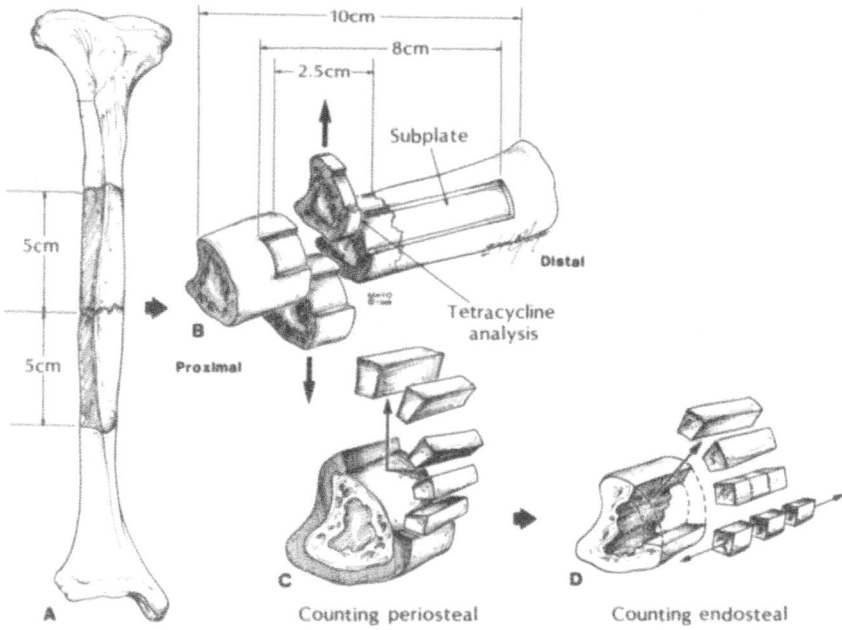

Figure 1. The tibial section was divided into segments for measurement of blood flow at 90 days. One section, a cross section of the entire tibia at the fracture site, was used for tetracycline-microradiographic analysis. (Reproduced with permission from Mayo Foundation).

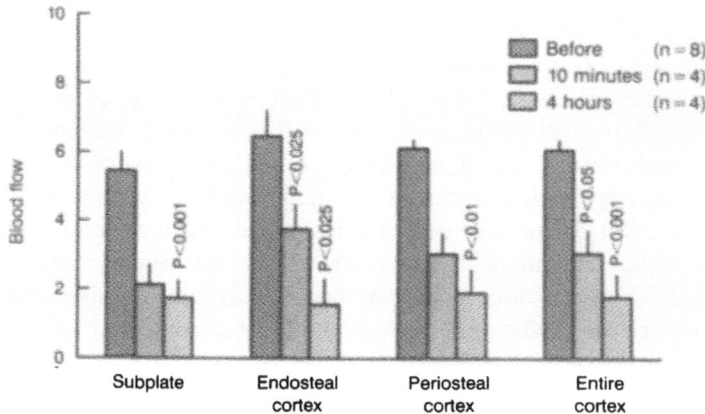

Figure 2. Blood flow (ml/min/100 g) to fracture without fixation. Subplate and periosteal blood flow is significantly less than bone prior to osteotomy. Bars indicate SEM. N is number of observations.

than perfect rotational stability offered by intramedullary nail fixation which may be countered by interlocking. On the other hand, an interlocking nail may prevent compression of fracture fragments. Tibial fractures, especially compound fractures, remain an area of concern and controversy. The complication of infection increases with the degree of injury and the degree of fixation (Gustilo et al., 1990), and has led many fracture surgeons to recommend external fixation for tibial fractures that have a high degree of probability for infection.

The purpose of our studies was to see if light could be shed on what these three forms of fixation might do to blood flow at timed intervals after an experimental fracture. An unsuspected bonus was the opportunity to compare bone mass, bone formation, and bone porosity at 90 days after fixation with these forms of fixation. Ninety days represents the endstage of intermediate fracture healing in a canine tibia and subsequent changes are principally long term remodeling. The adult dog was chosen because the canine bone remodeling resembles man and total metabolic rate in man and the canine are similar. Microspheres were chosen to measure blood flow since they appear to be the accepted benchmark for measurement of regional bone blood flow (Li et al., 1989). Bone remodeling was measured by combining microradiography and tetracycline labeling (Vanderhoeft et al., 1962). Blood flow was measured without fixation at 10 minutes and 4 hours after osteotomy, prior to fixation and again after fixation at 4 hours, 48 hours, 14 days, and 90 days. Common forms of fixation were used, i.e. intramedullary rod fixation (IMR) with reaming, external fixation (Sukhtian-Hughes, D. Howse, London England) (EF), plates (PL) which were DCP (Synthes Ltd., Wayne, PA). The IMR was a fluted titanium intramedullary nail (3M Orthopaedic Products, Minneapolis). The design of the study and the details of experimental methodology have been previously reported (Smith, 1990). The method of sample preparation for blood flow and histologic studies are illustrated in Figure 1. Essentially, we studied the bone at the fracture site, bone beneath the plate, termed the subplate region, and the endosteal and periosteal cortex of the fracture site.

OBSERVATIONS ON BLOOD FLOW

Experimental fracture of the midtibial region without fixation at 10 minutes and four hours produces a profound decrease in regional blood flow in all three regions we studied, the inner one-half or endosteal cortex, the outer one-half or periosteal cortex and the area where one would ordinarily place a plate for fixation. The immediate (10 minute) blood flow was 58 % of intact in the endosteal cortex dropping to 30 % by four hours. Figure 2 summarizes the changes in flow prior to osteotomy, at 10 minutes, and four hours after osteotomy.

The general observation was that blood flow was well maintained four hours after fixation. However, at four hours the PL subplate blood flow was significantly greater than the IMR subplate blood flow ($p < 0.05$). At 90 days, the PL subplate blood flow was greater than the subplate flow of the IMR or EF fixed tibial fracture. This difference is significant when comparing PL to IMR ($p < 0.02$) (Figure 3).
Endosteal cortical blood flows were profoundly affected by IMR fixation. At four hours IMR fixation blood flow was significantly less than EF or PL. At 14 days, endosteal blood flow was significantly less in IMR fixed tibias than EF fixed tibias (Figure 4). Periosteal flow was not significantly different in any of the fixation groups.

Table 1. Areas (cm^2) of regions on X-section at 90 days. Mean ± S.E.M.

Cortex			Medullary Canal			Periosteal Bone		
IMR	PL	EF	IMR	PL	EF	IMR	PL	EF
0.89	0.88	1.04	0.50	0.63	0.50	2.39	0.88	1.46
±0.11	±0.12	±0.14	±0.06	±0.09	±0.05	±0.54	±0.36	±0.62

No significant difference

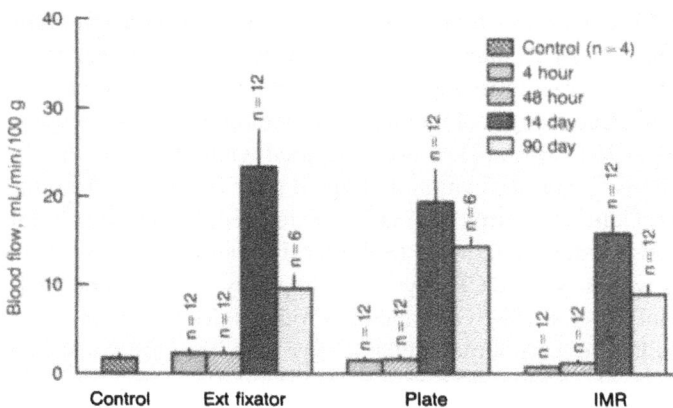

Figure 3. Blood flow to the subplate cortex at 4 hours, 48 hours, 14 days, and 90 days after fracture. In the tibia fixed with a PL, subplate cortex blood flow is significantly higher than IMR at 4 hours (p < 0.05) and 90 days (p < 0.02).

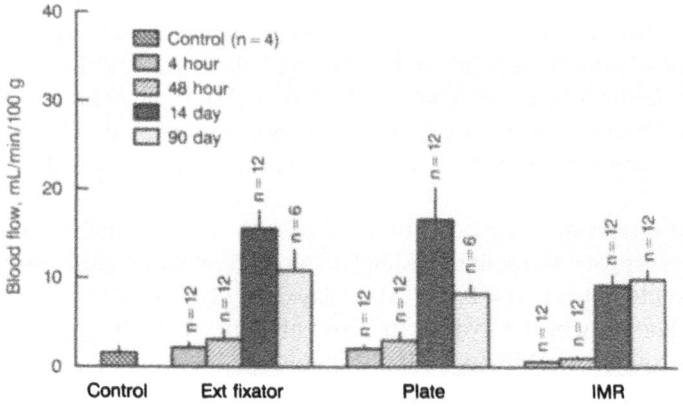

Figure 4. Endosteal cortex. At 4 hours, IMR fixation is significantly less than EF (p < 0.05) or PL (p < 0.01). At 14 days endosteal blood flow is less in IMR compared to EF (p < 0.025).

One area of concern with fixation of fractures is the effect on bone mass of the fixation device. The conclusion of this study and others that have emanated from our laboratory led us to conclude that at 90 days there is no significant difference between a PL, EF or IMR as regards bone mass. Table 1 indicates no significant change in cross-section area of cortex, medullary canal, or periosteal bone. Table 2 indicates that bone formation at 90 days is increased in the subplate region of tibias fixed with plates. This is consistent with increased blood flow in the subplate region at 90 days (Figure 3). There are trends, but not significant differences. The healing of fractures leads to changes in bone structure and mass. It has been postulated that this change in mass is secondary to alterations of mechanical factors by the fixation device (O'Sullivan et al., 1989). Increasing rigidity promotes bone formation but may produce porosity later on the remodeling process. We concluded that two factors are operative in changes in bone mass at 90 days: revascularization in response to vascular damage, and response to stability of fixation. Effects on bone mass beyond 90 days are areas of possible future study, however, such studies would be expensive and not easy to plan.

Table 2. Subplate Area. % of Subplate $\pm$ S.E.M.

Labeled			Unlabeled			Pore		
IMR	PL	EF	IMR	PL	EF	IMR	PL	EF
(N=10)	(N=5)	(N=5)						
38.5	57.4	39.1	41.3	28.0	48.7	20.1	14.6	12.2
$\pm$ 5.6	$\pm$ 10.0	$\pm$ 7.8	$\pm$ 7.3	$\pm$ 13.2	$\pm$ 9.0	$\pm$ 3.6	$\pm$ 5.1	$\pm$ 2.3

Trend: more bone formation, less old bone plate subcortex; no significant difference

MECHANICAL STUDIES ON OSTEOSYNTHETIC DEVICES

At the outset it was stated that blood flow is essential for initiation of fracture healing. It also is axiomatic that appropriate stability is essential to achieve fracture healing. For that reason in this same model, the canine tibia, we have gathered information on mechanical effects of common fracture devices.

A simulated tibial fracture was tested on a materials testing machine (MTS Systems, Minneapolis). After application of the fixation device bending force was applied in the anterior posterior and lateral planes as well as torque, compression and distraction. Relative rigidity is expressed as a percentage of the intact canine tibia. As might be expected a plate achieves overall rigidity. When parameters are meaned, external fixation in two planes equals a plate. Interlocking of an IMR enhances torsional rigidity, but no more than Orthofix external fixator (EBI, Verona, Italy). It may be that external fixation tends to prevent fracture sites from being held apart, especially if you compare each type of fixation when tested in distraction (Table 3).

Unfortunately, osteomyelitis is a consequence of fractures treated by the ever increasing plethora of methods of fixation (Gustilo et al., 1990). For this reason, external fixation was reintroduced as a form of fixation of tibial fractures, especially compound fractures. The hope was that an ideal form of external fixation would yield an ideal type of fracture healing. Furthermore, some as yet undefined manipulation of aspects of rigidity would deliver mechanical forces that would be favorable to the biologic process of fracture healing.

An alternative approach would be to view the fracture healing process as a three stage process. In arbitrary fashion, the initial stage lasts from zero to six weeks. During this period the stage is set for healing. Humoral and cellular processes are called into play to initiate the formation of callus. From six to twelve weeks a stage of intermediate fracture healing occurs wherein the callus is laid down periosteally and endosteally and biologic stability is achieved. In the canine, by 90 days, this process is completed. After 90 days bone remodeling reshapes the fracture site. Therefore, in a series of experiments it was possible to study dogs with bilateral tibiofibular osteotomies. Alignment of the fracture was maintained with an Orthofix external fixator. At 48 days testing for fracture stabilily was performed with a strain gauge device attached to the external fixator pins. If the fractures were equally stable, fixators were removed and an Aircast brace pressurized with aircells to 28.3 ± 1.0 mmHg was applied and a traditional cast applied to the other tibia (resin cast). All animals were sacrificed at 84 days and both tibias tested mechanically by torsional testing. All fracture sites were labeled during the study period with tetracycline (Vanderhoeft, 1962; Dale, 1989; Dale, 1992). The end result was more dense periosteal bone with less porosity, and a fracture that tested stronger in torque on the side treated with the pressurized brace (Dale, 1989; Dale, 1992).

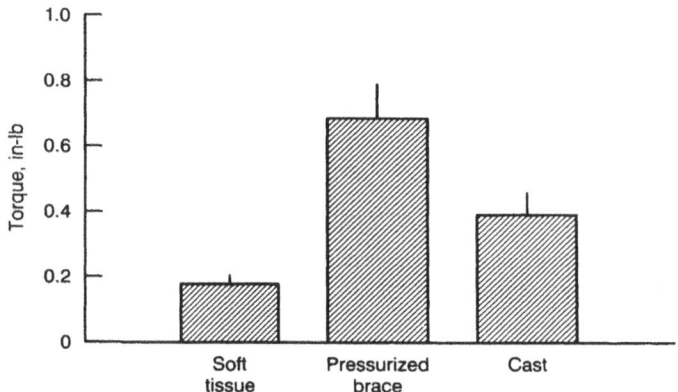

Figure 5. The stability of a pressurized brace and a conventional cast was determined in vitro. The pressurized brace provides significantly more stability than the conventional cast ($p < 0.005$). The soft tissue values are shown for interest. They have been subtracted from the pressurized brace and cast values.

The possible explanations for this phenomenon of the pressurized brace are: a more efficient form of control of torque; a compressive molding effect of the pressurized brace that produces a less porous but denser periosteal callus during the phase of callus formation; the enhancement of capillary filtration due to an elevation of osseous venous resistance that would elevate capillary hydrostatic pressure and increase perfusion of cells with substrate. Capillary filtration is mainly controlled by changes in pressure within the capillary influenced by Starling's law. Such changes in osseous venous resistance do enhance periosteal bone formation in puppies (Kelly and Bronk, 1990). At this time, evidence favors a mechanical factor, that is, a more efficient form of control of torque. Figure 5 indicates a significantly improved control of torque by a pressurized brace when compared to a traditional cast. Studies underway suggest that the level of pressure produced by the pressurized brace is an important factor in determining the response of periosteal bone during the period from 6-12 weeks.

Table 3. Fixation rigidity "in vitro" modeling. Percent intact bone volume

	Torque					
	Tight Rod	Interlocking Rod	Plate	1 Plane	2 Plane	EBI Fixator
Mean	26.3	66.6	131.5	60.7	166.6	67.9
S.D.	10.9	23.6	21.9	15.7	26.8	4.7
S.E.	4.5	11.8	8.9	6.4	10.9	2.1
N	6	4	6	6	6	5
	AP Bending					
Mean	27.1	31.1	116.1	121.4	114.3	48.1
S.E.	3.1	1.0	10.3	9.2	6.2	5.9
N	6	4	6	6	6	5
	Lateral Bending					
Mean	30.5	30.3	121.2	34.9	192.5	184.4
S.E.	3.8	2.3	26.6	3.4	12.2	38.6
N	6	4	6	6	6	5
	Distraction					
Mean	75.5	120.8	106.9	6.2	13.5	*
S.E.	2.4	27.7	5.3	1.2	0.5	
N	3	4	3	2	3	
	Compression					
Mean	78.3	66.4	111.2	75.2	78.4	100.0
S.E.	1.1	7.8	1.7	9.6	10.3	0.0
N	3	4	3	3	3	5
Mean of Observ.	47.5 (N=5)	63.0 (N=5)	117.4 (N=5)	59.7 (N=5)	113.1 (N=5)	100.1 (N=4)

* Not tested since fixator was dynamized.

ACKNOWLEDGMENT

These studies were supported by a grant from the National Institutes of Health, and from the Aircast Corporation, Summit, New Jersey, U.S.A.

REFERENCES

Akeson, W.H., SL-Y Woo, L. Rutherford, R.D. Coutts, M. Gousalves, P. Amiel: The effects of rigidity of internal fixation plates on long bone remodeling. Acta Orthop. Scand. 47: 241-249 (1976).

Blick, S.S., R.J. Brumback, A. Poka, A R. Burgess, N.A. Ebraheim: Compartment syndrome in open tibial fractures. J. Bone Joint Surg. 68A: 1348-1353 (1986).

Dale, P.A., J.T. Bronk, P.J. Kelly: Fracture healing with elevated venous pressure. Orthop. Trans. 13 (2): 462 (1989).

Dale, P.A., J.T. Bronk, M.E. O'Sullivan, E.Y.S. Chao, P.J. Kelly: A new concept in fracture immobilization: The application of a pressurized brace. In Press, Clin. Orthop. (1992).

Dankwardt-Lillestrom, G.: Reaming of the medullary cavity and its effect on diaphyseal bone. Acta Orthop. Scand. (Suppl) 128: 1-153 (1969).

Gustilo, R.B., R.L. Merkow, D. Templeman: The management of open fractures. J. Bone Joint Surg. 72A: 299-304 (1990).

Jacobs, R.R., B.A. Rahn, S.M. Perren: Effect of plates on cortical bone perfusion. J. Trauma 21: 91-95 (1981).

Kelly, P.J., J.T. Bronk: Venous pressure and bone formation. Microvasc. Res. 39: 364-375 (1990).

Kelly, P.J., R.H. Fitzgerald Jr, M.E. Cabanela, M.B. Wood, W.P. Cooney III, P.G. Arnold, G.B. Irons Jr.: Results of treatment of tibial and femoral osteomyelitis in adults. Clin. Orthop. 259: 295-303 (1990).

Li, G., J.T. Bronk, P.J. Kelly: Canine blood flow measurements estimated with microspheres. J. Orthop. Res. 7: 61-67 (1989).

O'Sullivan, M.E., E.Y.S. Chao, P.J. Kelly: The effect of fixation on fracture healing. J. Bone Joint Surg. 71A: 306-310 (1989).

Smith, S.R., J.T. Bronk, Kelly, P.J.: Effects of fracture fixation on cortical bone blood flow. J. Orthop. Res. 8: 471-478 (1990).

Trueta, J.: The role of vessels in osteogenesis. J. Bone Joint Surg. 45B: 402-418 (1963).

Uhtoff, W.H., SL-Y Woo, L. Rutherford, R.D. Coutts, M. Gousalves, P. Amiel: The effects of initial plates on post-traumatic remodeling and bone mass. J. Bone Joint Surg. 65B: 66-71 (1983).

Vanderhoeft, P.J., P.J. Kelly, L.F.A. Peterson: Determination of growth rates in canine bone by means of tetracycline-labeled patterns. Lab. Invest. 11: 714-726 (1962).

COMPARATIVE VASCULAR EVALUATION BY MRI OF AUTOLOGOUS AND BOVINE GRAFTS

J.V. CHRISTIANSEN, T. HINDMARSH, B. LEVANDER,
H. LOFGREN, T. OLSSON and F.P. REINHOLT

Länsjukhuset Ryhov
Jönköping, Sweden

More than 30 years ago Ralph B. Cloward introduced his method of anterior disc extirpation followed by fusion of the involved cervical spinal segments by using bone graft from the iliac crest (Cloward, 1958). To eliminate the postoperative pelvic pain emanating from the donor place, bovine grafts were introduced 15 years later (Taheri and Gueramy, 1972; Cartore and Fortuna, 1969) (Figure 1). Both methods are now well-known in clinical practice but there is still some dispute concerning the bovine grafts. One of the main questions is if there is a similar type of fusion after both procedures.

PATIENTS AND METHODS

Two series of patients undergoing Cloward anterior cervical fusion were analyzed. In the first retrospective series of 12 patients (29-67 years, average age 58 year), autologous bone grafts were used. In the other series of 20 prospectively studied patients (25-69 years, average age 47 years), bovine grafts (TM Unilab Surgibone) were implanted.
The bovine grafts were studied by MRI 3 days, 3, 6 and 12 months after the operative procedure, while the autologous grafts were examined up to 9 years postoperatively.
The MRI studies were performed on a 1.5-T superconducting magnet (Magneton Siemens Medical System). The contrast material was Gadolinium-DTPA (Shering AG, Berlin, Germany) administered at a dose of 0,1 mmol/kg body weight IV. Pre- and postcontrast T1-weighted sagittal and axial images with a slice thickness of 4 mm and with 1 mm gap were used.

RESULTS

In the group of autologous grafts the MRI study demonstrated homogeneous signal intensity in both involved vertebral bodies and in the interposed graft on T1-weigted

Bone Circulation and Vascularization in Normal and Pathological Conditions
Edited by A. Schoutens *et al.*, Plenum Press, New York, 1993

137

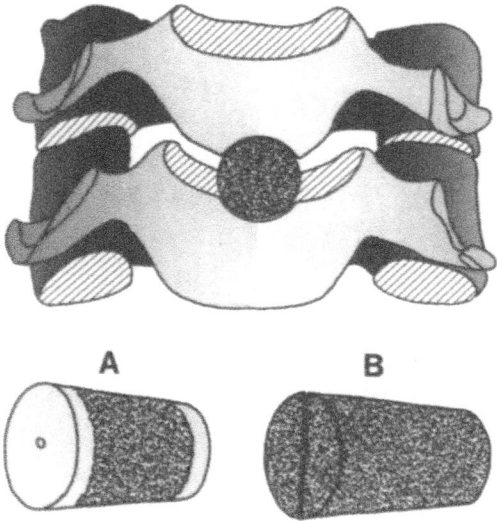

Figure 1. Illustration of the Cloward operation procedure. **A)** The autologous bone graft from the iliac crest. **B)** The bovine graft.

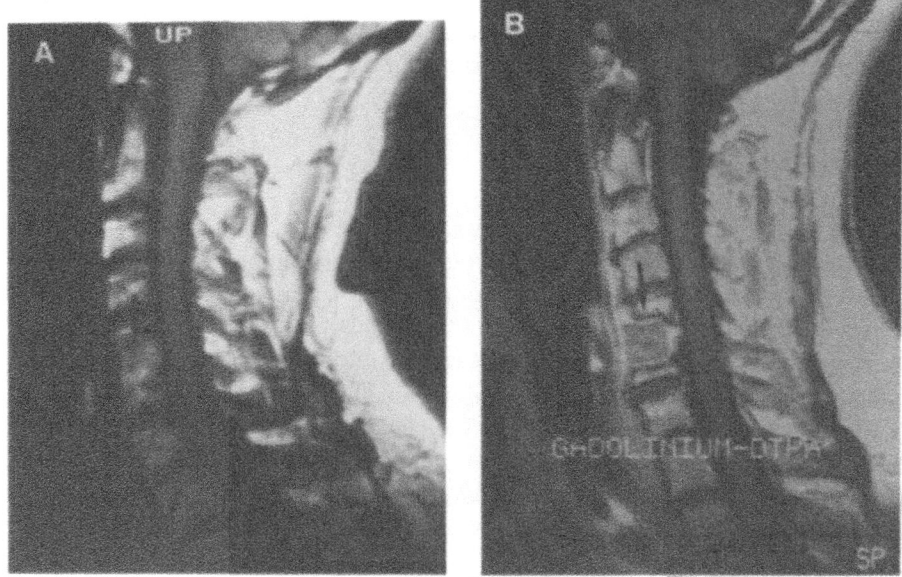

Figure 2. MRI 3 months postoperative after injection of gadolinium contrast. **A)** The implant from iliac crest with homogeneous signal intensity in both the involved vertebral bodies and in the interposed graft. **B)** The bovine graft with a slightly homogenous enhancement and with a border zone between the implant and the vertebral bodies showing marked enhancement (see arrows).

138

images. No enhancement was seen after Gd-DTPA contrast injection. This indicates that the circulation in the crista implant does not differ from the adjacent vertebral bodies (Figure 2A).

In patients receiving bovine bone grafts the MRI examination on Tl-weighted images showed a low signal intensity in the implant, but a normal signal in the adjacent vertebral bodies. After injection of Gd-DTPA the graft showed a slightly homogeneous enhancement but a rim of tissue in the border zone between the implant and the vertebral bodies showed marked enhancement. These MRI findings were seen from 3 months postoperatively onwards (Figure 2B). In two patients the bovine grafts have been extirpated 6 months after the implantation. Histopathological studies verified the vascularized fibrous fusion while central part of the graft showed multiple necrotic areas (Figure 3).

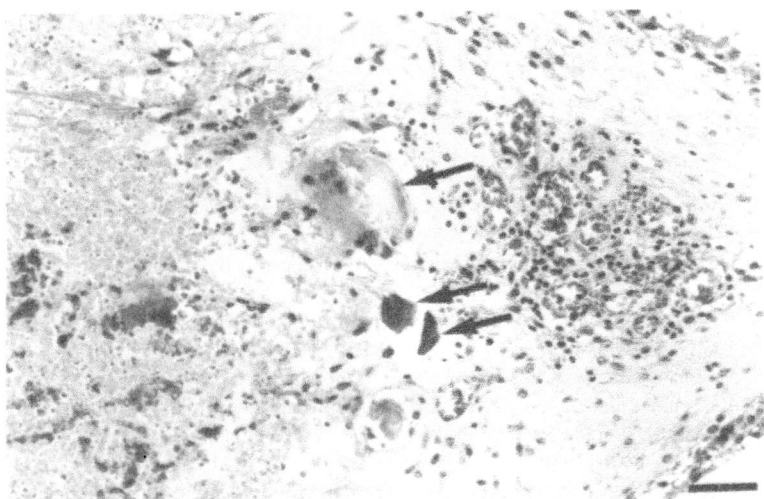

Figure 3. Light micrograph from border between necrotic bone/bone marrow (to the left) and fibrous connective tissue ingrowth (to the right). Small necrotic bone trabeculae are observed in the middle (arrows) some of them surrounded by resorbing multinucleated cells. Hematoxylin and eosin. Bar=50 μm.

Discussion

The autologous iliac graft showed a solid bone fusion after 3 months in all cases. A solid trabeculae fusion could be visualized on plain X-ray. This was confirmed on MRI which showed the same signal intensity in the graft as in the adjacent vertebral bodies.
The biodistribution of Gd-DTPA including time course of enhancement within epidural fibrotic connective tissue in the lumbar spine has been studied by Ross et al. (1989). Both clinical works and experimental works on dogs have been reported. The rapid contrast enhancement after IV injection on MRI imaging was correlated to morphological studies of tissue obtained at re-operation. Light and electron microscopy of human and dog epidural fibrotic connective tissue was performed to assess the vascularity and endothelial ultrastructure. A complementary study using a vascular injection with India ink was performed in dogs to define the microvascular anatomy of the same material. Hueftle et al. (1988) have studied the biodistribution of small molecule Gd-DTPA (mol wt, 600).

In general three things are necessary for contrast enhancement of any tissue: a vascular supply, a route for the diffuse of contrast material out of the vessels, and an interstitial space to receive the contrast material. The fibrotic connective tissue studied in the lumbar epidural compartment has all these three prerequisites as demonstrated by electron and light microscopy. Our study concerns bone grafts in the cervical spine but the biodistribution and kinetics of the contrast were agreement with previous reports (Stevens E. et al., 1962; Strich et al., 1985).

In conclusion, according to MRI findings, autologous iliac grafts develop a solid osseous fusion while bovine grafts exhibit a vascularized fibrous fusion after 3 months.

ACKNOWLEDGEMENT

This study was supported by grants from the Swedisch Medical Research Council, Karolinska Institutet and University College of Health and Care Jönköping, Sweden.

REFERENCES

Cartore B., A. Fortuna: Intersomatic fusion with calf bone in anterior surgical approach for treatment of myelopathy in cervical spondylosis. Acta Neurochir. 20: 59-62 (1969).

Cloward R.B.: The anterior approach for removal of ruptured cervical discs. J. Neurosurg. 15: 602-614 (1958).

Hueftle, M., M.T. Modic, J.S. Ross et al.: Lumbar spine: Postoperative MR imaging with Gd-DTPA. Radiology 167: 817-824 (1988).

Ross, J.S.: Gadolinium-DTPA-Enhanced MR imaging of the postoperative lumbar spine: Time course and mechanism of enhancement. American Journal of Neuroradiology 10: 37-46 (1989).

Stevens, E., B. Rossof, M. Werner, H. Spencer: Metabolism of the chelating agent diethylenetriamine pentaacetic acid (C 14-DTPA) in man. Proc. Soc. Exp. Biol. Med. 111: 235-2358 (1962).

Strich, G., P.L. Hagan, K.H. Gerber, R.A. Slutsky: Tissue distribution and magnetic resonance spin lattice relaxation. Effects of Gadolinium-DTPA. Radiology 154: 723-726 (1985).

Taheri Z. E., M. Gueramy: Experience with calf bone in cervical interbody spinal fusion. J. Neurosurg. 36: 67-71 (1972).

THE IMPORTANCE OF FLOW CONDUCTANCE OF CANCELLOUS BONE GRAFTS AS A CRITICAL FACTOR IN GRAFT INCORPORATION

P.W. HUI, P.C. LEUNG and A. SHER

Department of Orthopaedics and Traumatology
The Prince of Wales Hospital
The Chinese University of Hong Kong
Hong Kong

INTRODUCTION

The viability of any bone graft depends on the re-establishment of blood flow within the graft. If blood circulation is not re-established, it will necrose and then functions like a mechanical bridge only. We observed immediate "bleeding" from the cancellous surface of cortico-cancellous graft and tended to believe that to some extent cancellous bone could function like a blood conducting media containing numerous complex conducting channels.

The aim of this study is to test the hypothesis that flow conductance of processed cancellous grafts is a critical factor in graft incorporation and how the effects are reached: hematoma formation, local inflammatory response, formation of fibrovascular stroma and remodeling phase (Friedlaender, 1987). To achieve these goals we have designed a perfusion experiment to measure the flow conductance of cylindrical cancellous bone grafts. These grafts were obtained from porcine femoral heads and subsequently treated to reduce the risk of immunological response. Bone grafts of various conductance were then transplanted to the tibiae of rabbits to test the hypothesis stated above. Finally, histological study was carried out.

MATERIALS AND METHODS

Preparation of Porcine Bone Graft

Cylindrical specimens of 6 mm diameters were trephined from fresh porcine femoral head in a direction either parallel or perpendicular (anterior-posterior) to the femoral neck axis. Specimens of different lengths were obtained by cutting away the

Bone Circulation and Vascularization in Normal and Pathological Conditions
Edited by A. Schoutens *et al.*, Plenum Press, New York, 1993

141

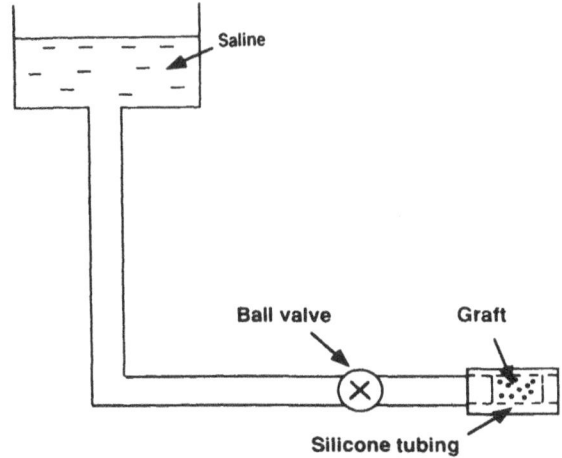

Figure 1. Perfusion apparatus for the measurement of flow conductance of cancellous bone graft.

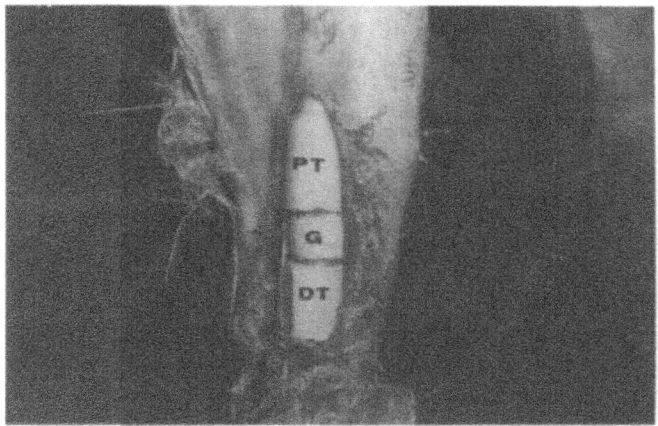

Figure 2. The graft (G) was held tightly between proximal tibia (PT) and distal tibia (DT) by a overlaying silicone tubing.

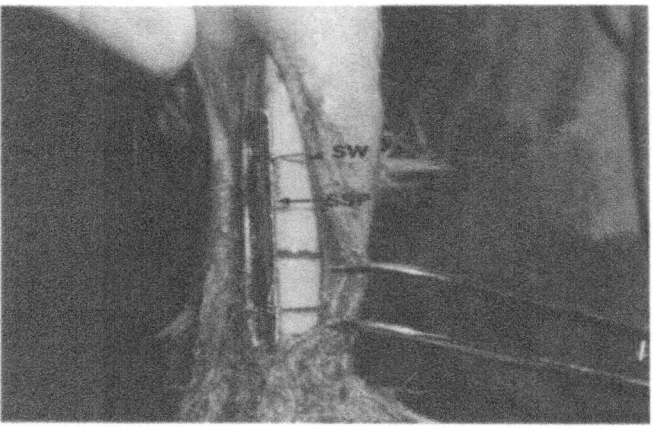

Figure 3. Stainless steel plate (SSP) and suture wire (SW) placed outside the silicone tubing were used as internal fixator.

portions containing cartilage and growth plates with a rotating diamond saw, defatting with acetone and then freeze-drying overnight (Karges et al., 1963). The treated bone grafts were porous in nature.

Blood Flow Conductance Measurement

To measure the flow conductance the bone graft was inserted into a tight fitting silicon rubber tubing which was then placed at the low pressure end of a perfusion apparatus (Figure 1). The pressure drop across the graft was determined by the height of liquid (normal saline) column and the induced flow was calculated by dividing the volume of the fluid perfusing from the graft by the perfusion time recorded with a stopwatch. The relation between these 2 parameters can be translated into a graph and the slope of the graph represents the conductance as will be shown in the results.

Rabbit Model

Bone grafts with various conductance were transplanted to the tibiae of white New Zealand rabbits of age 16 weeks to 18 weeks. After the rabbit had been anesthetized with 2.5 % pentobarbital an incision was made on the medial side of the left hind leg above the ankle to expose the tibia. A section of the tibia approximately the length of the graft was removed. The open ends of the silicone tubing containing the graft were slipped onto the osteotomized tibiae until the graft and tibiae were in close contact (Figure 2). Stainless steel plate and suture wire placed outside the silicone tubing were used as an internal fixator to stabilize the graft-tibia structure (Figure 3). Finally, the incision was closed with sutures and a Plaster of Paris cast was put on to provide further stability.

Histological Study

The rabbits were sacrificed 9 weeks after transplantation. The graft-tibia was removed and decalcified with formic acid. Histological sections stained with Hematoxylin and Eosin were examined using light microscopy. At the time of sacrifice special attention was given to the gross appearance of the grafts, in particular to any signs of blood clot formation.

Effect of Blood Flow Conductance on Graft Incorporation

Based on the histological findings and the gross appearances, we categorized the grafts into viable and nonviable ones. Subsequently, the relation between the fluid conductance and viability of graft was studied.

RESULTS

Figure 4 shows the typical results of flow conductance measurements. There was a linear relation between the pressure drop across the graft and the induced fluid flow. Flow conductance was derived by the slope of the regression lines and had different values depending on the physical characteristics of the grafts such as porosity, length and trabecular orientation.

Typical histological findings of different grafts with decreasing conductance were shown in figures 5 to 8. Figure 5a shows a graft with a conductance of 4.4×10^{-10} m^3/s/Pa which was one of the highest values obtained in the present study. There were unions at both ends of the graft even though the union at the distal junction was not a good as the proximal junction. At higher magnification (Figure 5b) we could see that new bones were formed within the inter-trabecular spaces while some of the trabeculae were resorbed. Occasionally, microvessels could be found within the newly formed bones.

A graft with conductance of 3.2×10^{-10} m^3/s/Pa was shown in Figures 6a and 6b. Histological findings were similar to the graft mentioned above except that union occurred only at the junction between proximal tibia and the graft.

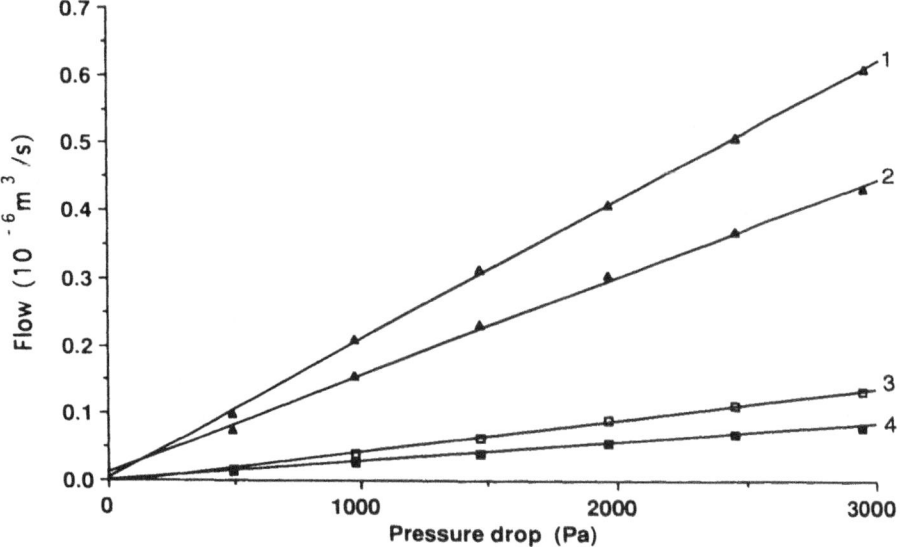

Figure 4. Typical results of perfusion experiment. Flow conductance = slope of regression line.
1) Length (L)=0.54 cm, porosity (p)=0.55, drilling direction (d)=parallel.
2) L=0.53 cm, p=0.40, d=parallel.
3) L=0.55 cm, p=0.42, d=perpendicular.
4) L=0.98 cm, p=0.39, d=parallel.

Figures 7a and 7b show a graft with conductance of 1.4×10^{-10} m^3/s/Pa. Again, only the proximal tibia was united with the graft. However, in contrast with the two previous grafts, many of the inter-trabecular spaces were not occupied with woven bone. Furthermore, the gross appearance of the graft at the time of its removal from the rabbit stained darkish red at the distal end of the graft indicating that blood clot might have been formed there.

Figures 8a and 8b show a graft with a conductance of 0.1×10^{-10} m^3/s/Pa which was one of the lowest values obtained in this investigation. There were no unions at both ends of the graft. We could not see any new bone or micro-vessels within the inter-trabecular spaces. Instead blood clots were seen in some of the pores. The trabeculae appeared to be intact. The gross appearance of the graft at the time of its removal stained darkish red along the entire graft.

Thirty three grafts were analyzed histologically and subsequently classified into either viable (fourteen) or nonviable (nineteen). Categorized as viable were those grafts

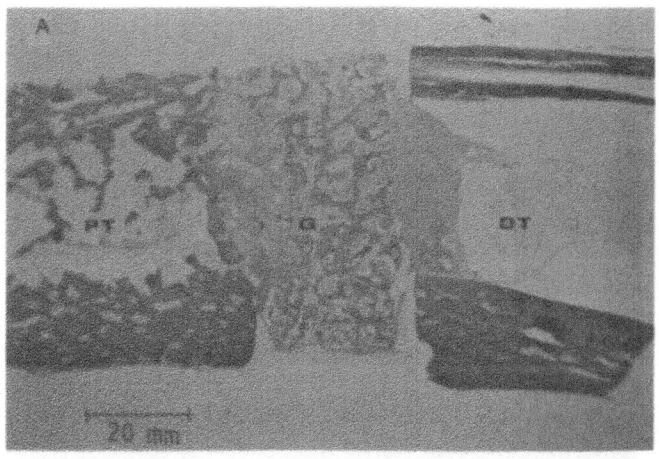

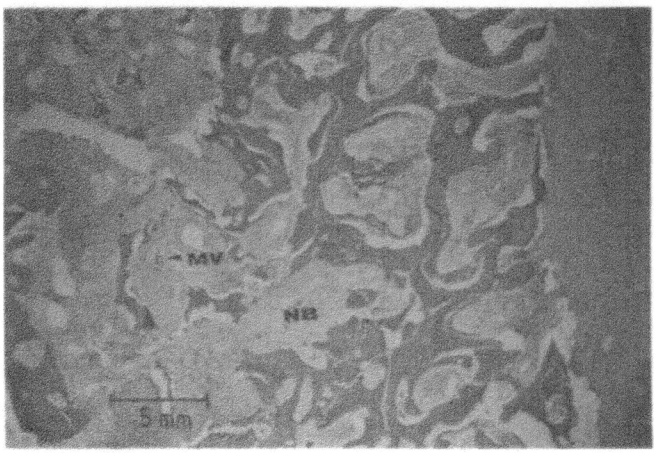

Figure 5. Histological sections of graft with conductance of 4.4×10^{-10} m³/s/Pa. A) PT: Proximal tibia, G: Graft, DT: Distal tibia. B) NB: New bone, MV: Micro-vessels.

with an union at at least the proximal tibia and with woven bone but no blood clots formed in the inter-trabecular spaces. Grafts not satisfying the above requirements were categorized as non-viable. The classification of each graft was plotted against its conductance as shown in Figure 9. It is obvious that there is a threshold between 1.4×10^{-10} and 1.6×10^{-10} m³/s Pa below which the grafts were non-viable and above which the grafts were viable. Only 5 out of the thirty three grafts did not accomodate this conclusion.

DISCUSSION

Flow conductance is defined as the slope of the linear flow-pressure drop relation. It is found to be useful to quantify the fluid conductivity of processed cancellous graft. Its value ranges from 0.05×10^{-10} to 13.4×10^{-10} m³/s/Pa depending on the physical properties of the graft.

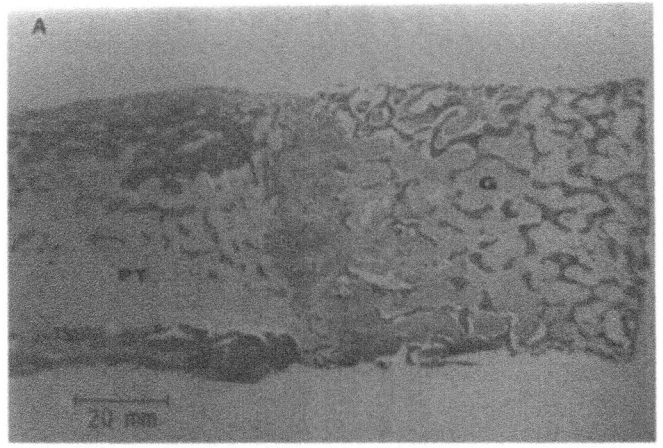

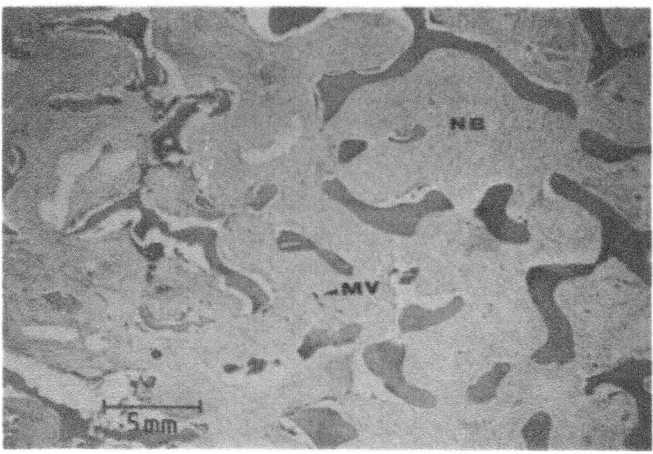

Figure 6. Histological sections of graft with conductance of 3.2×10^{-10} $m^3/s/Pa$. A) PT: Proximal tibia, G: Graft. B) NB: New bone, MV: Microvessels.

Results from the rabbit model support our hypothesis that incorporation of a graft is adversely affected by low conductance of the graft. In fact, non-union and blood clots were seen in most grafts below threshold (1.5×10^{-10} $m^3/s/Pa$) but not in those with a conductance above threshold. We propose that the sub-threshold conductance might increase clot formation during the inflammation stage or hinder clot removal near the end of inflammation stage. In-vivo blood flow measurement in a longitudinal study is definitely needed to verify the above proposal. The coagulated blood made revascularization difficult if not impossible (Burkhardt, 1983). The consequence of this is that new bone cannot be formed.

Cortico-cancellous bone grafts have been extensively used under different clinical situations to fill up bone defects. The assumption is that although there is no uniformity of the grafts, these will eventually survive and be incorporated. This study has shown that cancellous grafts function like fluid and blood conducting media and that the viability and subsequent incorporation of the graft is related to the flow conductance. Therefore it might be important to know the flow conductance of different types of grafts and the physical characteristics that affect their conductance. When a cortico-cancellous graft is found to conduct blood flow instantaneously and effectively it may be used as a perfect biological bridge for bone defects. Even vascularization using sophisticated microsurgical techniques might become unnecessary. In the more common situation chips of cancellous bone rather than block grafts are used. Further studies should look into the flow conductance of such chip packages packed with different degrees of pressure and the incorporation rate of these grafts in relation to their conductance values.

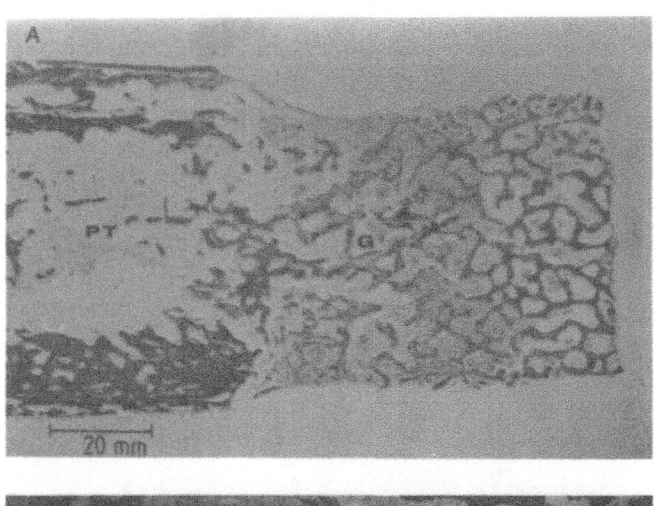

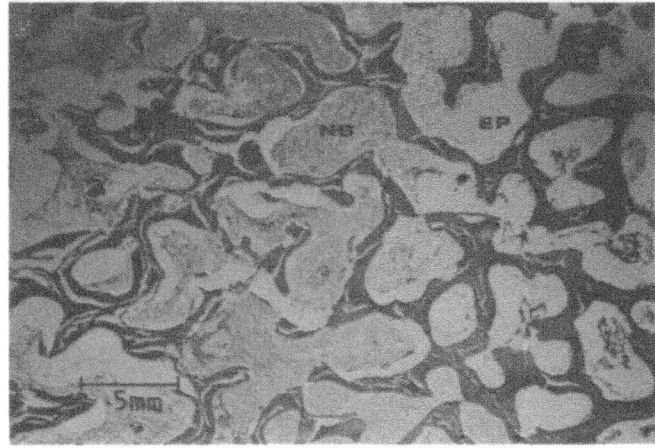

Figure 7. Histological sections of graft with conductance of 1.4×10^{-10} $m^3/s/Pa$. A) PT: Proximal tibia, G: Graft. B) NB: New bone, EP: Empty pore.

SUMMARY

Processed cancellous bone grafts are often used to fill up gaps resulting from bone loss under various conditions. The incorporation of the graft at its site of transplantation is essential. In this investigation we test the hypothesis that the fluid conductance of cancellous graft is a critical factor in its incorporation. Cylindrical cancellous bone grafts of 6 mm diameter were trephined from fresh porcine femoral heads in a direction either parallel or perpendicular to the femoral neck axis. The grafts were defatted, freeze-dried and stored until used. When the fluid conductance was measured, the graft was placed in a perfusion apparatus with normal saline as the perfusate. The pressure drop across the bone graft and the induced flow were measured. The conductance to fluid flow was calculated as the slope of the flow-pressure relation. Grafts of different flow conductance were transplanted into rabbits to replace segments of tibiae. Nine weeks after grafting, rabbits were sacrificed and sections of the decalcified grafts and adjacent tibiae were examined microscopically for evidence of new bone formation. Perfusion data indicated that fluid conductance ranged from 0.05×10^{-10} to 13.4×10^{-10} m^3/s/Pa. Data from rabbit model supported the hypothesis given above. A threshold conductance was found to be about 1.5×10^{-10} m^3/s/Pa below which blood clots were formed within the intertrabecular spaces of the graft. Thus, revascularization and new bone formation could not be attained. The clinical implication of this study is discussed.

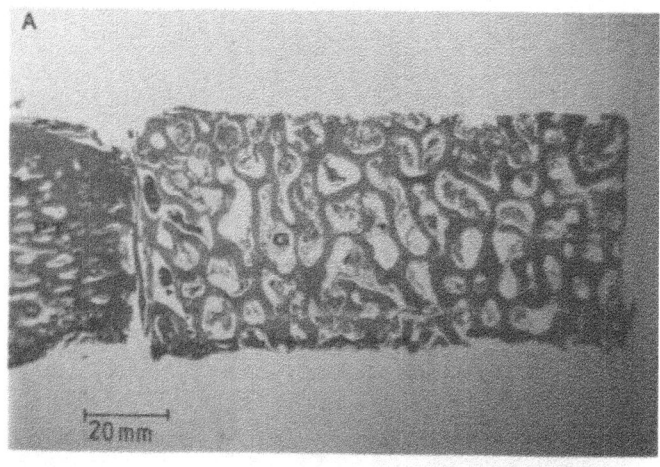

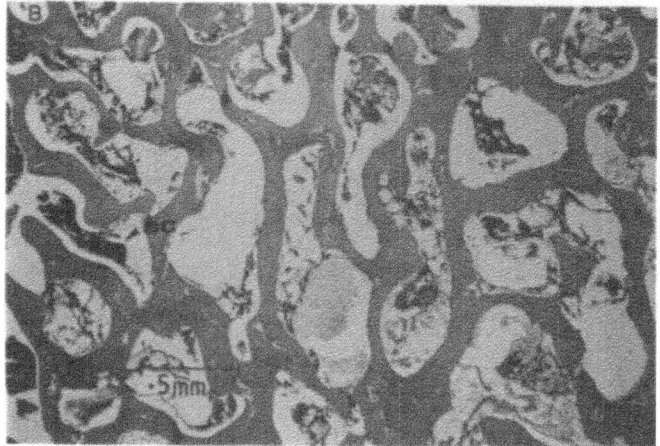

Figure 8. Histological sections of graft with conductance of 0.1×10^{-10} m^3/s/Pa. A) PT: Proximal tibia, G: Graft. B) BC: Blood clot.

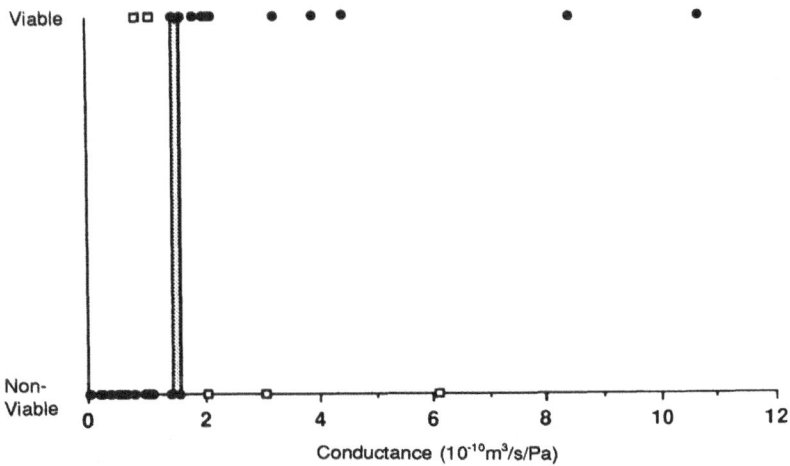

Figure 9. The dependence of graft incorporation on flow conductance. A threshold lies between 1.4×10^{-10} and 1.6×10^{-10} m^3/s/Pa. Open squares: grafts wrongly classified when flow conductance was used as the predictor.

ACKNOWLEDGEMENT

The authors would like to express their thanks to Mr. Simon Lee for his help in preparing bone specimen for histological study. This research is supported by a grant from Research Grants Committee, University and Polytechnic Grants Committee, Hong Kong.

REFERENCES

Burkhardt, H.: The biology of bone graft repair. Clin. Orthop. Rel. Res. 174: 28-42 (1983).

Friedlaender, G.E.: Bone grafts. The basic science rationale for clinical applications. J. Bone Jt Surg. 69A: 786-790 (1987).

Karges, D.E., K.J. Anderson, J.A. Dingwall and J. Jowsey: Experimental evaluation of processed heterogenous bone transplants. Clin. Orthop. 29: 230-246 (1963).

CIRCULATORY ASPECTS OF
BONE DISORDERS

MYELOGENOUS OSTEOPATHIES

R. BURKHARDT

Medizinische Klinik Innenstadt, der Universität München (Emeritus)
München, Germany

The synonyms of "Myelogenous Osteopathy" and "Haematic Osteodysplasia" were coined by Markoff (1942) and Gänsslen (1938) respectively. The authors were inspired from the phenomenon of haemopoietic hyperplasia resulting in osteodysplasia, most peculiarly represented in the brush-like deformation of the skull in children suffering from severe haemolytic disorders. In the following decennia however, the implications of the assumed common pathophysiology of bone marrow and bone stood aside from the clinical routine. This situation changed when it became clear that osteolytic and hypercalcaemic syndromes could be caused by interactions between the immunoreactive tissues of the bone marrow and osseous remodelling. This discovery was among the results of the complementation with histobiopsy of the routine radiologic analysis of the bone, and the exploration of experimental models of osteogenesis and haematopoiesis (Burkhardt, 1971; Mundy, 1974). This in short is the basis of the following review of some actual aspects of myelogenous osteopathies (MOPs).

GENETIC, ANATOMICAL AND FUNCTIONAL RELATIONS OF THE BONE AND MARROW TISSUES

The union of osteo- and haemopoiesis is a creation of developmental ingenuity, not an abnormality. Evidently the cancellous bone and its special capillarization, both offering widespread facilities for cellular and molecular exchange, are an almost ideal support for the vertebrate's highly specialized apparatus of blood cell renewal, as well as for mineral exchange, osseous nutrition, and repair. The common organization of these differently specialized tissues dates from the phyllogenetic step of lower vertebrates (frogs) in the form of a seasonal union (Burkhardt, 1980). Its developmental stages are rudimentally repeated during human embryogenesis and completed by its exclusive institutionalization within the truncal skeleton of the adult. Regarding its physiology we cannot fail to understand, why the disarray of one partner may immediately affect the other: the renewal and supply of bone and blood cells are allocated to one and the same site throughout life and the union of osteogenesis and haemopoiesis is established even in the experimental explantation of bone marrow cells. We assume therefore that the cancellous

Bone Circulation and Vascularization in Normal and Pathological Conditions
Edited by A. Schoutens *et al.*, Plenum Press, New York, 1993

153

Table 1. Pathologic interactions between bone marrow and bone cells.

BONE MARROW CELLS

Lymphocytes	Mastcells	Fibroblasts
Plasmacells	Endothelial Cells	Tumor Cells
Granulocytes	Macrophages	Megakaryocytes

STIMULATED BY

Traumatic lesion	Bacterial Toxin
Inflammation	Virus Toxin
Tumor Growth	

ACT UPON BONE CELLS

OSTEOBLASTS by

Growth Hormones
EGF = endothelial growth factor
FGF = fibroblast growth factor
TGF = tumor growth factor
IGF = insulinlike growth factor
BDGF = bone derived gr.factor
PDGF = platelet derived gr.factor

Facultative Stimulation with

Heparin   + Coupling Factor (?)
Cytokines e.g. Interleukin 1 (OAF)
Prostaglandins

OSTEOCLASTS by

Growth Hormones after Stimulation with

CSF = colony stimulating factor
Cytokines like Interleukin 1(OAF)
Prostaglandins (PGE2)
Plasminogen-Activator
TNFß = Lymphotoxin

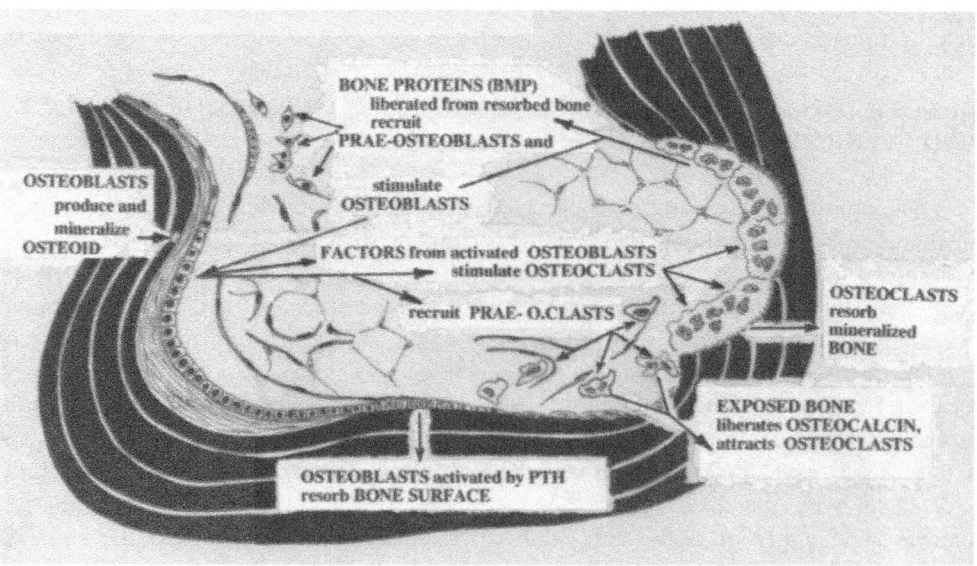

Figure 1. Bone remodelling cycle. Simplistic sketch. For explanation see text.

architecture of the bone including special capillarization are the indispensable requisites of efficient haemopoiesis. The same rules apply also to the abnormal conditions of repair after traumatic lesion and malignant haemoblastic proliferation or carcino-matous invasion: Consequently the overall anatomic characteristic of MOP is its localization within the truncal skeleton. As we know, the maintenance und functions of the bone depend of a lifelong equilibrium of the osseous remodelling cycle (Figure l). We assume that it follows roughly the indicated scheme: Osteoblasts are not merely thought to be responsible for the formative phase of the cycle by production and mineralization of the osteoidic bone matrix. They initiate also the resorptive phase by lysis of the trabecular collagenous coat after stimulation with the socalled "major bone hormones" like PTH and calcitriol; and they produce factors for the recruitment of pre-osteoclasts and stimulation of osteoclasts. These resorb bone and liberate bone matrix proteins like osteocalcin which in their turn close the cycle by recruitment and stimulation of osteoblast-cells (Burkhardt, 1992).

There are multiple ways of possible intervention with this cycle, open for the cells of the bone marrow (Table 1): not only various members of the immunoreactive cell family but probably also fibroblasts, endothelial, and tumor cells may interact with each other and with blood cells like granulocytes and platelets, when stimulated by the products of inflammation, necrosis, bacterial, viral or tumoral decay. The result is the cascadesque amplification of local factors, which may directly act upon the bone forming and resorbing cells, as is shown in a simplistic manner in the table. Numerous facts support the assumption that the pathogenic mosaic of MOP is composed of these elements. The problem is to disclose the mechanism which governs the composition, since our present knowledge of the local cellular organization is mostly based on in vitro experience. Another yet open question is the physiologic role of these events. The more we are learning about the fundamental interdependance of the bone marrow and bone tissues, the more the adjection "myelogenous" may loose its significance. From practical reasons however we are inclined to adhere to the old concept under the following, more liberal formula: MOPs are secondary disorders of the bone, caused by systemic distortion of all of those marrow tissues which are not normally and directly involved in osteogenesis.

DIFFFERENT PATHOGENIC MECHANISMS OF MOP IN RELATION TO THE RELEVANT PRIMARY DISORDERS OF THE BONE MARROW

Table 2 is a survey of several hypothetical pathogenic mechanisms of MOP of the disorders of the bone marrow in which they are most frequently observed. The pathogenic hypotheses are deduced from a synopsis of the available clinical, radiologic, histologic, and experimental data. Some special vascular aspects will be discussed in the last chapter. This survey is by no means complete. It may serve as a guideline for the gradual replacement and correction of our preliminary assumptions by future research. The following results and conclusions are of special interest: Non-bacterial myelitis due to immunologic processes is among the causes - if not the main cause - of the diffuse atrophy of haemopoiesis, manifest in aplastic anaemia. It shares the symptoms of systemic osteopenia and decreased osseous remodelling with the acute and chronic forms of malignant myeloproliferation. In aplastic anaemia however, the numbers of the venous bone marrow capillaries and especially trabecular sinusoids are diminished, whereas in myeloproliferation the venous and arterial capillaries are regularly increased. In the first case, the common capillary supply is reduced, in the second, augmented. Its balance,

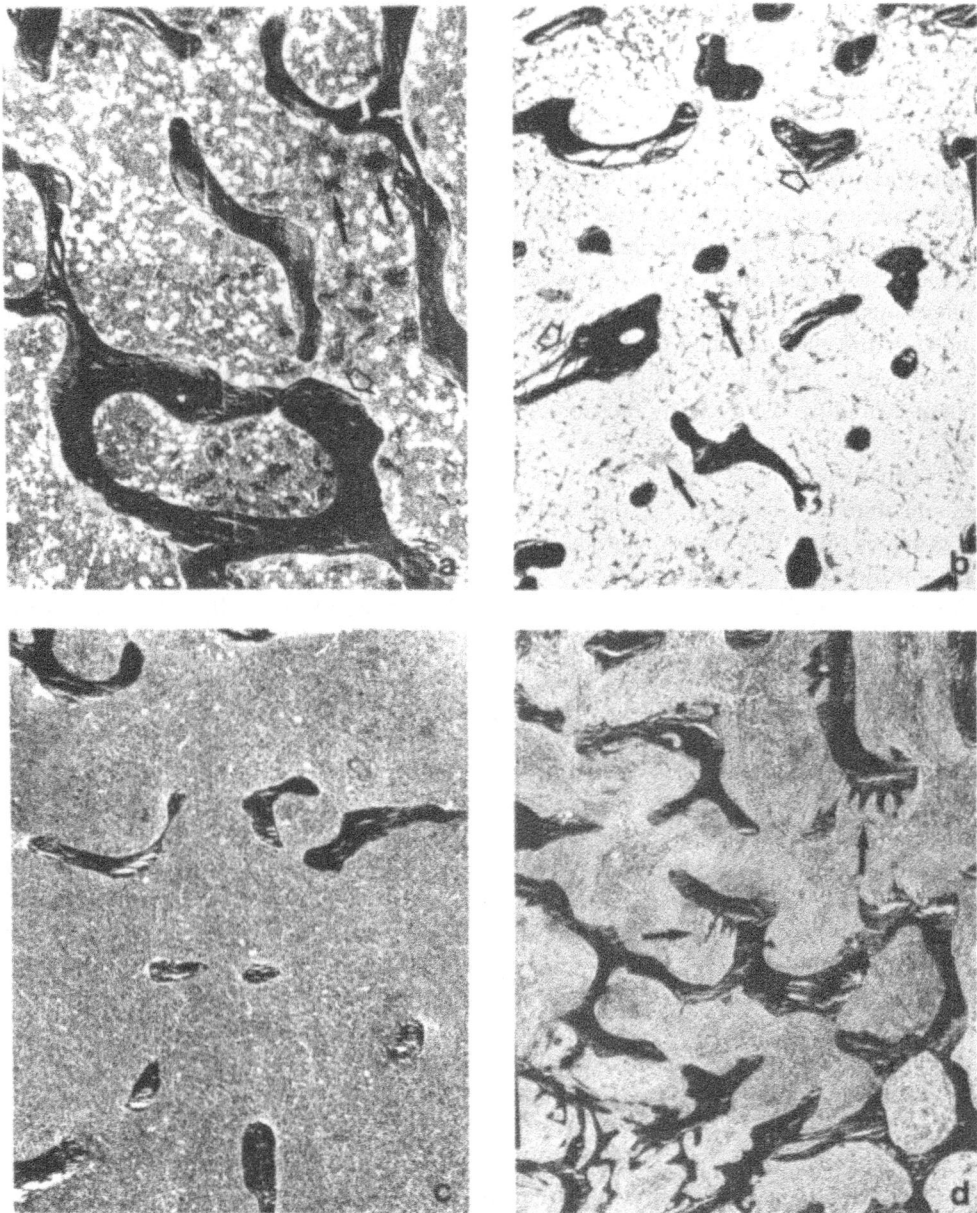

Figure 2. Gomori's stain. **a)** Normal, male 23 years. Normal trabecular framework and distribution of haemopoiesis and adipose tissue; note the association of marrow sinuses and haemopoiesis (dark spots, arrows); the flat sinuses covering wide areas of the trabecular surface (arrow) are hardly visualized at this magnification. **b)** Aplastic anaemia, inactive osteopenic picture: female patient, 25 years. Trabecular bone severely reduced; poor remodelling activity confined to paratrabecular sinuses (arrows); haemopoiesis replaced by adipose tissue; severe reduction of marrow sinuses (arrows). **c)** Chronic granulocytic leukaemia, male patient, 72 years. Trabecular bone reduced to an thin framework; granulocytic hyperplasia with numerous arterial capillaries totally replacing the fat cells. Absence of osteo-blastic-clastic activities. **d)** Malignant lymphoma (centroblastic-cytic type): male patient, 47 years. Irregular rarefaction and sprouts of newly formed woven bone. Zones of mesenchymal activation and osteoblastic-clastic activity (arrows) are clearly demarcated from the lymphoma occupying the central marrow spaces.

however, is disturbed in favour of the abnormal haemopoiesis, and factors, produced by the granulocytic hyperproliferation may suppress the osteogenesis (Gerwitz, 1981). This suppression may be among the reasons of its difference from osteodysplasia - not osteopenia - in the non-malignant haemoproliferation of severe haemolysis and iron deficiency. In these groups, the bone is supposed to provide additional marrow space for the compensatory haemopoiesis. There is hyperactive remodelling of the osseous trabeculae, extending into the long bones, and increased, especially venous, capillarization (Burk-

Table 2. Pathogenetic hypotheses of MOPs.

Primary Disease	Structural Pattern	Osseous Remodelling	Pathogenesis
Myelitis, Aplastic Anaemia	*osteopenia*	*decreased*	disturbance of microcirculation (immunol.dis.?)
Myeloproliferation chron./acute	*osteopenia*	*decreased*	requisition of microcirculation; suppression of osteogenesis
Haemolysis especially in growth period	*osteopenia*	increased	compensatory osteodystrophy; increase in venous capillaries
Osteomyelosclerosis	osteosclerosis	increased	mesenchymal stimulation by factors of megakaryocytes and platelets
Malignant Lymphoma, Plasmacytoma	osteolysis / sclerosis	increased	mesenchymal stimulation by factors of lymphocytes, plasmacells, macrophages
B.- marrow Infarction	cystic osteolysis / sclerosis	local increase	mesenchymal stimulation by factors of cell.necrosis
Metastatic Carcinoma	osteolysis / sclerosis	local increase	mesenchymal stimulation by factors of tumor cells

hardt, 1980). That the quality and not the mass of the cellular proliferation (as has been assumed formerly), is the pivotal element of the secondary osseous changes, is again illustrated by the special form of myeloproliferation in osteomyelosclerosis. Here the myeloproliferation and hypercapillarization are accompanied with osteosclerosis. The main reason of this distortion is the liberation of bone active factors from megakaryocytes, according to the work of Castro-Malaspina and Moore (1981), and our earlier proof of local fibrosclerosis in the regions of massive decay of these cells together with the interstitial dislocation of platelets (Burkhardt, 1971).

Similar osseous changes are caused by some special forms of malignant lymphoma, e.g. centrocytic lymphoma and single cases of myeloma. The mechanism is as yet unexplained. The majority of this group however, including Hodgkin's lymphoma, exhibits more or less characteristical osteolytic changes; Mundy et al. (1974) were the first to present the biochemical key to this subspecialty of MOP. The production and activation of osteoclast-activating factors (OAF) is probably not merely the domaine of the immune cell family. Not only Interleukin 1, the specification of the original OAF, but also other cytokines may be among the causes of osteoclastic osteolysis which is so frequently observed in other neoplastic invasive or metastatic processes of the bone marrow. A comparatively simple, focal model of the osteolytic-sclerotic action of cytokines from cellular necroses (TNF ß) is presented with the sclerosing bone cysts, consequences of the bone marrow infarction, best known in drepanocytosis. Various combinations of these factors including tumor angiogenetic factor (TAF) (Folkman, 1974) are at work in carcinomatous osteodysplasia (COD) (Burkhardt, 1982; Burkhardt, 1984). COD is classified under MOPs not alone from practical reasons - namely the radiological and histological similarity with other members of our group, especially non-Hodgkin's lymphoma and osteomyelosclerosis. Further arguments are its spread within the marrow space as consequence of the special capillary situation of the cancellous bones, and its partly mesenchymal pathogenesis: COD is not only produced by the products of the tumor cells themselves, but even more by stimuli from the stromal bone marrow cells. These are activated in the course of the mesenchymal reaction which follows the transgression of the neoplastic foci from the

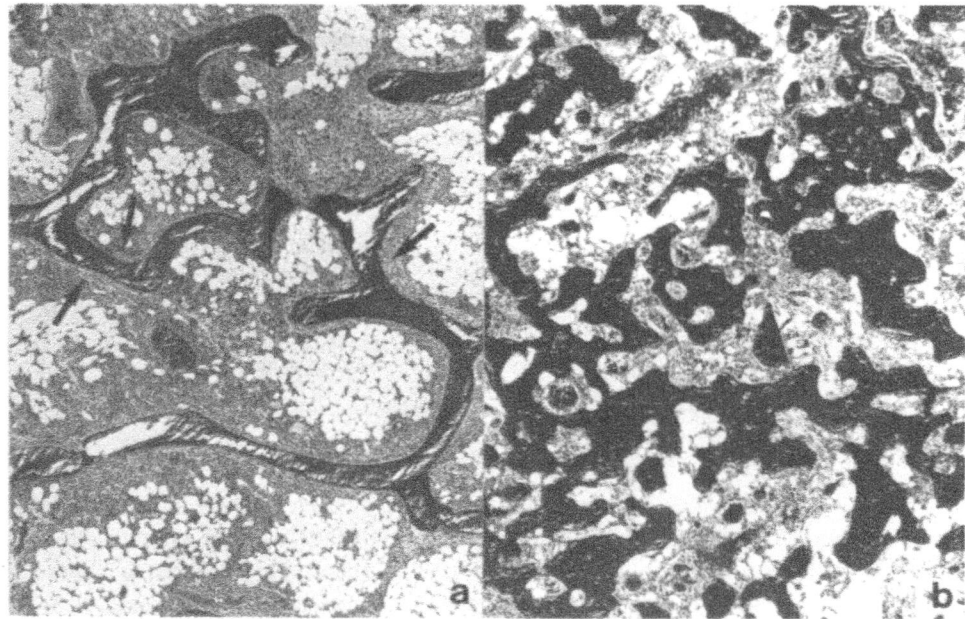

Figure 3. Gomori's stain. **a)** Plasmacytoma: female patient, 63 years. Irregular rarefaction of osseous trabecules, paratrabecular marrow densely infiltrated with plasmacytes, and richly capillarized. Poorly capillarized adipose tissue in the central marrow spaces. Extreme osteoblastic-clastic activity together with paratrabecular mesenchymal activation (arrows). **b)** Osteomyelosclerosis: male patient, 59 years. Irregular osteocondensation. Note the striking association of mesenchymal activation, increase in capillaries, and bone neoformation.

intra- into the extracapillary space (Burkhardt, 1984). The histomorphology of COD therefore is a most attractive model for almost all of the structural forms that MOP can adopt, and at the same time a mirror of the biologic activities of the metastatic cells which are responsible for their expression. In conclusion, the pathomorphologic shapes of MOPs are varied between systemic osteopenia or osteosclerosis, and the focal variants of osteolysis-osteocondensation. These conditions are to be exemplified with a few microscopic illustrations from iliac crest biopsies (Figures 2,3,4): In figure 2 the normal situation in the adult (a) is compared with the osteopenia of aplastic anaemia (b) and chronic granulocytic leukaemia (c), and the mixed osteosclerotic-lytic changes in malignant lymphoma (centrocytic-blastic type) (d). In figure 3 the active (lytic) osseous distortion of plasmacytoma (a) is opposed to the sclerotic osseous distortion of osteomyelosclerosis (b). Figure 4 shows the detail of the mesenchymal reaction in plasmacytoma (a), the normal ("active") osseous remodelling in a child (b), and the fibroblastic-osteoclastic changes in the case of figure 1 d.

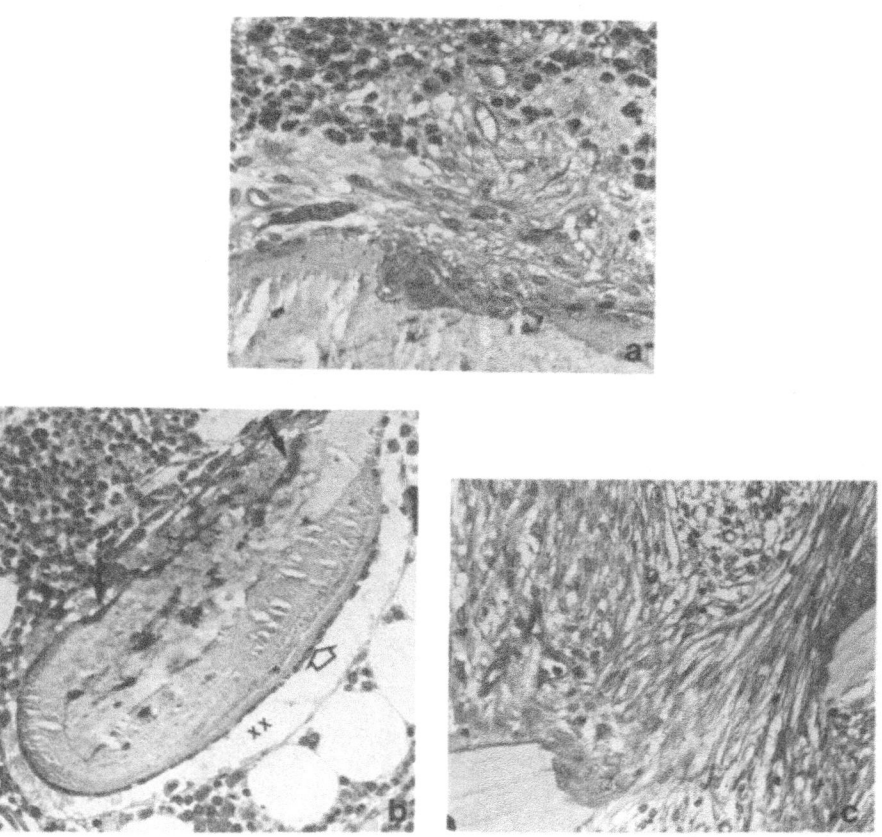

Figure 4. Gallaminblue- Giemsa stain. **a)** MOP in plasmacytoma; detail of figure 2a) From bottom to top: mature trabecular bone; osteoid seam; active osteoblasts and -clasts along the trabecular surface; zone of mesenchymal activation, rich in capillaries, pre-osteoblasts and -clasts; densely packed myeloma cells. No direct contact of myeloma cells and bone. **b)** Normal osseous trabeculae, haemopoiesis and fat cells: male child, 2 years. Active osseous remodelling (arrows); lower surface of the trabecule covered with flat osteoblasts (broad arrow), in continuity with the paratrabecular sinusoid. **c)** Active MOP in malignant lymphoma, detail of figure 2d): from bottom to top: bone; osseous remodelling; zone of activated mesenchyme rich in precursor cells and capillaries; neoplastic tissue.

MOPs are shaped directly by the specialized osteogenic cells, according to the stimulatory or inhibitory impulses from other cellular sources. Malignant cellular proliferation causes MOPs mostly by biochemical, not by physical interactions. They probably reflect the abnormal excesses of physiologic interactions between the bone and marrow tissues. This is the reason of their relative diagnostic insignificance on one hand, and of their relevance as clinical signposts of these interactions in the human on the other. MOPs show a more or less typical clinical profile only in a few of the corresponding bone marrow diseases, like the distended facial bones of the infantile haemolytic disorders, the spotty aspect of the drepanocytic marrow necroses, or the proximal femoral deformation of Gaucher's disease.

RANGE OF MOP AND DIFFERENTIAL DIAGNOSIS FROM OTHER GENERALIZED BONE DISEASES

MOPs are rated primarily in the field of internal medicine. Their importance is based on the following circumstances: frequency, differential diagnosis from other bone diseases, complications, and heuristical value for the understanding of the local pathophysiology of the bone.

Table 3. Frequency of MOP in the biopsy of the most relevant bone marrow disorders.

Primary Disease	Number of Biopsies	Incidence of MOP
Osteomyelo-sclerosis	543	100 %
Systemic Mastocytosis	166	100 %
Plasmacytoma	1.456	90 %
Metastatic Carcinoma	824	90 %
Sarcoidosis	114	85 %
Hodgkin's Disease	188	81 %
Aplastic Anaemia	155	70 %
Chronic Myeloproliferation	5.420	65 %
Malignant Lymphoma	3.376	35 %
Acute Leukaemia	895	17 %

This comparison (Table 3) reflects the relative frequency of the microscopic diagnosis of MOP among the relevant disorders of the bone marrow. These are osteomyelosclerosis, systemic mastocytosis, plasmacytoma, sarcoidosis, Hodgkin's disease, aplastic anaemia, metastatic carcinoma, malignant lymphoma, chronic myeloproliferation, and acute leukaemia. The microscopic diagnosis of MOP is made by the coincidence of osseous changes and primary bone marrow distortion, and by exclusion. The total number of the biopsies showing MOP was about twice as high as those from patients with other, mostly metabolic, bone diseases. The proportion of biopsies from all patients with osteopathies was 38 % of a total of 44,000 biopsies that were evaluated in our department in Munich between 1970 and 1985. These numbers of course indicate merely the relevance of MOP among the material of a biopsy department. Also for MOPs, the diagnostic scope of skeletal radiology differs widely from the specific microscopic image. In the case of systemic disorders however, the high specificity of the latter easily compensates for the narrow view. MOP therefore is primarily the domaine of histobiopsy. Table 4 shows the histologic differential diagnosis of the most frequent radiologic diagnoses of systemic bone diseases.

Table 4. Radiologic aspect and histologic differential diagnosis of MOP.

RADIOLOGIC ASPECT	HISTOLOGIC DIFFERENTIAL DIAGNOSIS
OSTEOPOROSIS	Osteomalacia / Hyperparathyroidism / Plasmacytoma / Mal.Lymphoma / Mastocytosis / Sarcoidosis / Diffuse Metastatic Carcinoma / Aplastic Anaemia / Haemoblastosis
PAGET's DISEASE	Osteomyelosclerosis / Metastatic Ca. / Hodgkin's Disease / Mastocytosis
HYPERPARA-THYROIDISM	Plasmacytoma / Paraneoplastic Osteodysplasia / Malignant Lymphoma / Metastatic Carcinoma
BONE TUMOR	Metastatic Carcinoma / Hodgkin's Disease / Gaucher's Disease / Histiocytosis X

Radiologic osteopenia may not only mean osteoporosis but also osteomalacia, hyperparathyroidism, plasmacytoma, malignant lymphoma, systemic mastocytosis, sarcoidosis, diffuse haemoblastic or carcinomatous spread, or systemic aplasia of the bone marrow. Paget's type of osseous changes is not unfrequently imitated by osteomyelosclerosis, metastatic carcinoma, and Hodgkin's disease. The radiologic image of hyperparathyroidism might be created by paraneoplastic parathyroidlike activity in the course of plasmacytoma, malignant lymphoma, and metastatic carcinoma; and foci of Gaucher's disease, Histocytosis X, metastatic carcinoma, and Hodgkin's disease can mimic primary bone tumours. The risk of being caught in the diagnostic trap is of course minimized by the haematologic and biochemical symptoms in most cases of MOPs. However situations are left when histobiopsy alone brings the decision. This is true especially in cases of systemic mastocytosis, sarcoidosis, Gaucher's disease, Histiocytosis X, metastatic carcinoma, and atypical Hodgkin's and non-Hodgkin's lymphoma. I cannot deal here with the special radiologic aspects, and the different expression of MOP in the growing and adult skeleton. These aspects and, as an example, the massive osteopenia caused by the rheumatic myelitis of childhood deserve special attention. MOPs can mimic radiologic osteoporosis, Paget's disease, hyperparathyroidism, and bone tumors. Therefore the non-invasive radiologic and the invasive biopsy techniques are suited to substitute, not to replace each other in the diagnostic evaluation of MOP.

MOP, SYMPTOMS, COMPLICATIONS AND THERAPY

MOPs may be accompanied by diffuse or localized bone pain, caused by the changes of the intramedullary pressure or by imminent bone fracture respectively. These symptomes are sometimes mistaken as rheumatism. The biochemical analysis, generally reflecting increased osseous remodelling, is insufficient to establish the correct diagnosis. Increase in calcitriol in the blood is observed in singular cases of sarcoidosis, and decrease in cases of neoplastic hypercalcaemia, which may be manifest also with suppression of the physiologic activity of the parathyroids. However, the most frequent reason of hypercalcaemia in MOPs is the production and activation of Interleukin 1 OAF) and prostaglandin E2 respectively (Gowen, 1985; Sato, 1988). Hypercalcaemia and spontaneous bone fractures are the main complications of MOPs, both deserving special therapy. In most cases however, MOP is not a therapeutic challenge of its own.

HISTOLOGIC RESULTS AND SPECULATIONS

Progress in experimental techniques brings more and more light into the hidden scene of the local tissue interactions. Clinical histology is the mirror which reflects the pathogenic results of these actions in the human. As has been shown, it offers a number of prominent features. To understand them, we have to realize that they are meaningful, however ambiguous. That is true especially of the microcirculatory changes of the bone marrow that were mentioned before.

For better characterization, the capillary changes were compared in 6 representative groups of MOP and 3 equal groups of other systemic diseases of bone, Paget's disease, hyperparathyroidism, and osteoporosis. Altogether 1,000 biopsies were evaluated, including 10 parameters of the bone and marrow qualities, and the counts of the bone marrow's arteries, arterioles, arterial capillaries, and sinusoids per 100 square millimeters/section of each case (Burkhardt, 1983) (Table 5).

It appeared that osseous remodelling was regularly increased only in the groups of plasmacytoma, Paget's disease, and hyperparathyroidism. The indices of osteoblasts and osteoclasts varied always equidirectionally, indicating regular coupling. Increase of the stromal cells coincided with osseous remodelling. Diminution of haemopoiesis and increase of the adipose tissue in the marrow space were correlated in osteoporosis, aplastic anaemia, plasmacytoma, and hyperparathyroidism, whereas in polycythaemia the reverse was true. Both components were diminished in osteomyelosclerosis, metastatic carcinoma, and Paget's disease, when fibrous changes replaced the marrow parenchyme. The most remarkable deviations from the normal were registered of the counts of the bone marrow vessels. This table, for sake of comprehension, was shortened to six groups with the extreme values of venous sinuses, arterial capillaries, and arterioles; the proportions of venous to arterial capillaries, and of these to the arterioles, are indicated inbetween.

Evidently, the groups differ markedly in the capillary counts. We have to compare these differences with the changes of those tissues, which may depend of the regular function of these vessels. Of course, we cannot expect to find plain correlations. Vascularization is the essential, but not unique, condition of structural organization (Burkhardt, 1984). Therefore

Table 5. Capillarization of the bone marrow in MOPs and other conditions.

	Venous Sinuses / 100 mm^2	VS : AC	**Arterial Capillaries** / 100 mm^2	VS : AC	**Arterioles** / 100 mm^2
Normal	1.700 910-2520	17 : 1	101 38-208	4 : 1	26 4 -81
Osteo - porosis	867 518-1312	12 : 1	75 43- 90	3 : 1	22 8 - 48
Aplastic Anaemia	1.000 335-1420	4 : 1	229 91-355	3 : 1	73 32-168
Polycyt- haemia	3.756 2490-4910	12 : 1	307 230-412	5 : 1	66 48-97
Granulo- cytic Leukaemia	2.460 1868-3112	2 : 1	1.056 410-1812	19 : 1	57 26-86
Osteo- myelo- sclerosis	2.073 1412-2938	1 : 1	2.126 1280-3340	12 : 1	183 125-250
Hyper- para- thyroidism	2.076 817-2996	4 : 1	596 330-1195	11 : 1	54 10-173

40 patients in each group; 20 males + 20 females.
According to Burkhardt R, Frisch B, Bartl R, Sommerfeld W, Mahl G, Jäger K, Schlag R, Hill W. (1984)

the results cannot be interpreted without reservation, including the general primate of the vascularization: Osteoporosis and aplastic anaemia, as we have shown earlier, are similar to each other from the correlation of the increase in adipose tissue and decrease of trabecular bone volume (Burkhardt, 1987). In osteoporosis as in senile osteopenia, the rarefaction of the osseous trabecules and the atrophy of the trabecular sinuses are correlated, osteopenia, however, is more accentuated than in aplastic anaemia. In both conditions the overall number of the sinuses is reduced - but, whereas the number of the arterial capillaries in osteoporosis is also reduced, it is increased in aplastic anaemia. This disproportion correlates with the depression of the haemopoiesis, the decrease alone of the sinuses in osteoporosis, does not. Correspondingly, polycythaemia shows an almost normal relation of sinuses' and arterial capillaries' numbers together with massive augmentation of both capillary sections. We therefore assume that the balance of the erythrocytopoiesis depends on the numbers of venous sinusoids in combination with a predominantly venous type of oxygenation of the bone marrow. This assumption may be true also for the anaemic conditions of granulocytic leukaemia and osteomyelosclerosis.

In both groups the strong increase of the venous sinuses is matched with the imbalance of the arterio-venous capillarization. In the group of hyperparathyroidism again, the same situation is given, however at a lower level. In contrast with the erythropoiesis, similar relations cannot be established for granulo- and megakaryocytopoiesis. Probably other factors of the milieu, and also the locomotion of the mother cells, and the integrity of the sinus walls, are more important for the generation and export of these elements. Considering the relations between the trabecular bone volume and capillarization in these groups, and in addition in Paget's disease, plasmacytoma, and metastatic carcinoma, we find a correlation between the diminution of the sinuses and osteopenia on one hand, and increase of stromal cells, and osseous remodelling on the other. The groups with osteosclerosis, like osteomyelosclerosis and Paget's disease, are those with the extreme values of arterioles and arterial capillaries.

In conclusion, disturbed microcirculation of the bone marrow is a regular consequence of those disorders, which most frequently are accompanied by MOPs . The reactive increase of the stromal marrow cells on one hand, and the disproportion of the venous capillarization in relation to the arterial capillary bed are among the principal conditions of these complications. These results encourage to consider the myelogenic aspects also of osteoporosis in addition to the search for its metabolic and hormonal clues. Secondary osteoporoses associated with rheumatic disorders are caused by immunologic processes (Burkhardt, 1992; Joffe, 1991; Rubin, 1988), which notoriously involve the capillary system. The study of MOPs has shed light on the close pathophysiological union of the bone and the bone marrow. This insight will certainly be fruitful for future research in osteology and haematology.

SUMMARY

Osseous and haemopoietic tissues are distinguished by their special performances and at the same time related by their genetic and structural organisation. Osteogenic trouble caused by haemopoietic dysfunction is called myelogenous osteopathy (MOP). Quasi as experiments of nature, MOPs convey us a message of the pathophysiologic interactions of the bone and marrow, which are most spectacular on the stage of cancellous bone. To read the message, radiology is a poor medium. We would need instead a new synthesis of histomorphology and molecular analysis at the cellular level. Accordingly, case controls including repeated biopsies and pursuit of the local factors of the structural organisation are among the promising guides to that purpose. At present the way is marked mostly by experimental data and biopsy snapshots, which need integration into clinical pathology.

From our own experience with about 44,000 biopsies of the human iliac crest the rate of MOP is 63 %. Therefore the symptoms of MOP are among the regular requisites of the differential diagnosis in every osteopathologic case. The diagnosis and therapy of MOP - not alone its hypercalcaemic manifestations - may become a clinical target of its own. The study of its pathopysiology offers new insights into the field of internal osteopathology in general, including the microcirculation of the bone marrow.

REFERENCES

Burkhardt, R.: Bone Marrow and Bone Tissue. Color Atlas of Clinical Histopathology. Springer Publ., Berlin Heidelberg New York 1971.

Burkhardt, R.: Myelogene Osteopathien. In: F. Kuhlencordt u.H.Bartelheimer (eds): Handbuch der inneren Medizin VI/IB. pp. 1057-118. Springer Publ., Berlin Heidelberg New York 1980.

Burkhardt, R., B. Frisch, R. Schlag, W. Sommerfeld: Carcinomatous osteodysplasia. Skel. Radiol. 8: 169-178 (1982).

Burkhardt, R.: Myelogene Osteopathien. Ther.Woche 33: 5794-5806 (1983).

Burkhardt, R., R. Bartl, B. Frisch, K. Jäger, G. Mahl, W. Hill, G. Kettner: The structural relationship of bone forming and endothelial cells of the bone marrow. In: J. Arlet, R.P. Ficat, D.S. Hungerford (eds): Bone Circulation. pp. 2-14. Williams and Wilkins Publ., Baltimore London 1984.

Burkhardt, R., G. Mahl, B. Frisch, R. Bartl, G. Kettner, M. Sund: Metastasen-Entwicklung im menschlichen Knochenmark in Abhängigkeit von Kapillaren und Tumor-Angiogenese. Verh. Dtsch. Ges. Path. 68: 316-321(1984).

Burkhardt, R., G. Kettner, W. Böhm, M. Schmidmeier, R. Schlag, B. Frisch, B. Mallmann, W. Eisenmenger, T. Gilg: Changes in trabecular bone, haematopoiesis and bone marrow vessels in aplastic anaemia, primary osteoporosis, and old age. A comparative histomorphometric study. Bone 8: 157-164 (1987).

Burkhardt, R.: Rheumatische und primäre Osteoporose. Arthritis Rheuma 2: 6-14 (1992).

Burkhardt, R.: Der Osteoblast - Schlüssel zum Verstandnis des Skelettorgans. Osteologie 1/2 (1992).

Castro-Malaspina, H., E.M. Rabellino, A. Yen, R.L. Nachman, M.A.S. Moore: Human megakaryocyte stimulation of proliferation of bone marrow fibroblasts. Blood 57: 781-787 (1981).

Folkman, J.: Tumour angiogenesis factor. Cancer Res. 34: 2109-2113 (1974).

Gänsslen, M.: Skelettveränderungen bei Blutkrankheiten. Münch. Med. Wochenschr. 27: 1048 (1938).

Gerwitz, G., A. Vignez, A. Stewart, B. Hoffmann: Production of a bone resorbing factor by myeloid leukemia cells. Blood, Suppl. l: 138-143 (1981).

Gowen, M., D. Wood, G.G. Russell: Stimulation of human bone cells in vitro by human monocyte products with interleukin-l activity. J. Clin. Invest. 75: 1223-1229 (1985).

Joffe, I., S. Epstein: Osteoporosis associated with rheumatoid arthritis; pathogenesis and management. Semin. Arthritis Rheum. 20: 256-272 (1991).

Markoff, N.: III. Die myelogene Osteopathie. Die normalen und pathologischen Beziehungen von Knochenmark zum Knochen. Erg. Inn. Med. Kinderheilkde 61: 132 - 342 (1942)

Mundy, G.R., R.A. Luben, L.G. Raisz, J.J. Oppenheim, D.N. Buell: Bone-resorbing activity in supernatants from lymphoid cell lines. New Engl. J. Med. 290: 867-871(1974).

Rubin, K., M. Ballow, R. Baron: Malignant osteoporosis and defective immunoregulation. J. Bone mineral Res. 3: 509-518 (1988).

Sato T., Y. Hakeda, E. Kurihara, et al.: Effect of prostaglandin (PG) on osteoclastic cell formation in vitro. Calcif. Tissue Int. Suppl.42: A 10 (1988).

CIRCULATORY ASPECTS OF BONE DISEASE IN ENDOCRINOPATHIES

A. SCHOUTENS

Service de Médecine nucléaire
Hopital Erasme
Université Libre de Bruxelles
Bruxelles, Belgium

Suppression or excess of insulin, calcitonin, thyroid and parathyroid hormones, estrogens, androgens, corticosteroids will modify bone metabolism while exerting either obvious and measurable, or surprisingly little effects on bone circulation under experimental conditions. Some hormone-like substances active on bone were never tested for circulatory changes, among them the very potent 1,25-Dihydroxyvitamin D3. It must be stressed that data in man are scarce and when available depend on the interpretation of the kinetics of bone seekers like ^{18}F-fluoride, a technique which awaits further validation.

DIABETES IN THE RAT

The suppression of insulin secretion induces drastic changes in bone metabolism and circulation in the rat. In man, such changes are less evident and it is so far generally believed that differences in bone growth, bone mineral content or bone metabolism in growing children or young adults, small as they are, could be secondary to more general consequences of insulin deprivation like the slower gain in weight of children (Mathiassen et al., 1990; Roe et al., 1991).

The development of diabetes in rat can be genetically determined (BB rats) or induced by streptozotocin. The effects of insulin deprivation on bone metabolism are similar in both models and the circulatory changes induced by streptozotocin can be reversed by insulin. A toxic effect of streptozotocin is thus excluded as the cause of the changes which include a cessation of bone growth, a reduction of bone formation by 50 % or more, a normal or low normal bone resorption and a resulting trabecular bone loss. Bone histomorphometry of diabetic rats compared to animals having not developed the disease show that the number of osteoblasts per osteoid surface can be less than 10 % of the values in controls, while the surfaces undergoing osteoclastic resorption are unchanged (Hough, et al., 1982; Wood et al., 1984).

Bone Circulation and Vascularization in Normal and Pathological Conditions
Edited by A. Schoutens *et al.*, Plenum Press, New York, 1993

167

Tibial blood flow (Microspheres trapping as the technique) was measured in streptozotocin-diabetic rats (Lucas, 1985, 1987). Blood flow is reduced to one third of controls at 14 days and is 50-60 % of control values 28 and 56 days after streptozotocin injection. Bone resistance to flow is 171 % of controls at 14 days (Figure 1). Yet bone vasoconstriction is submaximal and can be increased still further by noradrenaline infusion. Insulin treatment after streptozotocin injection will actually increase bone blood flow above control values. Circulation changes in diabetic rats are observed in small intestine, kidneys and bone but it is in this last organ that the amplitude of the changes is maximal. It is tempting and probably accurate to relate reduced bone blood flow on one hand and, on the other hand, reduced bone formation, reduced number of osteoblasts and supposedly reduced osteoblast proliferation.

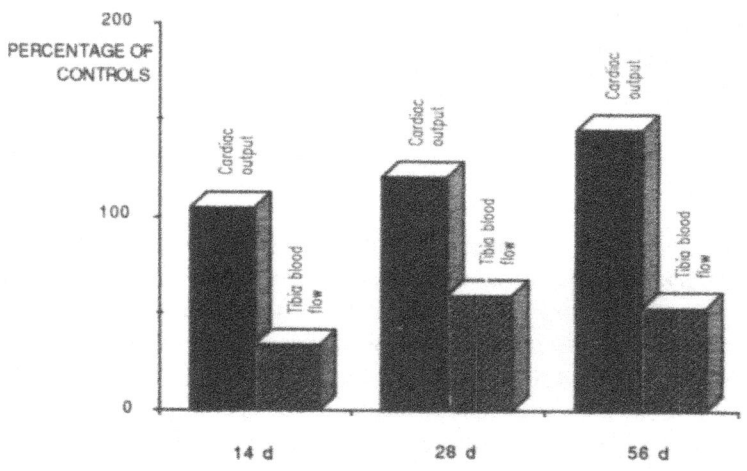

Figure 1. Low bone blood flow and normal or high cardiac output after 14 to 58 days insulin deprivation in the rat (After Lucas, 1987).

CALCITONIN IN EXCESS

Calcitonin was studied in two non comparable situations, one acute, the other related to induced bone metabolic changes.
In acute experiments on isolated canine tibia perfused at constant flow, Driessens and Vanhoutte (1979) observed an increased bone resistance to flow under calcitonin.On the other hand, rats treated with calcitonin for 30 days showed a reduced osteoclastic activity, an increased periosteal apposition while femoral blood flow (Microspheres trapping as the technique) was found increased (Schoutens et al., 1988). In this species, blood flow could be contemplated once again as related to osteoblastic activity and/or to osteoblastic proliferation (supposedly increased but not measured actually).
Bone blood flow (Kinetic interpretation of ^{18}F uptake as the technique) was measured in one single case of secretory medullary carcinoma of the thyroid and was found normal (Green et al., 1987).

HYPO- AND HYPERTHYROIDISM

Deprivation or excess of thyroid hormones induce a reduced or increased bone turnover in all mammals tested. In mature dogs, such manipulations lead to a wide range of bone tetracycline labeling indices. Bone blood flow was measured in these animals by the clearance of osteotropes, a technique now criticized. Notwithstanding it would seem that bone blood flow changes adapt to bone remodeling changes (Sim et al., 1970). The arteriovenous difference in O_2 remains unchanged and a fair relationship is found between O_2 consumption (^{85}Sr clearance x arteriovenous difference in oxygen content) and bone tetracycline labeling (Figure 2).

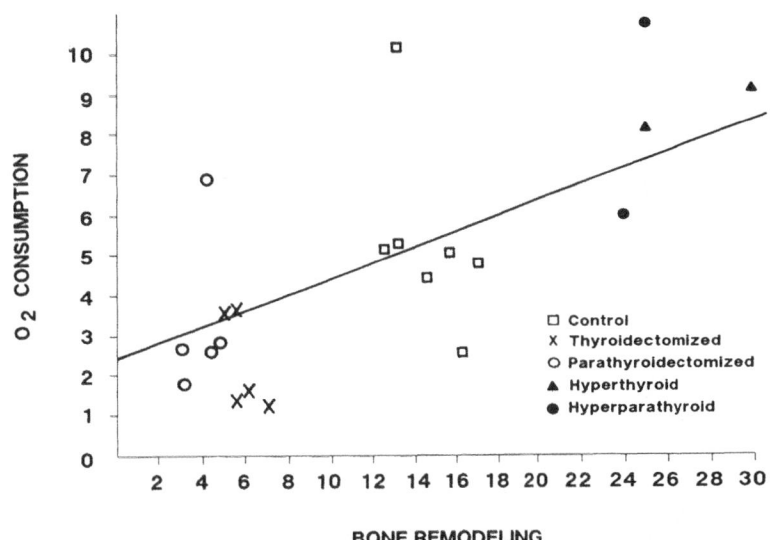

Figure 2. Relationship of oxygen consumption to bone remodeling in mature dogs under conditions of hypo- to hyper-thyroidism or parathyroidism (Sim and Kelly, 1970, with permission Journal of Bone and Joint Surgery).

HYPO- AND HYPERPARATHYROIDISM

Parathyroidectomy and excess of parathyroid hormones in adult dogs lead to lower and higher blood flows than control values (Sim et al., 1970).
Thyroid and parathyroid hormonal manipulations concur in the relationship between O_2 consumption and bone remodeling illustrated by figure 2.

In humans, bone blood flow (^{18}F kinetics as the technique) is normal in primary hyperparathyroidism (Green et al., 1987) (Figure 2 of Peter's paper, p 180).

CALCIUM DEFICIENT DIET WITH OR WITHOUT VITAMIN D DEFICIENCY

Bone metabolic changes associated with calcium deficiency are generally considered as resulting from secondary hyperparathyroidism. However a close scrutiny of the initial events, hours or days after having put mature 120-day-old rats on a calcium deficient diet shows that the induction of a high bone turnover is not dependent on the presence of parathyroid hormone in excess. Nor does it rely on an increased serum level of 1,25 Dihydroxyvitamin D3. The number of osteoblasts increase markedly in the metaphysis of long bones through enhanced cell proliferation, the calcium accretion rate and the single

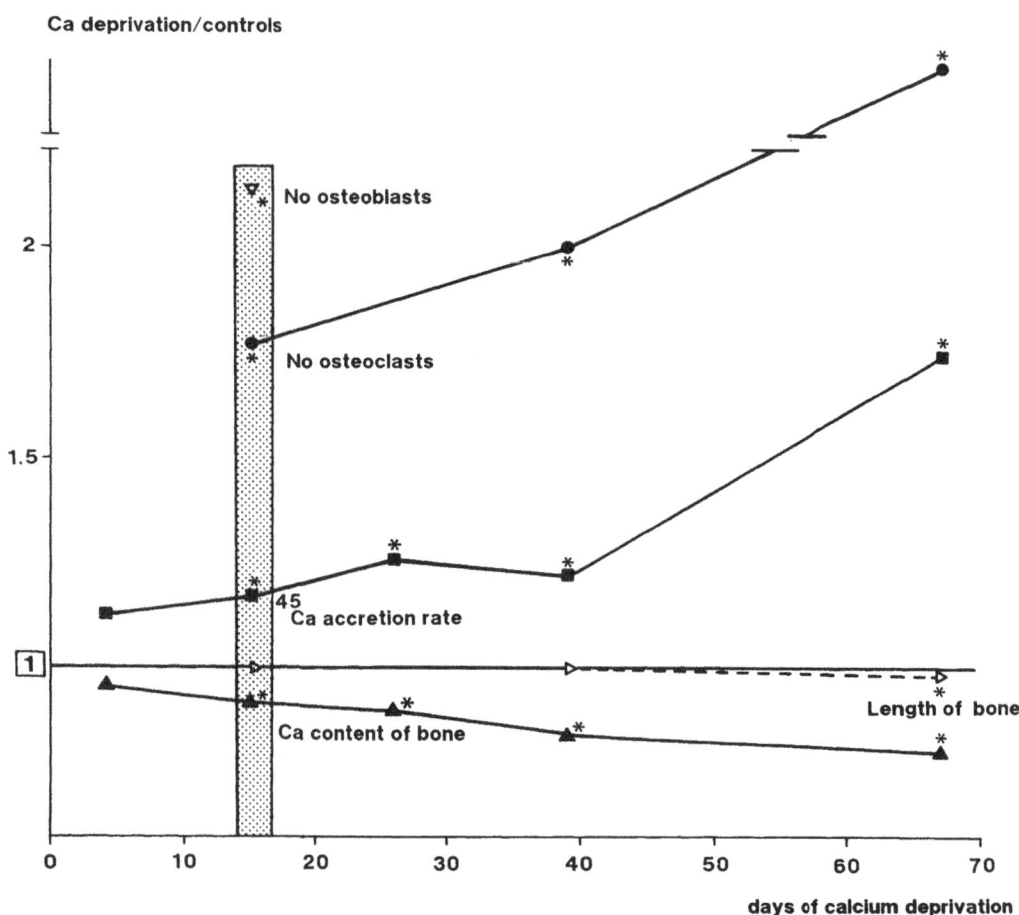

Figure 3. Calcium deprivation in the diet in mature female rats. As compared to controls, high indices of bone turnover, loos of bone calcium, no change in bone blood flow (Egrise, unpublished).

tetracycline labeling increase, but the double labeling (bone formation) decreases. Serum osteocalcin is high, osteoid tissue is being deposed in excess. Thus a great density of osteoblasts are present and active but prove uncompetent in deposing calcified bone (Egrise, personal communication). The osteoclastic surface increases as well and osteopenia progresses rapidly (Figure 3).

Rockoff et al. (1969) demonstrated in immature rats that the femoral uptake of ^{86}Rb over the 30 first seconds after injection is high between 3 and 9 days of dietary calcium deficiency and normal thereafter, indicating a rapid, significant and time dependent increase in the cardiac fraction of cardiac output to bone. We were unable to document any significant change in bone blood flow (Microspheres trapping as the technique) in tibiae and femurs of female adult 120-day-old rats and consider the model of calcium deficiency in the rat as departing from the general assumption that flow rate and osteoblast number/activity are closely related.

VITAMIN D DEFICIENCY

Fourteen out of twenty patients with osteomalacia had high values of bone blood flow (18F kinetics as the technique). All patients were studied before treatment; the cause of osteomalacia, vitamin D deficient diet or vitamin D malabsorption, was not stated (Green et al., 1987)

OVARIECTOMY

Ovariectomy in the rat seems to be a reliable model for human postmenopausal osteoporosis. Bone resorption and bone absorption are both high. Bone loss affects mostly trabecular bone, is initially rapid and will subside temporarily thereafter to resume into a phase of slow bone loss. Even in mature 120 days old rats, growth is still active and increases under estrogen depletion. Density of osteoblasts on trabeculae is 35 % higher than in controls and in vitro proliferation of osteoblast-like cells is markedly increased in primary and secondary cultures (Egrise et al., 1992).

Bone blood flow (Microspheres trapping technique) is increased from 28 to 84 days postovariectomy and then decreases to control values. Blood flow increase is observed both in metaphysis and in diaphyseal cortical bone and is thus not related to the intensity of bone loss. There seems to be a parallelism between bone blood flow changes and those of bone formation and turnover as monitored by ^{45}Ca accretion rate, serum alkaline phosphatase and osteocalcin (Figure 4).

ORCHIDECTOMY

Orchidectomy in post-pubertal 55 days old rats leads to a decrease in tibial length, an increase in osteoclastic bone resorption, a brief increase at 31 days of the calcium accretion rate to be followed by a decrease in bone formation at 86 and 120 days, and to a loss in calcium content of long bones (Figure 5). Bone blood flow (Microspheres trapping technique) increases threefold with a peak at 31 days, concomitant with the rapid increase in the number of tibial metaphyseal osteoclasts and with the transient increase in calcium accretion rate. BBF is low normal at 86 and 120 days (Schoutens et al., 1984). Testosterone replacement therapy, initiated after the initial blood flow increase, inhibits the negative effect of castration on bone growth and bone mineral.

One-year-old rats have no measurable bone growth anymore. Orchidectomy at this age leads to a decrease in long bone volume and bone calcium content, to a thinning of the cortical width, and after an initial increase in calcium accretion rate to a depressed

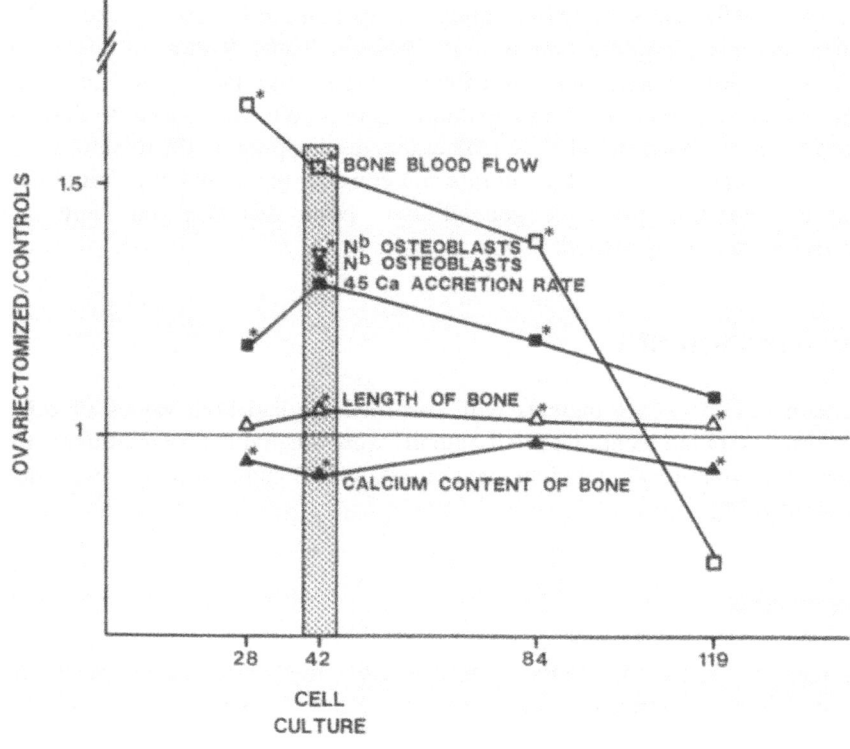

Figure 4. Ovariectomy in the rat. As compared to cotntrols, bone turnover is high, length of bone increased and bone calcium content low. Bone blood flow is high, up to 84 days postovariectomy (Egrise et al., 1992).

formation rate. The number of osteoclasts increases. Bone blood flow (Microspheres trapping technique) is increased throughout the 120 days observation period (Figure 5) (Verhas et al., 1986).

CORTICOIDS

Treatment with corticoids leads in mammals to fatty replacement of bone marrow and to a marked decrease in bone formation rate. Corticoids in excess is a well known risk factor in man for the development of osteonecrosis, with the femoral head as first target. Ischemia was proposed as the main mechanism resulting in the death of bone marrow and bone cells. And indeed important modifications of bone vascularization can be documented (Arlet et al., 1991). However, treatment of adult geese with high doses of corticosteroids for 5 months did not result in altered bone blood flow (Microspheres trapping technique) (Bouteiller et al., 1980, 1983). It could not be concluded that osteonecrosis is not related to an altered bone circulation because these animals did not develop osteonecrosis or even fatty replacement of bone marrow. The latter is unexpected as the species is worldwide reknown for the easiness with which fatty liver can be induced by forcefeeding; apparently bone defends itself far better than does the liver against overfeeding and corticoids.

Wang (1990) treated rabbits with weekly doses of 12 mg methylprednisolone for 12 weeks. Femoral head blood flow (Microspheres trapping technique) is unchanged after 6 weeks. But at 10 weeks, blood flow is decreased, at a moment when bone marrow composition is altered. This late circulatory change can be prevented by antilipid agents.

These results suggest that corticoids have no direct or immediate effect on bone blood flow, whereas they reduce osteoblast proliferation through a primary mechanism. With corticoids in excess, we do not confirm the parallel behavior of bone formation rate and bone blood flow, which was suggested in situations like experimental insulin deprivation. However, considering the scarcity of data, we probably need additional experiments to better describe the effects of corticoids on bone vascularization and bone blood flow.

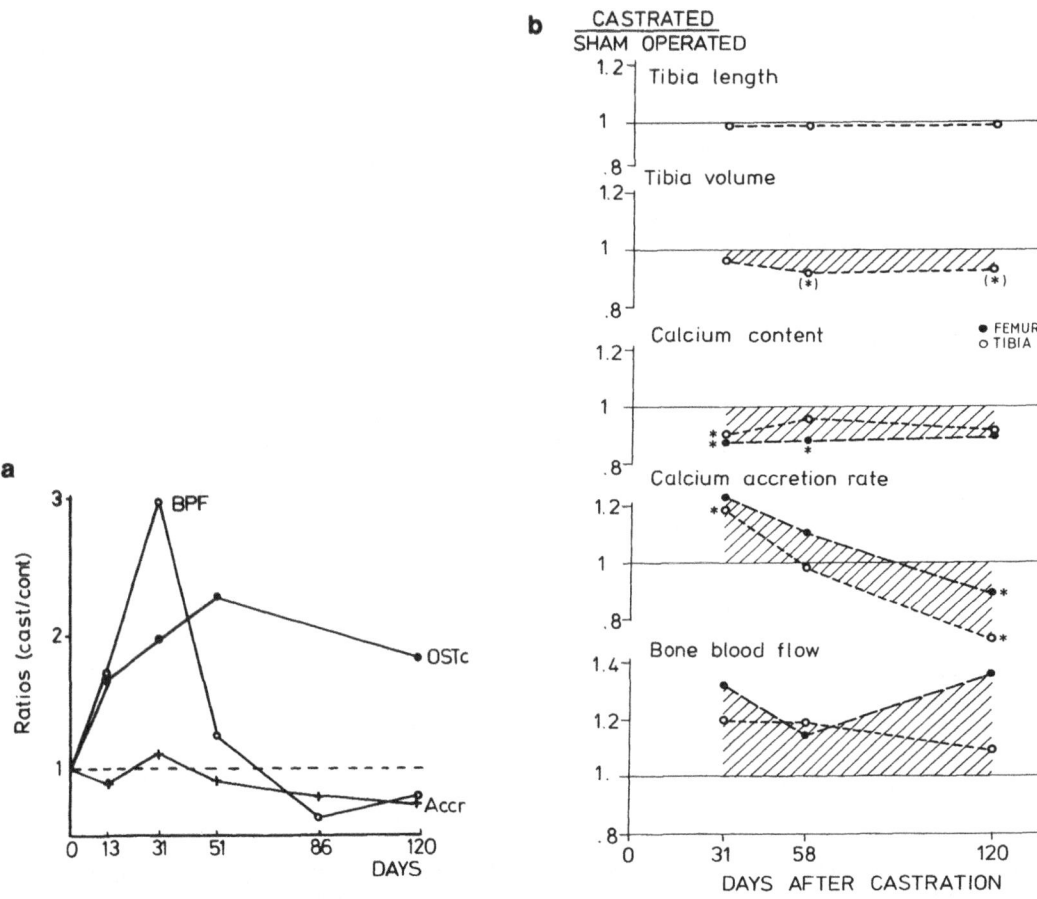

Figure 5. a) Orchidectomy in postpubertal 55-day-old rats. Day O is day of castration. Evolution of the ratios of castrates over controls for bone plasma flow, number of osteoclasts and ^{45}Ca accretion rate of bone (Schoutens et al., 1984, with permission Scand. Univ. Press). **b)** Orchidectomy in 1-year-old rats. Evolution of the ratios of castrates over controls for the bone parameters (Verhas et al., 1986, with permission Springer Verlag).

DISCUSSION

In many instances, bone blood flow changes would accomodate parallel changes in the rate of bone formation. However, even in the small number of situations presented, this parallelism is twice at fault, once in calcium deficiency where an important osteoblast activity, however not resulting in bone formation, is not accompanied by a significant or lasting increase in bone blood flow; and also in hypercorticism where bone formation is markedly depressed in most species, with no change in bone blood flow in geese and de-layed reduction in rabbits.

Thus far, most experimental work and thinking were aimed at understanding which advantages or disadvantages for bone metabolism result from bone blood flow changes. It was suggested that the last could help in stabilizing the arteiovenous difference in O_2. The question is whether bone does need that stability when one considers how low bone O_2 dif-ference is, actually two times less than the mean for the whole body, being only equalled by the kidney. Alternative hypotheses were presented for the role of bone blood flow, the most interesting thus far trying to relate microanatomy, microcirculation, cytokin production, blood flow and bone metabolism. By working on these relationships, the interest shifted to-wards a search for the trigger at tissue level able to increase bone blood flow. It was pro-gressively accepted, at least implicitely, that bone blood flow changes could be the conse-quence of significant tissue adaptative mechanisms, instead of conditioning them.

In the hypothesis that validated techniques for measuring bone blood flow in humans should become available, the question is still open whether they could help in the diagnosis of bone metabolic diseases and in the monitoring of their treatment.

REFERENCES

Arlet, J., M.Laroche, R.Soler, M.Thiechart, M.T.Pieraggi, B.Mazieres: Histopathology of the vessels of the femoral heads in specimens of osteonecrosis, osteoarthritis and algodystrophy. ARCO News Letter . 3,2: 108-112(1991).

Bouteiller, G., J.Arlet, A.Blasco, F.Vigoni, A.Elefterion: Is osteonecrosis of the femoral head avascular ? Bone blood flow measurements after long-term treatment with corticosteroids. Metab. Bone Dis. Relat. Res. 4: 313-318(1983).

Bouteiller, G., A.Blasco, F.Vigoni, J. L.Decamps, P.Puel, H.Regis, J. Arlet: Regional femur blood flow measurements with radiolabelled microspheres in fatty liver-induced or corticosteroid treated animals: preliminary report. Biomedicine 33: 119-123(1980).

Driessens, M., P.M.Vanhoutte: Vascular reactivity of the isolated tibia of the dog. Am. J. Physiol. 236: H904-908(1979).

Egrise, D., D.Martin, P.Neve, A. Vienne, M.Verhas, A. Schoutens: Bone blood flow and in vitro proliferation of bone marrow and trabecular bone osteoblast-like cells in ovariectomized rats. Calcif.Tissue Int. .50: 336-341(1992).

Green, J.R., J. Reeve, M. Tellez, N. Veall, R. Wootton: Skeletal blood flow in metabolic disorders of the skeleton Bone 8: 293-297(1987).

Hough, S., J.E.Russell, S.L.Teitelbaum, L.V.Avioli LV: Calcium homeostasis in chronic streptozotocin-induced diabetes mellitus in the rat. Am.J.Physiol. 242: E451-456(1982).

Lucas, P.D.: Effects of streptozotocin-induced diabetes and noradrenaline infusion on cardiac output and its regional distribution in pithed rats. Diabetologia 28: 108-112(1985).

Lucas, P.D.: Reversible reduction in bone blood flow in streptozotocin-diabetic rats. Experientia 43: 894-895(1987).

Mathiassen, B., S. Nielsen, J. Ditzel, P. Rodbro: Long-term bone loss in insulin-dependent diabetes mellitus. J. Int. Med. 227: 325-327(1990).

Rockoff, S.D., A. Bravo, H. Kaye, R.P.Spencer: Rapid increase in fraction of cardiac output to bone in experimental calcium deficiency. Calcif.Tissue Res.3: 17-29(1969).

Roe, T.F., S. Mora, G.Costin, F.Kaufman, M.E.Carlson, V.Gilsanz: Vertebral bone density in insulin-dependent diabetic children. Metabolism 40: 967-971(1991).

Schoutens, A., M.Verhas, M.L'Hermite-Baleriaux, M.L'Hermite, A.Verschaeren, N.Dourov, M.Mone, A. Heilporn, A.Tricot: Growth and bone haemodynamics responses to castration in male rats. Acta Endocrinol. 107: 428-432(1984).

Schoutens, A., M.Verhas, N. Dourov, P. Bergmann, F.Caulin, A.Verschaeren, M.Mone, A. Heilporn: Bone loss and bone blood flow in paraplegic rats treated with calcitonin, diphosphonate, and indomethacin Calcif.Tissue Int. 42: 136-143(1988).

Sim, F.H., P.J. Kelly: Relationship of bone remodeling, oxygen consumption, and blood flow in bone. J.Bone Joint Surg.[Am] 52: 1377-1389(1970).

Verhaeghe, J., E.Van Herck, W.J.Visser, A.M.H.Suiker, M.Thomasset, T.A.Einhorn, E. Faierman, R. Bouillon: Bone and mineral metabolism in BB rats with long-term diabetes. Diabetes 39: 477-482(1990).

Verhaeghe, J., W.J.Visser, T.A.Einhorn, R.Bouillon: Osteoporosis and diabetes: lessons from the diabetic BB rat. Horm.Res.34: 245-248(1990).

Verhas, M., A.Schoutens, M.L'Hermite-Baleriaux, N. Dourov, A. Verschaeren, M. Mone, A. Heilporn: The effect of orchidectomy on bone metabolism in aging rats. Calcif.Tissue Int. 39: 74-77(1986).

Wang, G.J., R.E.Fechner, J.P.O'Hara, N.Stoffel, W.G.Stamp: Improvement of femoral head blood flow in steroid-treated rabbits. In: J. Arlet, B. Mazières (eds.): Bone circulation and bone necrosis, pp. 395-398. Springer Verlag, Berlin 1990.

Wood, R.J., L.H.Allen, F.Bronner: Regulation of calcium metabolism in streptozotocin-induced diabetes. Am.J.Physiol. 247: R120-123(1984).

BONE TURNOVER IN OSTEOPOROSIS

A. M. PETERS

Department of Diagnostic Radiology
Hammersmith Hospital
London, United Kingdom

INTRODUCTION

Skeletal turnover is the result of bone remodelling and is achieved through the activity of microscopic bone remodelling units consisting of osteoclasts and osteoblasts respectively resorbing and laying down new bone. These basic multicellular units (BMU) are active for periods of 3-6 months, and, at any one time, about 10^5-10^8 are active throughout the skeleton. Normally, slightly more bone resorption occurs compared with bone replacement, so that with advancing age, there is a slow progressive loss of bone mass. Osteoporosis may be the result of an abnormal imbalance between formation and resorption within the BMUs.

Several techniques are available for assessing skeletal turnover, ranging from simple biochemical measurements to complex histomorphometric analysis and radio-calcium kinetic studies. Of considerable interest to nuclear medicine specialists are techniques for measuring bone mineral content from dual energy X-ray absorptiometry and the uptake and retention by the skeleton of several radiopharmaceuticals, particularly ^{99m}Tc-diphosphonates, radio-calcium, radio-strontium and fluorine-18. Skeletal blood flow (SBF) is also of interest, in particular, techniques for measuring SBF and appropriate forms in which to express it, and its relationship with the uptake of bone-seeking radiopharmaceuticals. In this review, some of these techniques will be discussed in relation to osteoporosis.

SITES AND KINETICS OF SKELETAL DIPHOSPHONATE UPTAKE

^{99m}Tc-diphosphonate (MDP) is deposited in bone at the interface of new bone and osteoid. It is not taken up by the osteoblast, but reflects osteoblastic activity. Its retention by the skeleton has been widely used as an index of bone turnover. This was popularised by Fogelman et al. (1978, 1980) and is expressed as the whole body count rate at 24 hours as a fraction of the count rate at 5 minutes after intravenous injection of ^{99m}Tc-MDP.

Bone Circulation and Vascularization in Normal and Pathological Conditions
Edited by A. Schoutens *et al.*, Plenum Press, New York, 1993

177

Since no losses have occurred by 5 minutes, the count rate at this time represents 100 % of the dose. Patients with high rates of bone turnover, such as hyperparathyroidism and osteomalacia, retain abnormally high percentages of the injected dose (Fogelman et al., 1978). Diphosphonates are analogues of pyrophosphate, and localise at active sites of bone remodelling. Their rates of uptake, therefore, reflect the activity of the BMUs on a whole body basis. The whole body retention (WBR) at 24 hours, provided the extravascular space is cleared of tracer and there is no urinary retention, reflects the competition during the preceding 24 hours between the kidneys and skeleton for the available MDP. The popularity of this technique, and its subsequent large literature, is based on its simplicity and the large numbers of patients referred to nuclear medicine units for bone scanning with these agents. In order to validate MDP as a marker of bone turnover, it is clearly necessary to have as much information as possible on its precise mechanism of uptake and the influence of other variables, such as bone capillary permeability and skeletal blood flow, on its uptake.

^{99m}Tc-MDP is a small hydrophilic solute. With a molecular weight of 261, its kinetics of transfer across bone capillaries between plasma and bone interstitial fluid are similar to those of ^{51}Cr-EDTA and sucrose transfer across similar vascular beds possessing a continuous endothelium. Thus, unidirectional (instantaneous) extraction efficiency is about 50 % (McCarthy et al., 1980) and since average bone blood flow, when expressed in units of ml per 100 g per min, is similar to resting muscle, the permeability surface area product for MDP in bone capillaries is similar to the value for muscle. McCarthy and Hughes (1989) have measured the various transport constants of ^{99m}Tc-MDP in the model illustrated in figure 1. Starting from their estimated values of k1 (assumed to be equal to k2) and k, one can draw the following general conclusions concerning the kinetics of skeletal ^{99m}Tc-MDP uptake. 1) Clearance is minimally influenced by bone blood flow, especially at low baseline levels of flow; 2) clearance is minimally influenced by capillary permeability, except at high levels of bone transport constant (k), and high blood flows; and 3) clearance is influenced by the bone transport constant, k (presumably reflecting the availability of bone surface area available for the tracer and bone metabolic activity), especially at high levels of capillary permeability.

Clearly, the relationship between bone blood flow and MDP uptake is not linear, and this has been previously demonstrated using continuous infusion of ^{81m}Kr and radiolabelled microspheres by Lavender et al. (1979). The steady state extraction efficiency of MDP by bone is an important determinant of MDP clearance. Extraction efficiency is a function of the transport constants k1 and k, and is apparently increased in pathological lesions containing woven bone (Galasko, 1975).

CLINICAL MEASUREMENT OF BONE TURNOVER

The clinical measurements of skeletal blood flow, MDP clearance and MDP steady state extraction efficiency by bone are difficult. The most widely used clinical method for measuring skeletal blood flow is the ^{18}F method of Wootton et al. (1976). This deconvolutional approach relies on an extraction efficiency of 100 % of ^{18}F by bone. It is important to appreciate that this is the unidirectional, rather than the steady state, extraction fraction, and requires an endothelial permeability surface area product well in excess of that of blood flow. The technique gives blood flow as a percentage of total blood volume per minute. It is slightly reduced in osteoporosis (Reeve et al., 1988), increased in osteomalacia and Paget's disease, and unchanged in hyperparathyroidism

(Tellez et al., 1983) (Figure 2). SBF correlates significantly with histomorphometric indices of bone apposition in patients with osteoporosis (Reeve et al., 1988). One of the difficulties of expressing SBF in this form is that it is difficult to compare with SBF expressed in more conventional units of ml per 100 g per min. The latter is itself conceptually difficult, since conventional blood flow tracers, like ^{133}Xe (Lahtinen et al., 1981), ^{15}O labelled water (Martiat et al., 1987) and antipyrine (McInnis et al., 1980) have a distribution volume in bone which is smaller than total bone volume. If bone mass is reduced in osteoporosis, SBF, although reduced in absolute terms, may be increased in terms of perfusion. Regional bone blood flow is even more difficult to measure. The most promising approach currently is the use of positron emission tomography and ^{18}F (Hawkins et al., 1992).

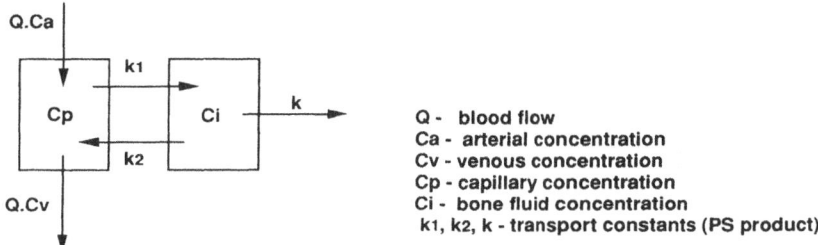

Q - blood flow
Ca - arterial concentration
Cv - venous concentration
Cp - capillary concentration
Ci - bone fluid concentration
k1, k2, k - transport constants (PS product)

Figure 1. Compartmental model for bone uptake of hydrophilic solutes, such as ^{99m}Tc MDP. The fundamental equations are based on the Fick Principle:

$$E_{pi} = \frac{k_1 . C_p}{Q . C_a} \tag{1}$$

$$E_{pb} = \frac{k . C_i}{Q . C_a} \tag{2}$$

where E_{pi} and E_{pb} are extraction fractions of tracer from plasma to bone fluid and plasma to bone, respectively. But, at steady state

$$k_1 . C_p = k . C_i + k_2 . C_i = Q . E_{pi} . C_a \tag{3}$$

Rearranging and substituting for $Q . C_a$,

$$E_{pb} = \frac{k}{k + k_2} . E_{pi} \tag{4}$$

E_{pi} can be calculated for different values of Q from the Crone-Renkin equation if k_1 is known. Note that k_1, k_2 and k have units of ml per min, and are equivalent to permeability surface area products rather than fractional rate constants.

Notwithstanding an extra-renal soft tissue uptake of about 5 %, ^{99m}Tc-MDP plasma clearance is equal to the sum of renal and skeletal clearances. ^{99m}Tc-MDP clearance is difficult to measure from blood sampling, as the clearance curve is multiexponential (Caniggia and Vattimo, 1980). Whereas the terminal slope of, say, a ^{99m}Tc-DTPA

179

clearance curve is monoexponential with a rate constant reflecting GFR/ECF volume, the MDP clearance curve approaches a very slow terminal exponential which reflects the slow reflux of ^{99m}Tc or ^{99m}Tc-MDP from bone to plasma. MDP clearance should, therefore, be based on the blood clearance curve from which the asymptote representing this reflux has been subtracted. Alternatively, and more easily, the MDP clearance can be derived from dynamic whole body or regional skeletal counting with the whole body counter, gamma camera or probe counter. The count rate by any of these methods reaches an asymptote or slow terminal phase when the tracer is cleared completely from plasma. The rate constant with which the asymptote is approached is a close reflection of MDP clearance normalised for its distribution volume (the ECF).

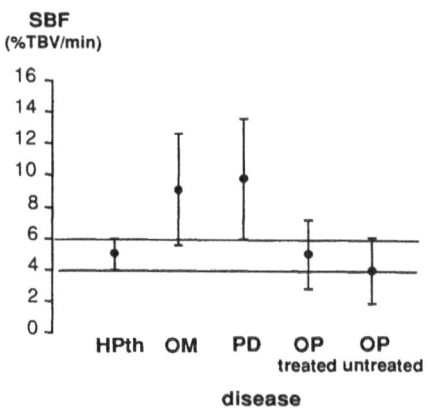

Figure 2. Skeletal blood flow (per cent total blood volume per min) determined using the technique of Wootton et al. (1976) in several bone diseases: HPth-hyperparathyroidism; OM-osteomalacia; PD-Paget's disease; OP-osteoporosis. Mean values ± 1 SD. The horizontal lines enclose the mean value ± 2 SD in normal subjects.

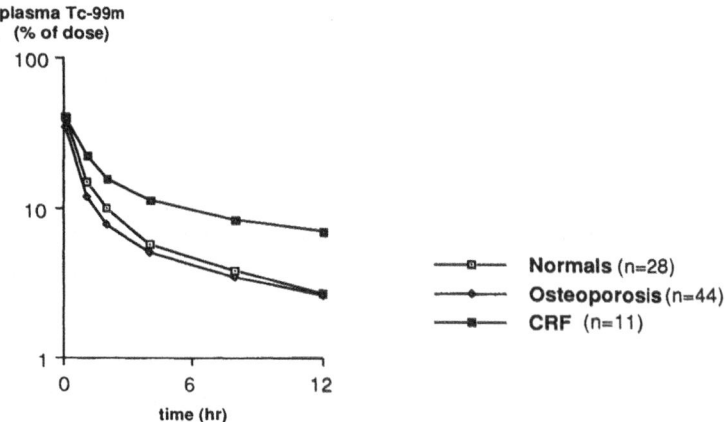

Figure 3. Composite plasma ^{99m}Tc MDP disappearance curves (error bars not shown for clarity) in normal subjects, osteoporosis and chronic renal failure (CRF). Note the logarithmic vertical axis. From Caniggia and Vattimo, Calcif. Tissue Int. 30: 5-13 (1980), with permission.

180

Fractionation of total MDP clearance between skeletal and renal clearances is probably not essential for the evaluation of bone disorders in which skeletal clearance is grossly elevated, and in which, therefore, variations in renal clearance have little impact. However, in diseases, particularly osteoporosis, in which skeletal clearance is only modestly abnormal, a correction for renal clearance is essential, especially in view of the normal reduction in renal function associated with ageing. This can be tackled in several ways: Nisbet et al. (1984) performed a ^{51}Cr-EDTA plasma clearance simultaneously with a ^{99m}Tc-MDP plasma clearance and found that normally the plasma concentration ratio of the two tracers remained essentially constant at about 1.4 (EDTA/MDP) up to 6 hours after injection. In conditions associated with increased bone turnover, the ratio, as expected, increased. No patients with osteoporosis were included in this study.

Another way is to fractionate the normalised MDP clearance, obtained from dynamic whole body or regional counting, between renal and skeletal clearances using the 24 hour whole body retention value. Thus, if MDP clearance is 0.012 min^{-1} (Molloi et al., 1989) and WBR is 40 %, renal clearance would be approximately 0.0072 min^{-1} (which compares favourably with the normal value for ^{51}Cr-EDTA clearance) and skeletal clearance would be approximately 0.0048 min^{-1}. This approach has the attraction that no blood sampling is required. In patients having imaging doses of ^{99m}Tc-MDP, their renal clearance can be measured from dynamic gamma camera imaging over the initial 3 minutes after injection, analogously to the measurement of individual kidney GFR using ^{99m}Tc-DTPA (Piepsz et al., 1977; Peters et al., 1988).

CLINICAL APPLICATIONS OF TURNOVER MEASUREMENTS

How have these techniques been applied to osteoporosis, and what has been their value, a) in telling us about the fundamental pathology of osteoporosis and b) in the management of patients with osteoporosis with respect to diagnosis, prognosis and follow-up? Most groups have reported a modest increase in ^{99m}Tc-MDP WBR in osteoporosis, although some have reported a decrease. The contribution of renal clearance to the WBR measurement has often not been taken into account and this is a serious omission, particularly in studies attempting to correlate bone turnover with age. Thomsen et al. (1986) measured WBR as a function of age, showing it to increase with age. The fact that aged-matched men did not show such an increase suggests that deteriorating renal function was not entirely responsible for the increasing WBR. Davie et al. (1987) also observed an increased whole body retention in osteoporosis, even after allowing for age-related deterioration in renal function. They noted a correlation between serum alkaline phosphatase and WBR in normal subjects, but not in patients with osteoporosis. WBR was also shown by Martin et al. (1983) to correlate with alkaline phosphatase in metabolic bone disease but not in osteoporosis. WBR also correlates with urinary hydroxyproline in several metabolic bone diseases (Martin et al., 1983; Mosekilde et al., 1987). This correlation is also seen in osteoporosis, although for any given range of hydroxyproline values, WBR is less in osteoporosis (Martin et al., 1983). Martin et al. (1983), working with the closely related compound ^{99m}Tc-pyrophosphate, demonstrated that although the WBR increased in ageing, thereby confirming the findings of Thomsen et al. (1986), it was reduced in patients with osteoporosis compared with age-matched controls. Importantly, they showed no correlation between WBR and the number of crush fractures in patients with osteoporosis. In other words, increased MDP uptake as a result of fracture does not apparently interfere with WBR as an index of bone turnover in osteoporotic bone. Mosekilde et al. (1987) compared WBR with biochemical indices as a predictor of radiocalcium accretion rates in bone. Alkaline phosphatase and urinary hydroxyproline were better predictors in patients grouped under the general heading of metabolic bone disease (with a wide range of WBR values) and in osteoporosis, even when renal clearance was allowed for in the WBR measurement.

One of the most interesting of the many contributions of Fogelman et al. to this field is the finding that WBR is elevated in oophorectomised women not receiving oestrogen replacement therapy (Fogelman et al., 1980). Indeed, they noted a significant inverse correlation between oestrogen dose and WBR. They also recorded a significant relationship between bone loss as recorded on bone densitometry and WBR, a correlation also noted by Caniggia and Vattimo (1980). The latter authors, although, like many others, finding an elevated WBR in osteoporosis, recorded a plasma MDP clearance which was similar in controls and patients with osteoporosis. In contrast, patients with chronic renal failure, in whom the renal component of MDP plasma clearance is effectively missing, had a slower rate of clearance. This confirms that impaired renal clearance is not the cause of an increased whole body retention in osteoporosis.

If we accept that WBR is increased in osteoporosis, then skeletal MDP clearance must be increased. As skeletal blood flow, is, if anything, reduced, skeletal MDP extraction efficiency is likely to be increased. This could be the result of an increase either in k1 or k (figure 1) or both. A possibility, as suggested by Martin et al. (1983), is that skeletal extraction of MDP is promoted in osteoporosis by a larger surface area of trabecular bone available for tracer exchange, consequent upon the more porous bone of osteoporosis. In other words, increased whole body retention is simply an incidental consequence of osteoporosis, rather than implying any fundamental change in bone turnover. The alternative view is that MDP retention is increased as a result of increased bone turnover in this condition.

REFERENCES

Caniggia A., A. Vattimo: Kinetics of technetium- 99m-tin-methylene- diphosphonate in normal subjects and pathological conditions: a simple index of bone metabolism. Calcif. Tissue Int. 30: 5-13 (1980).

Davie M.W.J., J.M. Britton, M. Haddaway, I.W. McCall: ^{99m}Tc-MDP retention in osteoporosis: relationship to other indices of bone cell activity and response to calcium and vitamin D therapy. Eur. J. Nucl. Med. 13: 462-466 (1987).

Fogelman I., R.G. Bessent, J.G. Turner, D.L. Citrin, I.T. Boyle, W.R. Greig: The use of whole body retention of ^{99m}Tc-diphosphonate in the diagnosis of metabolic bone disease. J. Nucl. Med. 19: 270-275 (1978).

Fogelman I., R.G. Bessent, H.N. Cohen, D.M. Hart: Skeletal uptake of diphosphonate. Method for prediction of postmenopausal osteoporosis. Lancet 2: 667-670 (1980).

Galasko C.S.B.: The pathological basis for skeletal scintigraphy. J. Bone Joint Surg. [B] 57: 353-359 (1975).

Hawkins R.A., Y. Cho, S-C. Huang, C.K. Hoh, M. Dahlbom, C. Scheipers, N. Satyamurthy, J.R. Barrio, M.E. Phelps: Evaluation of the skeletal kinetics of fluorine-18-fluoride ion with PET. J. Nucl. Med. 33: 633-642 (1992).

Lahtinen T., E.M. Alhava, P. Karjalainen, T. Romppanen: The effect of age on blood flow in the proximal femur in man. J. Nucl. Med. 22: 966-972 (1981).

Lavender J.P., R.A.A. Khan, S.P.F. Hughes: Blood flow and tracer uptake in normal and abnormal canine bone: comparison with ^{85}Sr microspheres, ^{81m}Kr and ^{99m}Tc-MDP. J. Nucl. Med. 20: 413-418 (1979).

Martiat Ph., A. Ferrant, M. Cogneau, A. Bol, C. Michel, J. Rodhain, J.L. Michaux, G. Sokal: Assessment of bone marrow blood flow using positron emission tomography: no relationship with bone marrow cellularity. Br. J. Haematol. 66: 307-310 (1987).

Martin P., A. Schoutens, O. Manicourt, P. Bergman, M. Fuss, M. Verbanck: Whole body and regional retention of ^{99m}Tc-labeled pyrophosphate at 24 hours: physiological basis of the method for assessing the metabolism of bone in disease. Calcif. Tissue Int. 35: 37-42 (1983).

McCarthy I.D., S.P.F. Hughes: Multiple tracer studies of bone uptake of ^{99m}Tc-MDP and ^{85}Sr. Am. J. Physiol. 256: H1261-1265 (1989).

McCarthy I.D., S.F.P. Hughes, J.S. Orr: An experimental model to study the relationship between blood flow and uptake for bone-seeking radionuclides in normal bone. Clin. Phys. Physiol. Meas. 1: 135-143 (1980).

McInnis J.C., R.A. Robb, P.J. Kelly: The relationship of bone blood flow, bone tracer deposition and endosteal new bone formation. J. Lab. Clin. Med. 96: 511-522 (1980) .

Molloi S., R. Mazess, H. Bendsen, M. Wilson: Whole body and regional retention of ^{99m}Tc labeled diphosphonates with a whole body counter: a study with normal males. Calcif. Tissue. Int. 44: 322-329 (1989).

Mosekilde L., C. Hasling, P. Charles, F.T. Jensen: Biphosphonate whole body retention test: relations to bone mineralisation rate, renal function and metabolic bone disorders. Eur. J. Clin. Invest. 17: 530-537 (1987).

Nisbet A.P., S. Edwards, C.R. Lazarus, J. Malamitsi, M.N. Maisey, G.D. Mashiter, P.J. Winn: Chromium-51 EDTA/technetium-99m-MDP plasma ratio to measure total skeletal function. Br. J. Radiol. 57: 677-680 (1984).

Peters A.M., I. Gordon, K. Evans, A. Todd-Pokropek: Background in ^{99m}Tc-DTPA renography evaluated by the impact of its components on the measurement of individual kidney glomerular filtration rate. Nucl. Med. Commun. 9: 545-552 (1988).

Piepsz A., A. Dobbeleir, F. Erbsmann: Measurement of separate kidney clearance by means of ^{99m}Tc-DTPA complex and a scintillation camera. Eur. J. Nucl. Med. 2: 173-177 (1977).

Reeve J., M. Arlot, R. Wootton, C. Edouard, M. Tellez, R. Hesp, J.R. Green, P.J. Meunier: Skeletal blood flow, iliac histomorphometry, and strontium kinetics in osteoporosis: a relationship between blood flow and corrected apposition rate. J. Clin. Endocrinol. Metab. 66: 1124-1131 (1988).

Tellez M., R. Wootton, J. Reeve: Skeletal blood flow measured with ^{18}F in patients with osteomalacia and hyperparathyroidism. Eur. J. Nucl. Med. 8: 299-302 (1983).

Thomsen K., A. Gotfredsen, C. Christiansen: Bone turnover in healthy adults measured by whole body retention and urinary excretion of ^{99m}Tc MDP. Normalisation by bone mass. Scand. J. Clin. Lab. Invest. 46: 587-592 (1986).

Wootton R., J. Reeve, N. Veall: The clinical measurement of skeletal blood flow. Clin. Sci. Mol. Med. 50: 261-268 (1976).

Wootton R., J. Reeve, E. Spellacy, M. Tellez-Yudilevich: Skeletal blood flow in Paget's disease of bone and its response to calcitonin therapy. Clin. Sci. Mol. Med. 54: 69-74 (1978).

BONE BLOOD FLOW AND SPACEFLIGHT OSTEOPENIA

R. M. DILLAMAN and R. D. ROER

Center for Marine Science Research and
Department of Biological Sciences
University of North Carolina at Wilmington
Wilmington, USA

INTRODUCTION

Spaceflight osteopenia remains one of the most serious impediments to long term exposure to microgravity. Much of the research on its cause has been focused on the absence of weightbearing in the bones of the lower extremities. The etiology, however, has remained elusive. There is mounting evidence that the alterations in blood flow, also associated with weightlessness, are partially responsible for the loss of bone mineral during prolonged spaceflight.

Recent work in our laboratory and others has clearly demonstrated that there exists an extravascular flow of interstitial fluid (ISF) through the mineralized matrix of haversian and nonhaversian bone. Such a flow of ISF has been shown to be capable of the perfusion of osteocytes deep within the bone matrix, and exceeds the ability of other putative mechanisms for nutrient and oxygen supply to bone, such as diffusion and convective flow due to bone flexion.

There is, moreover, ample evidence of an intimate relationship between blood flow in bone and the type of cardiovascular and hemodynamic alterations observed during spaceflight. It is, therefore, likely that these alterations will affect the physiological state of the bones, particularly those of the lower extremities where cardiovascular effects of spaceflight are most pronounced.

FLUID MOVEMENT IN BONE COMPARTMENTS

Bone is a complex organ whose basic physiology can only be understood when all its compartments and their interrelationships are defined and understood. These compartments include: the vascular elements, the extravascular space and interstitial fluid (ISF), the bone lining cells and the mineralized matrix and its constituent cells, the osteocytes. The movement of gases, ions and molecules into and out of the various

Bone Circulation and Vascularization in Normal and Pathological Conditions
Edited by A. Schoutens *et al.*, Plenum Press, New York, 1993

185

compartments of bone is important, not only for the normal growth and development of bone, but for it to perform its role in calcium homeostasis.

The extravascular perfusion of the bone matrix is critical to the maintenance of aerobic metabolism of osteocytes and to the exchange of bone mineral with the plasma. Consequently, a number of studies have focused on the pattern and rates of extravascular flow through the mineralized portion of bone. Using such markers as thorotrast, ferritin and horseradish peroxidase in a variety of experimental animals (mice, rats, chicks, cats and dogs), it was determined that the osteoblastic layer was not able to appreciably restrict the flow of materials into the mineralized portion of the bone (see Dillaman et al., 1991 for review). The markers invariably showed up in the osteocytic lacunae and mineralized matrix within minutes of injection. In the most comprehensive study to date, Montgomery et al. (1988) examined the movement of ferritin into the haversian bone of dogs. Perfusion of ferritin into the tibial nutrient artery resulted in appearance of ferritin in most of the haversian capillaries at time zero; by 5 minutes the number of ferritin capillaries decreased markedly. At all times between zero and 25 minutes, ferritin could be seen in the extravascular compartment of bone. A similar time frame of marker movement for horseradish peroxidase was observed in the femur and tibia of the rat (Dillaman et al., 1991). Peroxidase was localized histochemically within endosteal bone and throughout most osteocytic lacunae within 5 minutes of injection, and began to disappear within 4 hours. In both studies, the marker progressively moved outward from the vessels in the endosteal region and returned to the circumvascular region near the periosteum. The resulting halo pattern seen in cross sections of the diaphysis was very similar to that seen in the woven bones of chicks (Dillaman, 1984) and argues against the outward movement by diffusion. Montgomery et al. (1988) also described two apparent extravascular pathways in the haversian bone, referring to them as "perivascular prelymphatics" and "matrix prelymphatic channels". These observations indicate a unidirectional centrifugal flow of ISF from the endosteal toward the periosteal surface. Montgomery et al. (1988) and Dillaman et al. (1991) maintain that the driving force for this flow is the pressure gradient that exists between the endosteal and periosteal surface (Brookes, 1971).

A computer model describing the vascular and extravascular flow patterns within portions of the femur of the rat has been used (Dillaman et al., 1991), employing the fluid dynamics modeling program, Fluent (Creare, Inc.), and anatomical data derived from studies of rat bone vasculature. The model, given reasonable estimates of physical parameters within the bone, predicts a pattern of extravascular flow which corresponds to the observations derived from the marker studies described above, including entry of the marker into the endosteal surface, exit of the marker from the vessels near the endosteum, and return of the marker to the vessel canals toward the periosteum. Despite a relatively low permeability assigned to the mineralized matrix, the initial results of the modeling procedure predict a substantial flow of ISF through the matrix, with the absolute rates of flow within these compartments being determined by the total blood flow to that portion of the bone. However, the model further predicts that the vasculature architecture and the relation of the porosity fields to the vessels are primary factors dictating the pattern and relative magnitudes of vascular and extravascular flow.

SPACEFLIGHT OSTEOPENIA

Osteopenia and alterations in calcium homeostasis have been a consistent consequence of spaceflight. Calcium and phosphorus loss in man during spaceflight was noted in the Gemini VII flight and in some early Soviet flights, but it was the Skylab program that first offered accurate determination of calcium metabolism (Whedon, 1982). Results indicated a gradual rise in urine calcium over the first two to four weeks, leveling off at values 60-100 % greater than preflight values. This significant negative calcium balance persisted throughout the 60 and 84 day duration of the Skylab 3 and Skylab 4 flights. Fecal calcium excretion, at the same time, showed a decrease in the first few weeks of flight followed by a gradual rise to levels higher than preflight values for the remainder of the flight. Both urinary and fecal calcium decreased to restore positive balance immediately following the termination of the flight. The conclusion was that the negative calcium balance constituted a hazard in flights of 9-12 months or more. This was based on the presumption that negative calcium balance would be reflected in decreased bone mass. In fact, radiographic densitometry and photon absorptiometry (Rambaut et al., 1975; Vogel et al., 1977) have demonstrated a decrease in bone mass as a result of spaceflight. However, the decrease in bone mass was not uniformly distributed throughout the skeleton. Rather, the os calcis showed significant decreases in bone density while the radius and ulna showed no such loss. Photon absorption measurements have shown a 19 % loss of bone in the tibia after long term Soviet missions. The localized nature of bone loss serves to further emphasize the hazards of long-term spaceflight in that a weak link is created and is associated with load-bearing bones. These data, in addition to the lack of change in the calcicotropic hormones (PTH, calcitonin, 25-hydroxycholecalciferol) have led to the assumption that the load-bearing nature of bones may, in itself, contribute to the loss of minerals during spaceflight (Whedon, 1982).

EXPERIMENTAL MODELS FOR SPACEFLIGHT OSTEOPENIA

Human Studies

Several ground-based models have been used to simulate and test the effects of hypogravity on calcium loss and bone demineralization. In man, the situation most closely reflecting the effects of hypogravity is prolonged bed rest (Greenleaf, 1991). Under such conditions one also encounters elevated urine and fecal calcium, resulting in a net negative calcium balance along with increase in hydroxyproline excretion, all of which suggest bone resorption (van der Weil et al., 1991). Furthermore, measurements of bone density have revealed the same pattern of decreased bone mass in the os calcis but not in the radius (Donaldson et al., 1970). Interestingly, Arnaud et al. (1988) have reported, in a 6° head-down bed rest study of 11 men, a significant increase in bone density of the head, indicating a redistribution of bone mineral. Effects of bed rest have been shown to be most effectively ameliorated by quiet standing for at least 3 hours per day, while exercise in the supine position only aggravated negative calcium balance and urinary excretion of hydroxyproline (Arnaud et al., 1986).

Animal Studies

Rats subjected to spaceflight have exhibited defects in bone metabolism similar, in many respects, to those observed in astronauts. Results from Cosmos 782 and 936 demonstrated a 47 % and a 43 % decrease in tibial periosteal bone formation (Morey and Baylink, 1978). Along with slight increases in medullary area, the decrease in bone formation was characterized by arrest and reversal lines that were apparent after post-flight recovery. These lines presumably represented that region that was the mineralizing front at the end of the flight and whose character was sufficiently altered so that subsequent mineralization post-flight rendered them visible in histological sections (Morey and Baylink, 1978). Asling (1978) examined medullary bone turnover in rats and noted that the spaceflight group had reduced numbers of trabeculae and that those trabeculae were much shorter than controls, only rarely reaching into the secondary spongiosa. More extensive analyses of bone parameters were made in conjunction with the Cosmos 1129 flight (Simmons, 1981; Wronski et al., 1980), Cosmos 1667 (Vico et al., 1991) and Cosmos 1887 (Földes et al., 1990). Periosteal bone formation rate in the tibia and humerus was seen to be reduced by 30 % - 40 % and the fractional area of trabecular bone was significantly lower than controls. Concomitant with the reduction of medullary bone was an increase in the fractional area of the marrow cavity. Since the numbers of osteoclasts were essentially normal at the end of the flight, and since ultrastructural analyses of osteoclasts indicated that they were less than fully active (reduced development of ruffled border), it was concluded that osteoclastic resorption was not the major cause of decreased bone mass (Matthews, 1981). Evidence from rats flown aboard the 7 day Spacelab 3 Shuttle flight indicated an increase in fragility, a reduction in osteocalcin, and a reduction in calcium/hydroxyproline in the humeri (Patterson-Buckendahl et al., 1987). While seven days appeared to be insufficient for the development of histomorphologic changes in cortical bone and vertebrae (Wronski et al., 1987), alterations were noted in the epiphyseal plates (Montufar-Solis and Duke, 1988).

The most widely used animal model for simulating the effects of spaceflight on bone has been the tail-suspended rat model first described by Morey-Holton and Wronski (1981). Tail suspension provides both an unloading of the hindlimbs and a cephalad shift in blood and extravascular fluid. Tail-suspended rats display many of the changes in bone observed in spaceflight. Globus et al. (1986) further showed a marked decrease in tibial bone growth between one and two weeks of suspension, followed by resumption of growth from weeks two to four. Nonetheless, a difference in bone mass in the tibia and lumbar vertebrae persisted. Roer and Dillaman (1990) demonstrated that tail-suspended rats grew at the same rate as controls and remained in calcium balance throughout the three week period of suspension. Nonetheless, there was a decrease in the wet weight and dry weight of the femora and tibiae in suspended compared to controls. Humeri and radii/ulnae demonstrated no difference in body-weight specific wet weight and a significant increase in body-weight specific dry weight. Skulls and mandibles showed a significant increase in both dry and ash weights in suspended animals compared to controls (Figure 1). These data, as those of Arnaud et al. (1988), imply a redistribution of bone mineral from the posterior to the anterior as a function of posture.

FLUID SHIFTS AND REDISTRIBUTION

One of the major physiologic responses to hypogravity is the redistribution of body fluids (Charles and Bungo, 1991). Under altered postures (bed rest and/or head-down tilt) or in the absence of gravity, blood and ISF volumes in body regions are shifted from the distribution found under normogravic orthostasis. Both prolonged antiorthostasis and spaceflight induce a cephalad shift in blood and ISF. These shifts are manifested in a decrease in leg girth, decreased leg tissue thickness and volume of up to 30 %, as well as facial edema and congestion (Sandler, 1982). These changes also alter blood pressure and hemodynamics. Venous pressure in the legs decreases while that in the arms increases such that these pressures were found to be equal during flight aboard Salyut 6 (Kozerenko et al., 1981).

Horizontal bed rest and antiorthostatic hypokinesia (AOH) simulate these particular hemodynamic effects of spaceflight (Sandler, 1982). As one might expect, decreases in fluid volume are more pronounced in regions normally inferior to the level of the heart during orthostasis and, consequently, thoracic volume may increase, albeit transiently (Grundy et al., 1991). Thus, changes in microcirculation during AOH show increased blood flow in the conjunctiva and edema of the sclera, a decrease in capillary flow in the skin folds of the toes, but relatively little alteration of finger skin fold capillary patterns (Chernukh et al., 1980). The magnitude of the fluid volume shift depends upon the conditions and duration (Fortney et al., 1991) of the study, but in all cases exceeds the 600-700 ml shift observed in subjects upon arising from a supine posture at 1 g. These values can be compared to inflight measurements demonstrating cephalad fluid shifts of 1-4 liters (Sandler, 1982; Nicogossian et al., 1983).

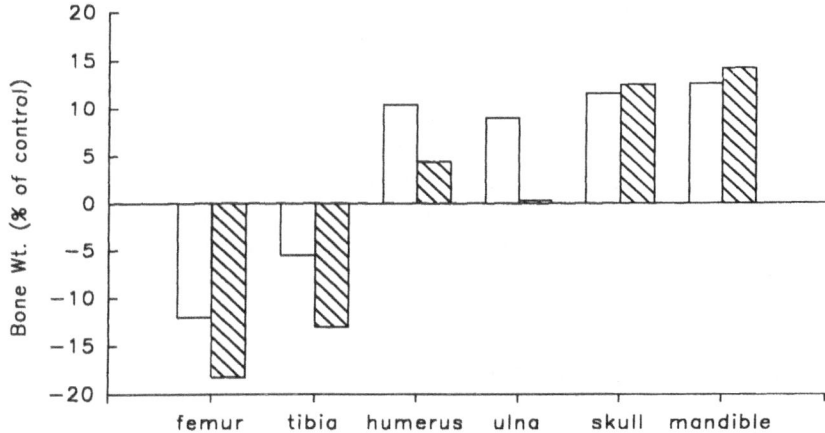

Figure 1. Dry weight (open bars) and ash weight (hatched bars) of bones from juvenile rats which were tail-suspended for 3 weeks., as compared to bone weights of nonsuspended littermates (data from Roer and Dillaman, 1990).

We have used a chronically implanted blood flow probe to measure femoral artery flow in rats before, during and after tail-suspension. These data demonstrate significant reductions in mean flow rates on all days throughout the duration of suspension, which reached magnitudes as high as 42.7 % ± 17.9 % (n = 13), 43.4 % ± 23.4 % (n = 5) and 45.4 % ± 19.8 % (n = 7) (mean ± S.D.) on days 1, 3 and 5 of suspension respectively. Mean flows increased upon return to normal, orthostatic posture, and despite the pattern described above, attained values not significantly different from pre-suspension rates.

These observations demonstrate the immediate and sustained effect of tail-suspension on femoral artery flow. It would seem evident that a decrease in hindlimb blood flow of this magnitude would have profound consequences on the perfusion of hindlimb bones. Immediate reductions in blood flow to the hind limbs and to the hindlimb muscles and bones occur with tail suspension due to the effects of an altered gravitational vector on vascular and extravascular volumes and pressures (Hargens et al., 1992).

While the hindlimb musculature may be capable of limited, local autoregulation of blood flow in the face of reduced plasma volume and femoral artery flow, it is unlikely that the hindlimb bones possess such capability. Bone vascular smooth muscle undergoes α-adrenergic mediated vasoconstriction in response to norepinephrine, methoxamine and endothelin-1 (Brinker et al., 1990) causing an increase in bone vascular resistance and a decrease in bone blood flow (Cochrane and McCarthy, 1991). The bone vasculature, however, shows a distinct lack in sensitivity to vasodilatory agents such as ATP (Cochrane and McCarthy, 1991), nitroglycerine, acetylcholine, and 8-bromo-cyclic GMP (Brinker et al., 1990).

As stated above, computer modeling of ISF flow in bone suggests that decreased bone blood flow should result in decreased extravascular perfusion of the bone matrix (Dillaman et al., 1991). These predictions were supported by Handley and coworkers (1990) who demonstrated that perfusion of the metatarsal bone of the isolated cow forefoot was dependent upon the perfusion rate of the ulnar artery. Decreased blood flow in the rat, in response to vasoactive agents, also resulted in a decrease in the rate of ^{85}Sr clearance from bone, which also indicates a reduced extravascular perfusion (Cochrane and McCarthy, 1991).

The consequence of the reduced hindlimb blood flow due to tail suspension in the rat on extravascular perfusion was recently demonstrated. Dillaman et al. (1991) injected HRP into suspended and control rats, and then sacrificed at intervals following injection for histochemical determination of HRP distribution. At 5 min. after injection, HRP was evident along the endosteal margins of the femur in control rats, but not in the bones of the suspended animals. The difference was even more marked at 15 min. postinjection. Washout of HRP from the bones of suspended animals was also slower; little HRP evident in control femurs by 4 h postinjection, while suspended rats still displayed significant amounts of HRP in the endosteal bone.

MECHANISMS OF TRANSDUCTION

As recently reviewed by Dillaman et al. (1991), evidence abounds that alterations in the perfusion of bone result in changes in mineral dynamics, deposition and resorption of bone. The means by which flow can be transduced into an effect at the level of bone cells remains to be addressed. Recent suggestions of the basis for transduction mechanisms can be grouped into three categories: metabolic factors, electrical signals and shear stress.

Since both bone lining cells and osteocytes require nutrients and produce metabolites, a decrease in the rate of bone perfusion would be expected to reduce the supply of oxygen to the cells (particularly those more distant from the endosteal surface and vessel canals) and cause a build-up of such metabolites as CO_2, lactate, acid and nitrogenous wastes. Such changes are evident in subchondral bone in response to simulated joint effusion in rabbits (Kofoed and Lindenberg, 1986) and to arterial tamponade of the canine knee (Holm et al., 1990), which induce a decrease in tissue pO_2 and pH, and an increase in pCO_2 and lactate. Osteotomy of the femoral shaft of dogs also results in an interruption of endosteal blood flow and induces tissue hypoxia and hypercapnia, with a subsequent reactive hyperemia from the periosteal circulation (Kofoed et al., 1985).

Ischaemia and hypoxia have been shown to affect bone cells, both in terms of proliferation and phenotypic expression. Ischaemia induces necrosis in osteocytes after as little as 2 h (James and Steijn-Myagkaya, 1986; Usui et al., 1989). Less severe decreases in pO_2 have also been shown to decrease DNA content in cultures of proliferative and hypertrophic zone chondrocytes and to inhibit ^{35}S incorporation into glycosaminoglycans (Clark et al., 1991). Deren and coworkers (1990) demonstrated that maximal expression of alkaline phosphatase phenotype occurred in periosteal cell cultures when they were incubated at normal capillary oxygen tensions. Cultures of rat calvarial bone cells demonstrated an increase in mitotic rate when cultured at low pO_2 (≤ 9 % O_2) whereas they increased alkaline phosphatase activity, incorporation of ^{35}S into proteoglycans and ^{14}C-proline into collagen at higher pO_2 (≥ 13 % O_2) (Brighton et al., 1991).

Decreases in oxygen tension arising from decreased perfusion may also directly affect acid/base regulation. Decreased pH, resulting from metabolic acidosis, has numerous deleterious effects on bone formation and mineralization. Even a slight lowering of pH has been shown to decrease collagen synthesis in chick calvarial bone cells in culture (Lenz and Ramp, 1989). Metabolic acidosis in vivo or decreased bicarbonate concentrations in vitro lead to loss of bone calcium, probably from carbonate stores (Bushinsky and Lechleider, 1987). Metabolic acidosis in vivo also increases bone resorption and depresses bone formation (Barzel and Jowsey, 1969; Kraut et al., 1986). Fluid flow, per se, may also affect cell responses either through electrical effects or shear stresses. Shear stresses of 6 and 24 dynes/cm^2 caused a 9- and 20-fold increase (respectively) in the rate of prostaglandin E_2 synthesis by rat calvarial osteoblasts in culture (Reich and Frangos, 1991). These cells also demonstrated increased cAMP (Reich et al., 1990) and inositol triphosphate (Reich and Frangos, 1991) levels with shear, which could be partially inhibited by PGE2 antagonists.

IN VITRO MODELS

One approach to separating the direct and indirect effects of alterations of various parameters on bone formation is to study these parameters in an isolated bone system in vitro. For example, in vivo, pH shifts and elevated lactate concentrations occur simultaneously. Using an in vitro system, one might adjust pH while holding lactate concentrations constant and, conversely, alter lactate concentrations at constant pH.

We have developed a system for the perfusion culture of the embryonic chick calvaria that allows us the type of in vitro analysis described above. The isolated calvaria is placed between two halves of a milled, cylindrical plexiglass chamber, sandwiched by silicone washers to effect a water-tight seal. The inflow port is attached to a peristaltic pump that supplies a precise flow of culture medium. The medium enters a 0.13 ml chamber and then perfuses across an 11 mm^2 region of the calvaria. After entering a 0.1 ml downstream chamber, the medium is collected in a fraction collector for subsequent analysis. Preliminary experiments have demonstrated that bones may be maintained for at least 3 weeks, during which time they show marked growth in the perfused region (Moore, 1990). Furthermore, calvaria have been prelabeled with ^{45}Ca three days prior to

Table 1. Calcium exchange in perfused chick calvariae upon reduction of flow from 15.0 μl/min. to 2.5 μl/min. Initial pH of both buffered and unbuffered media was 7.40

	Total % ^{45}Ca released	Final pH
Buffered Medium	14.17 ± 2.64	7.37 ± .020
Unbuffered Medium	17.94 ± 1.44	7.21 ± .027

placed in the chambers, and exchange of ^{45}Ca with the medium monitored (Jenkins, 1993). Results (Table 1) indicate that reducing the flow of pH 7.4 culture medium across the calvaria from 15 µl/min. to 2.5 µl/min. significantly increases the exchange of calcium with medium that is unbuffered as compared to medium that is buffered to pH 7.4. The probable mechanism for the increase in exchange is the significant drop in pH also noted in the table, which is likely due to anaerobic metabolism. The use of such an isolated organ culture system permits one to address the question of whether or not perfusion effects can be transduced at the bone cell level and have a significant effect on mineral dynamics.

SUMMARY

Spaceflight osteopenia is a serious problem whose cause(s) must be understood so that an effective countermeasure can be developed. Of the possible causes for this condition, hemodynamic alterations are a likely candidate. Future research should not only continue to correlate changes in cardiovascular and hemodynamic factors with bone loss, but should more directly explore the possibility that bone might respond directly to blood flow.

ACKNOWLEDGEMENTS

The authors wish to acknowledge the excellent technical support of Daniel M. Gay. This work was supported by grant NAG 2-391 from NASA. This is contribution #77 from the Center for Marine Science Research.

REFERENCES

Arnaud, S.B., M.R. Powell, J. Vernikos-Danellis, P. Buchanan: Bone mineral and body composition after 30 day head down tilt bedrest. J. Bone Mineral Res. 3: S119 (1988).

Arnaud, S.B., V.S. Schneider, E. Morey-Holton: Effects of inactivity on bone and calcium metabolism. In: J. Vernikos, H. Sandler (eds.): Inactivity: Physiological Effects, pp. 49-75. Academic Press, San Diego 1986.

Asling, C.W.: Histological studies on tibial bone of rats in the 1975 COSMOS-782 flight. Part 2. Microradiographic study of cortical bone. NASA TM-78525: 291-307 (1978).

Barzel, U.S., J. Jowsey: The effects of chronic acid and alkali administration on bone turnover in adult rats. Clin. Sci. 36: 517-524 (1969).

Brighton, C.T., J.L. Schaffer, D.B. Shapiro, J.J.S. Tang, C.C. Clark: Proliferation and macromolecular synthesis by rat calvarial bone cell grown in various oxygen tensions. J. Orthop. Res. 9: 847-854 (1991).

Brinker, M.R., H.L. Lipton, S.D. Cook, A.L. Hyman: Pharmacological regulation of the circulation of bone. J. Bone Joint Surg. 72A: 964-975 (1990).

Brookes, M.: The Blood Supply of Bone. Butterworths, Ltd., London 1971.

Bushinsky, D.A., R.J. Lechleider: Mechanism of proton-induced bone calcium release: Calcium carbonate dissolution. Am. J. Physiol. 253: F998-F1005 (1987).

Charles, J.B., J.W. Bungo: Cardiovascular physiology in space flight. Exper. Gerontol. 26: 163-168 (1991).

Chernukh, A.M., O. G. Gazenko, B.M. Fedorov, T.N. Krupina, P.N. Aleksandrov, D.I. Shagal, N.YE. Pan ferova T.M. Proskurina: Effect of hypokinesia on blood microcirculation. NASA TM-76308 (1980).

Clark, C.C., B.S. Tolin, C.T. Brighton: The effect of oxygen tension on proteoglycan synthesis and aggregation in mammalian growth plate chondrocytes. J. Orthop. Res. 9: 477-484 (1991).

Cochrane, E., I.D. McCarthy: Rapid effects of parathyroid hormone (1-34) and prostaglandin E2 on bone blood flow and strontium clearance in the rat in vivo. J. Endocrinol. 131: 359-365 (1991).

Deren, J.A., F.S. Kaplan, C.T. Brighton: Alkaline phosphatase production by periosteal cells at various oxygen tensions in vitro. Clin. Orthop. Relat. Res. 252: 307-312 (1990).

Dillaman, R.M.: Movement of ferritin in the 2-day-old chick femur. Anat. Rec. 209: 445-453 (1984).

Dillaman, R.M., R.D. Roer, D.M. Gay: Fluid movement in bone: Theoretical and empirical. J. Biomechanics 24: 163-177 (1991).

Donaldson, C.L., S.B. Hulley, J.M. Vogel, R.S. Hattner: Effect of prolonged bed rest on bone mineral. Metabolism 19: 1071-1084 (1970).

Foldes, I., M. Rapcsak, T. Szilagyi, V.S. Oganov: Effects of space flight on bone formation and resorption. Acta Physiol. Hung. 75: 271-285 (1990).

Fortney, S.M., K.H. Hyatt, J.E. Davis, J.M. Vogel: Changes in body fluid compartments during a 28-day bed rest. Aviat. Space Environ. Med. 62: 97-104 (1991).

Globus, R.K., D.D. Bikle, E. Morey-Holton: The temporal response of bone to unloading. Endocrinology 118: 733-742 (1986).

Greenleaf, J.E.: Physiology of Prolonged Bed Rest. ARC-12241, NASA, Washington D . C . 1991.

Grundy, D., K. Reid, F.J. McAardle, B.H. Brown, D.C. Barber, C.F. Deacon, I.W. Henderson: Transthoracic fluid shifts and endocrine responses to 6° head-down tilt. Aviat. Space Environ. Med. 62: 923-929 (1991).

Handley, R.C., T. Essex, J. Pooley: Laser Doppler flowmetry and bone blood flow in an isolated perfused preparation. J. Med. Engin. Technol. 14: 201-204 (1990).

Hargens, A.R., D.E. Watenpaugh, G.A. Breit: Control of circulatory function in altered gravitational fields. Physiologist 35: S80-S83 (1992).

Holm, I.E., H. Ewald, J. Bulow, C. Bünger: Vasoactive substances in subchondral bone of the dog knee. J. Orthop. Res. 8: 205-212 (1990).

James, J., G.L. Steijn-Myagkaya: Death of osteocytes. Electron microscopy after in vitro ischaemia. J. Bone Joint Surg. 68B: 620-624 (1986).

Jenkins, B.C.: Mobilization of Calcium from Chick Calvariae in an Organ Perfusion System. M.S. Thesis, University of North Carolina at Wilmington 1993.

Kofoed, H., S. Lindenberg: Effect of simulated joint effusion on subchondral haemodynamics and metabolism. Injury 17: 274-276 (1986).

Kofoed, H., E. Sjontoft, S.O. Siemssen, H.P. Olesen: Bone marrow circulation after osteotomy. Blood flow, pO2, pCO2, and pressure studied in dogs. Acta Orthop. Scand. 56: 400-403 (1985).

Kozerenko, O.P., A.I. Grigoriev, A.D. Egorov: Results of investigations of weightlessness effects during prolonged manned space flights onboard Salyut-6. Physiologist 24: 549-554 (1981).

Kraut, J.A., D.R. Mishler, F.R. Singer, W.G. Goodman: The effects of metabolic acidosis on bone formation and bone resorption in the rat. Kidney Int. 30: 694-700 (1986).

Lenz, L.G., W.K. Ramp: Effects of pH on cultured osteoblast-like cells. J. Bone Min. Res. 4: S421 (1989).

Matthews, J.L.: Quantitative analysis of selected bone parameters. Supplemental report 3B: Mineralization in the long bones. NASA TM-81289: 199-228 (1981).

Montgomery, R.J., B.D. Sutker, J.T. Bronk, S.R. Smith, P.J. Kelly: Interstitial fluid flow in cortical bone. Microvas. Res. 35: 295-307 (1988).

Moore, L.B.: The Design and Testing of a Perfusion System for Bone Organ Culture. M.S. Thesis, University of North Carolina at Wilmington 1990.

Morey, E.R., D.J. Baylink: Inhibition of bone formation during space flight. Science 201: 1138-1141 (1978).

Morey-Holton, E., T.J. Wronski: Animal models for simulating weightlessness. Physiologist 24: S45 (1981).

Nicogossian, A ., S . L. Pool, P. C . Rambaut: Cardiovascular responses to spaceflight. Physiologist 26: 578-580 (1983).

Patterson-Buckendahl, P., S.B. Arnaud, G.L. Mechanic, R.B. Martin, R.E. Grindeland, C.E. Cann: Fragility and composition of growing rat bone after one week in space flight. Am. J. Physiol. 252: R240-R246 (1987).

Rambaut, P.C., M.C. Smith, JR., P.B. Mack, J.M. Vogel: Skeletal response. In: R.S. Johnson, L.F. Dietlein, C. Berry (eds.): Biomedical Results of Apollo, pp. 303-322. NASA SP-368, Washington D.C. 1975.

Reich, K.M., J.A. Frangos: Effect of flow on prostaglandin E2 and inositol triphosphate levels in osteoblasts. Am. J. Physiol. 261: C428-C432 (1991).

Reich, K.M., C.V. Gay, J.A. Frangos: Fluid shear stress as a mediator of osteoblast cyclic adenosine monophosphate production. J. Cell. Physiol. 143: 100-104 (1990).

Roer, R. D., R. M. Dillaman: Bone growth and calcium balance during simulated weightlessness in the rat. J. Appl. Physiol. 68: 13-20 (1990).

Sandler, H.: Cardiovascular responses to weightlessness and ground-based simulations. In: N. Longdon (ed.): Zero-g Simulation for Ground-Based Studies in Human Physiology, with Emphasis on the Cardiovascular and Body Fluid Systems, pp. 107-146. European Space Agency, Paris 1982.

Simmons, D.J.: Adaptation of the rat skeleton to weightlessness and its physiological mechanisms. Results of animal experiments aboard the Cosmos 1129 Biosatellite. Physiologist 26: 565-568 (1981).

Usui, Y., K. Kawai, K. Hirohata: An electron microscopic study of the changes observed in osteocytes under ischemic conditions. J. Orthop. Res. 7: 12-21 (1989).

Van Der Weil, H.E., P. Lips, J. Nauta, J.C. Netelenbos, G.J. Hazenberg: Biochemical parameters of bone turnover during ten days of bed rest and subsequent mobilization. Bone Mineral 13: 123-129 (1991).

Vico, L., V.E. Novikov, J.M. Very, C. Alexandre: Bone histomorphometric comparison of rat tibial metaphysis after 7-day tail suspension vs. 7-day spaceflight. Aviat. Space Environ. Med. 62: 26-31 (1991).

Vogel, J.M., M.W. Whittle, M.C. Smith, Jr., P.C. Rambaut: Bone mineral measurement experiment MO 78. In: R.S. Johnston, L.R. Dietlein (eds.): Biomedical Results from Skylab, pp. 183-190. NASA SP-377, Washington, D.C. 1977.

Whedon, G.D.: Changes in weightlessness in calcium metabolism and in the musculoskeletal system. Physiologist 25: S41-S44 (1982).

Wronski, T.J., E. Morey-Holton, W.S.S. Jee: Cosmos 1129 spaceflight and bone changes. Physiologist 23: 579-582 (1980).

BONE VASCULARIZATION IN ARTHRITIS

Cody BÜNGER and Ebbe Stender HANSEN

University of Aarhus
Orthopaedic Hospital
Aarhus, Denmark

INTRODUCTION

Studies of the normal knee have demonstrated that the hemodynamic impact of joint effusion on the juxtaarticular tissues depends on the anatomical relationship between the joint capsule and the compartment of interest and the level and duration of the effusion pressure. During physiological use the normal intra-articular pressure (IAP) of the joint fluctuates from slightly subatmospheric to markedly negative values (Jayson and Dixon, 1970).

INFLUENCE OF INCREASED IAP ON JUXTAARTICULAR HEMODYNAMICS IN NORMAL JOINTS

In the presence of joint effusion IAP will show increased values by joint motion and muscular contraction during function. The impact of elevated joint pressure and juxta-articular intraosseous pressures was first investigated by Arnoldi (1979) in order to define the pathogenetic role for osteoarthrosis. Within the knee joint the tissue compartments most affected by an increased IAP are the capsule, the patella and distal femoral condyles. In dogs, joint pressures down to 0.9 kPa may cause elevation of intraosseous pressure (IOP), but changes in juxtaarticular regional bone blood flow (RBF), intraosseous blood gas tension and lactate have been reproduced at IAP' s below 50 mmHg or at IAP' s below 50 % of mean arterial pressure (MAP) (Bünger, 1987; Ewald et al., 1989; Holm et al. 1990). Acute elevation of IAP above 75 % of MAP does cause decline in subchondral RBF, intraosseous pO_2, whereas pCO_2 and lactate do increase as a sign of a shift into anaerobic metabolism. Accordingly ^{99m}Tc-diphosphonate uptake in juxtaarticular bone evaluated by scintimetry was also significantly depressed (Hansen et al., 1989a, 1989b). Significant differences exist between the susceptibility of hip and knee joint to IAP elevation. The hemodynamics of the hip being most susceptible is probably due to a more intraarticular located vascular supply and venous drainage. Also between species we found notable

Bone Circulation and Vascularization in Normal and Pathological Conditions
Edited by A. Schoutens *et al.*, Plenum Press, New York, 1993

195

differences. The pig knee does not show any change in RBF or IOP during joint tamponade (Hansen et al., 1992). This must also be explained by different vascular anatomy. Moreover the subchondral bone in pigs was more vascular with higher RBF and vascular volume rates.

EXERCISE AND JUXTAARTICULAR HEMODYNAMICS IN NORMAL JOINTS

Quantitative data on RBF changes during exercise are available in dogs and pigs which had instrumentation for long duration tracer microsphere techniques. Interestingly both in immature and mature cancellous bone a uniform vasoconstrictive response is elicited during exercise. This results in a more than 50 % reduction in subchondral RBF in both hip and knee joints during treadmill exercise. This is in contrast to a twofold increase in RBF of the joint capsule, where a beta-adrenergic response to exercise probably dominates (Bünger, 1987, 1990). In case of intraosseous vascular lesions the reduction of bone blood flow induced by exercise might be critical.

SUBCHONDRAL HEMODYNAMICS IN ARTHRITIS

The classical static conception of hemodynamic changes in arthritis is the elevation of IOP, as a sign of venous engorgement. This can be further visualized by intraosseous phlebography, which shows dilated veins with prolonged clearance of contrast media. However this analysis provides a poor 3-dimensional resolution of the vascular events, which can be better documented experimentally by scanning electron microscopy (SEM) (He et al., 1990).

In rabbits, immobilization of the knee in extension resulted in substantial vascular changes characterized by dilatation and fusion of sinusoids, leakage of contrast material through the sinusoid wall and development of shunts (He et al., 1990). In carragheenan induced chronic arthritis the quantitative regional assessment of hemodynamics showed a mean unchanged RBF and elevated vascular volumes (VV) measured by ^{125}I-fibrinogen and ^{51}Cr-red blood cells, and a prolonged transit time of blood components (Bünger, 1984). Using refined tissue sampling techniques a significant intracompartmental variation in all hemodynamic parameters was shown in experimental canine gonarthritis (Bünger, 1987; Hansen et al., 1991). The subchondral condylar bone was more hyperaemic in arthritis in contrast to the more central bone where blood flow remained largely unchanged. The vascular volumes were uniformly enlarged, both in the central and peripheral bone, whereas the calculated mean transit time was significantly prolonged only in the central bone (Table 1).

The corresponding bone metabolism illustrated by ^{99m}Tc-Diphosphonate uptake correlated positively with blood flow and plasma volume and negatively with red cell volume in a nonlinear fashion (Hansen et al., 1991). The vascular pathophysiology responsible for intracompartmental flow gradients in arthritis was investigated using the hypothesis that A-V shunts will redirect blood flow within the compartments. Simultaneous blood flow analysis with 15 μm and 50 μm microspheres in the same arthritis model and during simulated joint tamponade at 75 % of MAP showed unaffected RBF 50/RBF 15 ratio in the total bone compartment (Hansen et al., 1990). This provided evidence of precapillary vasodilation. This indicates an intraosseous vasoregulation which shifts the embolization site of 50 μm spheres downsteam when the central and peripheral

Table 1. Hemodynamics of the knee in experimental chronic arthritis. (Hansen et al., 1991)

	Blood flow ml/l00g/min		Vascular volume ml/l00g		Transit time seconds	
	A	C	A	C	A	C
Synovialis	67	3	2.3	1.7	4	50
	(19)	(1)	(0.2)	(0.2)	(1)	(12)
Subchondral bone	15	10	7.5	5.9	40	45
	(2)	(2)	(1)	(0.9)	(9)	(10)
Central condyle	6	7	8.3	6.5	120	85
	(0.5)	(1)	(1)	(1)	(35)	(20)

A = arthritis; C = Controls; (SD); N = 8

compartments are compared. The presence of A-V shunts could not be confirmed. During exercise, redistribution of RBF from all cancellous bone structures was measured both in the hip and the knee joint, irrespective of arthritis. Blood was redirected to the active muscles. The redistribution is not influenced by anatomical location, presence of increased venous outlet resistance or synovial inflammation (Bünger, 1987). It is suggested that the metabolic demands of subchondral bone and synovial tissue are not met during exercise, since significant hyperaemia is present in these locations after exercise. These data point to an aggravation of the inflammatory process by exercise in carragheenan induced arthritis.

ARTHRITIC BONE LOSS IN RELATION TO HEMODYNAMIC CHANGES

Both local and generalized osteoporosis accompanies rheumatoid arthritis (Sambrook et al., 1990). The local periarticular bone loss is mainly trabecular and amounts 20 %. Steroid treatment may enhance the bone loss further up to 30 % (Luckert BP et al., 1990).

The mechanism of arthritic bone loss is thus multifactorial. Histomorphometric data have demonstrated an increased bone turnover in subchondral bone in chronic experimental arthritis (Holm et al., 1985; Hauge et al, 1992). At cellular level the bone resorption is activated by an inflammatory release of cytokines, like Interleukin-1 and -6 and Tumor necrosis factor (TNF), which work in cooperation with prostaglandins and leukotrienes. During steroid treatment the bone formation is probably depressed due to inhibition of PG's release. Experimentally carragheenan induced arthritis results in a loss of total bone density by CT of 20 % and a 25 % reduction in trabecular bone volume when compared with a reference of uninflamed knees (Søballe et al., 1991; Hauge et al., 1992). The importance of this uniform inflammatory bone loss and the hemodynamic characteristics, particularly in relation to the enzymatic degradation of arachidonic acid by the cyclooxygenase and leukotriene pathways, have been investigated recently in dogs (Hansen et al., 1990; Hansen et al., 1992). Cyclooxygenase inhibition by Naproxen in the carragheenan model resulted in reduced joint effusion volume and pressure. Capsular and subchondral hyperaemia are almost normalized along with juxtaarticular vascular volumes. The ^{99m}Tc-Diphosphonate uptake turned normal and trabecular bone loss was partially prevented (Hauge et al., 1992). This suggests that cyclooxygenase inhibition protects against hemodynamic and metabolic changes in juxtaarticular bone in arthritis (Hansen et al., 1992). Inhibition of Leukotriene B4 formation by 15-hydroxy-eicosatetraenoic acid (15-HETE), an endogenous inhibitor does only result in mild changes of the synovial inflammation. No effect can be detected with regards to synovial and subchondral blood flow or bone metabolism as measured by ^{99m}Tc-Diphosphonate uptake (Hansen et al., 1990).

CONCLUSION

In arthritis early qualitative and quantitative changes are present in subchondral bone. The microcirculation is characterized by an increased permeable surface area between blood and bone with prolonged transit of blood. These vascular changes may be due to both inflammatory vasodilation and venous stasis due to efflux inhibition, but it may also indirectly reflect a hypermetabolic state of bone with net trabecular bone loss. Both vascular and metabolic changes of subchondral bone can be prevented by cyclooxygenase inhibition. Exercise training acutely outweighs the hemodynamic changes in arthritis and causes a major reduction of bone blood flow, which may aggravate the disease process. Future research should aim to describe the relationship between inflammatory activation factors and early vascular reactivity in normal and arthritic bone to define whether the vascular events play a primary role in juxtaarticular bone loss.

REFERENCES

Arnoldi, C.C., I. Reinmann, S. Bach Christensen, S. Mortense: The effect of increased intraarticular pressure on juxtaarticular bone marrow pressure. IRCS. Medical Science 7: 471 (1979).

Bjurholm, A.: Neuroendocrine peptides in bone. Thesis. ISBN. 91-7900-875-5, Stockholm, 1979.

Bünger, C.: Hemodynamic of the juvenile knee. Joint effusion and synovial inflammation studied in dogs. Acta Orthop. Scand. 58: 1-104 (1987).

Bünger, C., E. Tøndevold, J. Bülow, J. Hjermind: Microcirculation of the juvenile knee in chronic arthritis. Clin. Orthop. 110: 294-302 (1986).

Bünger, C., K. Sølund, F. Joyce, E. Stender Hansen, H.M. Schrøder, E. Bünger: Total and segmental blood flow of the femoral head epiphysis in coxitis at rest and during exercise. In: B. Mazières and J. Arlet (eds): Bone circulation and bone necrosis, pp 137-141. Springer Verlag, 1990.

Ewald, H., I. Holm, J. Bülow, C. Bünger: Effect of indomethacin on regulation of juxtaarticular bone blood flow during joint tamponade. An experimental study in puppies. Scand. J. Clin. Lab. Invest. 49: 273-278 (1989).

Hansen, E.S., T.B. Henriksen, I. Noer, C. Bünger: Hemodynamic effects of knee-joint tamponade. ^{99m}Tc-Diphosphonate scientimetry in growing dogs. Acta Orthop. Scand. 60: 549-554 (1989a).

Hansen, E.S., I. Noer, T.B. Henriksen, V.E. Hjortdal, K. Søballe, C. Bünger: The influence of synovial effusion on juxtaarticular ^{99m}Tc-Diphosphonate uptake in arthritis of the immature dog knee. J. Orthop. Rheumatol. 2: 31-38 (1989b).

Hansen, E.S., K. Fogh, V.E. Hjortdal, T.B. Henriksen, I. Noer, H. Ewald, T. Herlin, K. Kragballe, C. Bünger: Synovitis reduced by inhibition of leukotriene B4. Carragheenan-induced gonarthritis studied in dogs. Acta Orthop. Scand. 61: 207-212 (1990).

Hansen, E.S., K. Søballe, T.B. Henriksen, V.E. Hjortdal, C. Bünger: ^{99m}Tc-Diphosphonate uptake and hemodynamics in arthritis of the immature dog knee. J. Orthop. Res. 9: 191-202 (1991).

Hansen, E.S., V.E. Hjortdal, S. He., Kjølseth, K. Køj, K. Søballe, C. Bünger: Arteriovenous shunting is not associated with venous congestion in bone. Knee tamponade studied with 15 μm microspheres in dogs. Acta Orthop. Scand. 62: 268-275 (1991).

Hansen, E.S., S. He, V.E. Hjordal, D. Kjølseth, K. Søballe: Distribution of blood flow in normal and arthritic joints. Role of arteriovenous shunting studied in growing dogs with 15 μm and 50 μm sized microspheres. Am. J. Physiol. 262: H38-H46 (1992).

Hansen, E.S., S. He, K. Søballe, D. Kjølseth, T.B. Henriksen, V.E. Hjortdal, C. Bünger: ^{99m}Tc-Diphosphonate uptake and hemodynamics in experimental arthritis: Effect of naproxen in the canine Carragheenan injection model. J. Orthop. Res. 10: 647-656 (1992).

He, S., X. Zhenhua, E.S. Hansen, C. Bünger: Microvascular morphology of bone in artrosis. Scanning electron microscopy in rabbits. Acta Orthop. Scand. 61: 195-200 (1990).

Hauge, E., F. Mesen, E.S. Hansen: Preventive-action of narpoxen on Carragheenan induced arthritis in immature dogs. A histomorphometric study. Trans. Orthop. Res. Soc. 18: (1993) in press.

Holm, I., C. Bünger, F. Melsen: A histomorphometric analysis of subchondral bone in juvenile arthropathy of dog knee. APMIS sect. A 93: 290-304 (1985).

Holm, I., H. Ewald, J. Bülow, C. Bünger: Vasoactive substances in subchondral bone of the dog knee. J. Orthop. Res. 8: 205-212 (1990).

Jayson, M.I.V., A. St J. Dixon: Intraarticular pressure in rheumatoisd arthritis of the knee. Pressure changes during joint use. Ann. Rheum. Dis. 29: 401-408 (1970).

Luckert, B.P.: Glucocorticoid-induced osteoporosis: Pathogenesis and management. Ann. Intern. Med. 112: 352-364 (1990).

Sambrook P.N.: Rapid periarticular bone loss in rheumatoid arthritis. Arthritis Rheum. 33: 615-622 (1990).

VASCULAR ASPECTS IN DEGENERATIVE JOINT DISORDERS

Carl C. ARNOLDI

Rigshospitalet
Orthopaedic Surgery
Copenhagen, Denmark

INTRODUCTION

"Degeneration" seems to be the end stage of a number of joint affections and in most degenerative disorders all structures of the joint are affected (synovium and synovial fluid, subchondral bone and cartilage, the fibrous capsule and the muscles involved in movements and stabilisation). Joint degeneration may be described as a process during which highly specialized structures for some reason or reasons either disintegrate or are replaced by more primitive tissues, less well adapted to the function of the joint, movement within the limits of stability, and less resistant to the mechanical stresses inherent in normal joint function.

Tissue degeneration and death are generally, if not always, caused by circulatory dysfunction, and the effect of interruption of arterial supply to part of an organ is well known as infarction. The conception of degenerative osteoarthritis as a primarily vascular disorder is an old one, and until the appearance of the work by Harrison, Schajowicz and Trueta (1953) it was generally accepted that deficiency of arterial supply to the femoral head played a dominant role in the pathogenesis of osteoarthritis of the hip joint. These authors showed, however, that in this disorder the arterial supply to the head was intact, sometimes even increased, while the intraosseous veins and sinusoids showed signs of dilation and stasis. Since then an impressive amount of work has been performed, dealing with the changes of veins and venules in the various structures of osteoarthritic joints, and, later, also in joints with non-traumatic femoral head necrosis, and with the causes and consequences of impeded venous drainage. At the moment, the consensus seems to be that changes in venous drainage conditions seem to be a universal phenomenon in manifest osteoarthritis, regardless of location. However, opinions still differ as regards the place of these changes in the pathogenesis of the disorder. As regards non-traumatic femoral head necrosis most authors are still of the opinion that arterial blockage is of major importance.

This paper will deal with some clinical and experimental findings in degenerative osteoarthritis and non-traumatic osteonecrosis, and attempt to correlate them in a synthesis.

Bone Circulation and Vascularization in Normal and Pathological Conditions
Edited by A. Schoutens *et al.*, Plenum Press, New York, 1993

201

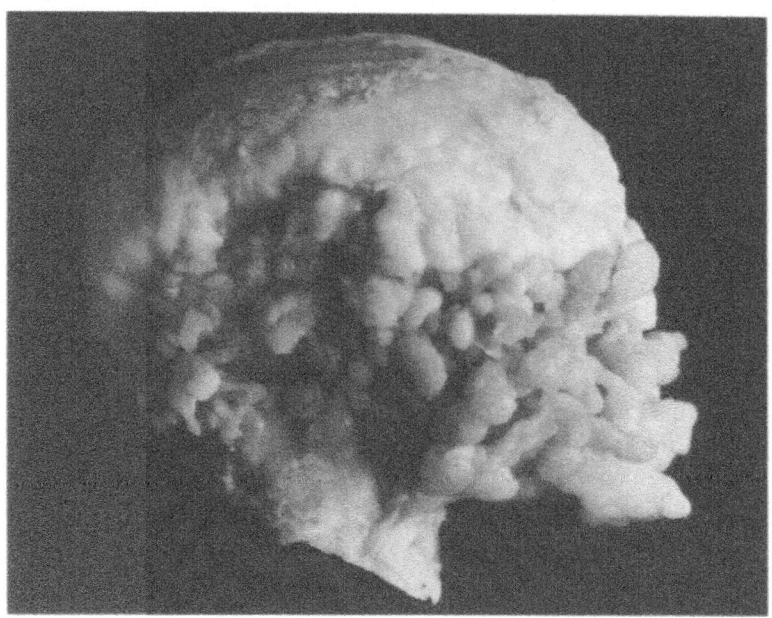

Figure 1. Femoral head from a 71-year-old woman with osteoarthritis of the hip joint. Proliferative osteoarthritis.

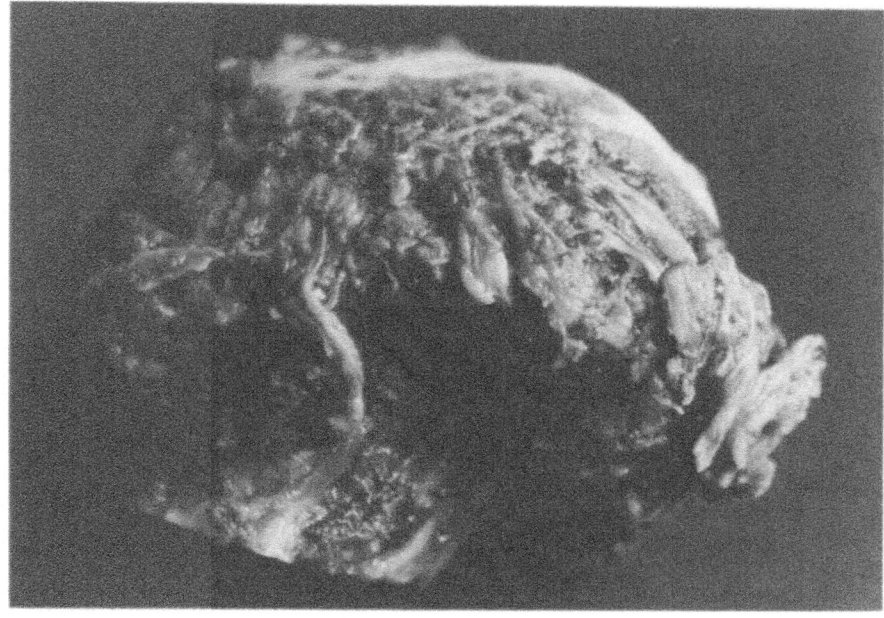

Figure 2. Femoral head from a 71-year-old man with osteoarthritis of the hip joint. Fibrous synovitis.

FINDINGS BY VARIOUS METHODS OF INVESTIGATION

Clinical Examination

The main clinical symptoms and signs of degenerative joint disorders are pain and restriction of movement. Pain at rest may be a very early symptom, but in the majority of cases it is not prominent in the initial stage. Pain on loading and joint movement is typical, and in the hip, pain on forced inward rotation is usually among the first signs of a disorder. Correspondingly, in the hip the rotational movements are the first to become restricted, whereas the patient may have a reasonably good flexion-extension movement even with grave radiological changes. The typical fixed contracture of the hip joint is in flexion, adduction and outward rotation, the position in which the joint capsule is maximally relaxed and the lumen of the joint cavity has its maximum capacity.

Synovium and Synovial Fluid

Figures 1 and 2 show different stages of synovial involvement. In Figure 1 the joint was distended by an excess of synovial fluid. The synovium is bulky, moist and oedematous. I use the term "proliferative synovitis". Figure 2 represents a long standing synovial disorder. The joint was dry and the bulky oedematous synovium has been replaced by fibrous scarlike tissue (Arnoldi et al., 1980).

Cartilage.

It is characteristic of degenerative osteoarthritis that the signs of cartilage degeneration are symmetrical, i.e. both opposing joint surfaces are involved. Further characteristics will be mentioned in the section on histology.

Intraosseous Phlebography

Figure 3 shows the typical difference between the unaffected and the osteoarthritic (or osteonecrotic) femoral head and trochanteric region. The contrast injected into the bone marrow of the normal living femoral head leaves immediately to extraosseous veins, and it is impossible to visualize more than a tiny area with minute vessels at the site of injection. On the osteoarthritic side the extraosseous veins are either very poorly filled or invisible, whereas the contrast fills the intraosseous veins far down in the femoral shaft. Figure 4 shows that, whereas the contrast has disappeared from all veins on the normal side within a few minutes, it remains in the bone marrow of the affected side, in our experiments up to more than one hour (Arnoldi et al., 1972).
It is interesting to note that the non-filling of veins of normal bone marrow is characteristic only for living bone. Injection of but a few ml of contrast into cadaveric bone will fill the entire intraosseous venous network of the femur.

Intraosseous Pressure Measurements

Figure 5 shows pressure tracings at rest in the supine position from the trochanteric region and the femoral vein from a patient with unilateral osteoarthritis of the hip joint. the pressure is highest and the pulsatile excursions greatest on the osteoarthritic side.

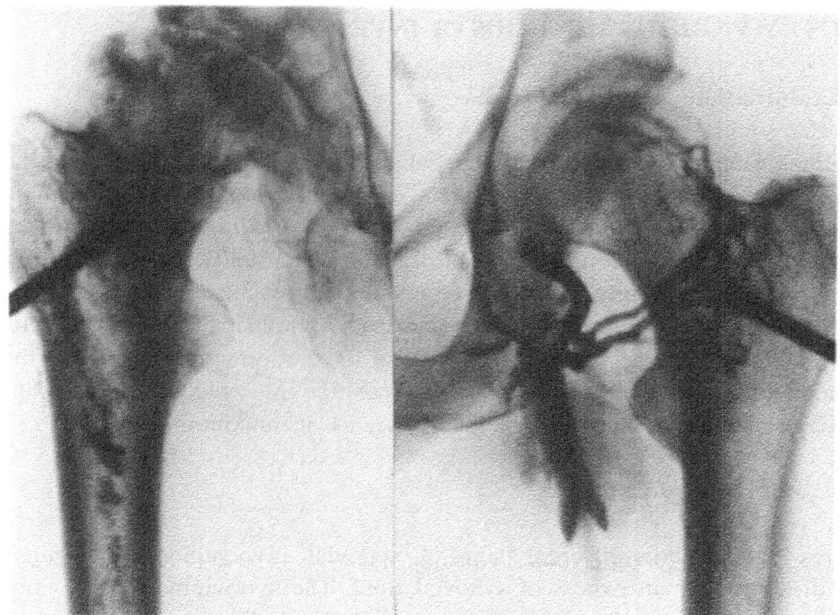

Figure 3. Intraosseous phlebographs from normal left hip and from right hip with severe osteoarthritis, exposed 30 sec after injection of contrast material. In the normal hip the contrast-mixed blood leaves the intraosseous space without noticeable filling of intraosseous vessels. Drainage takes place through superior retinacular veins to gluteal veins and via two medial circumflex veins to femoral and obturator veins, as well as through inferior retinacular veins, emptying into the trunk of the medial circumflex vein. The extraosseous veins are well filled and the arrangement of valves visible. - In the right hip the contrast is collected in large, tortuous intraosseous vessels, extending from the injection site distally into the diaphysis. No extraosseous veins visible.

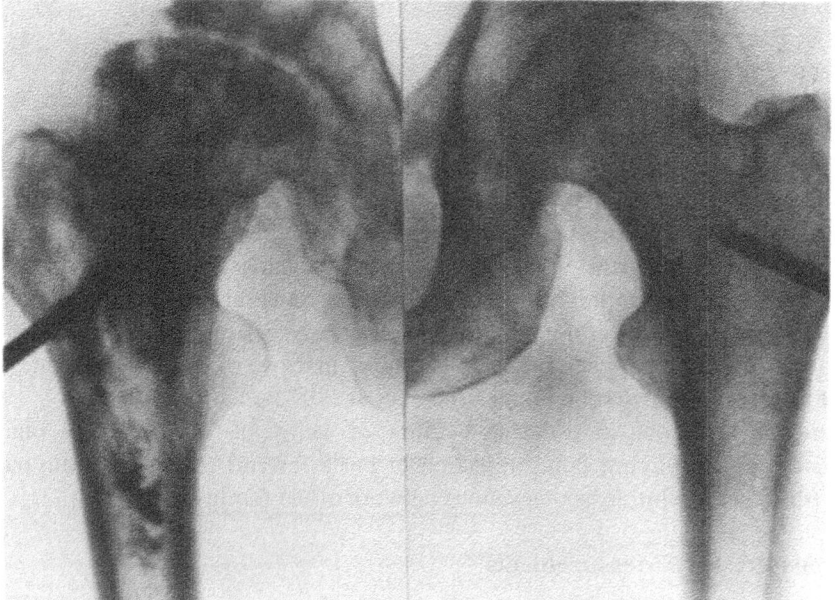

Figure 4. Intraosseous phlebographs from the same patient as in Figure 3, exposed 30 min after the injection. In the left hip there is no sign of contrast material in extra- or intraosseous veins (complete evacuation was noted 3 min after the injection). In the arthritic hip large quantities of contrast remain in the intraosseous space. No filling of extraosseous veins.

This is a characteristic finding in osteoarthritis of the hip and knee, confirmed my many other authors. Figure 6 shows that the high pressure is confined to the bone marrow and not due to extracapsular venous hypertension. The figure also shows another characteristic finding, i.e. the intramedullary pressure is highest near the articular surface (Arnoldi et al., 1972). An intraosseous pressure more than 40-50 mmHg above the extraosseous venous pressure is correlated to the symptom "rest pain" (Arnoldi et al., 1971; Lynch, 1974).

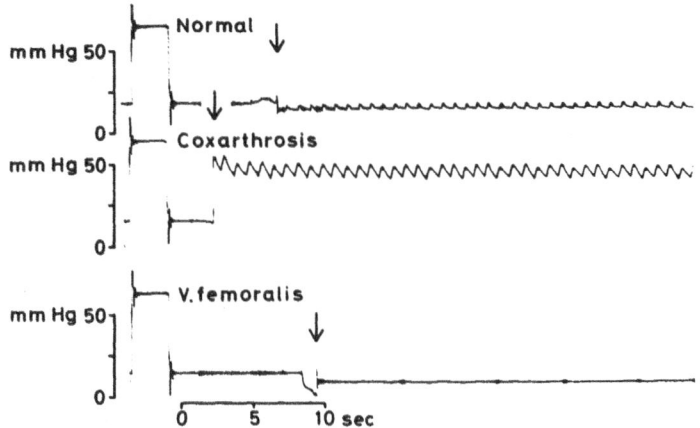

Figure 5. Pressure recordings from the femoral trochanteric region and femoral vein from patient with unilateral osteoarthritis of the hip joint.

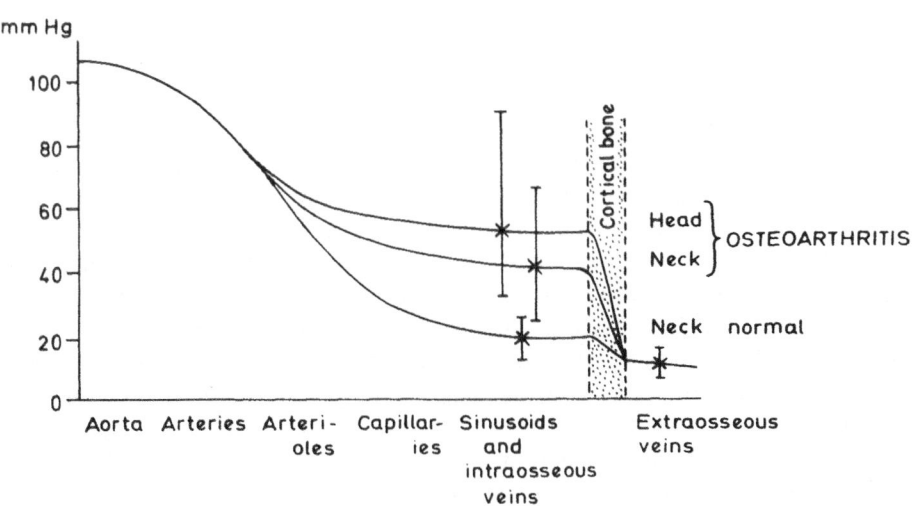

Figure 6. Mean systemic blood pressure and pressures (mean) from normal trochanteric region, from the trochanter and femoral head of osteoarthritic joints, and from the femoral vein. The values arranged as Landis's curves.

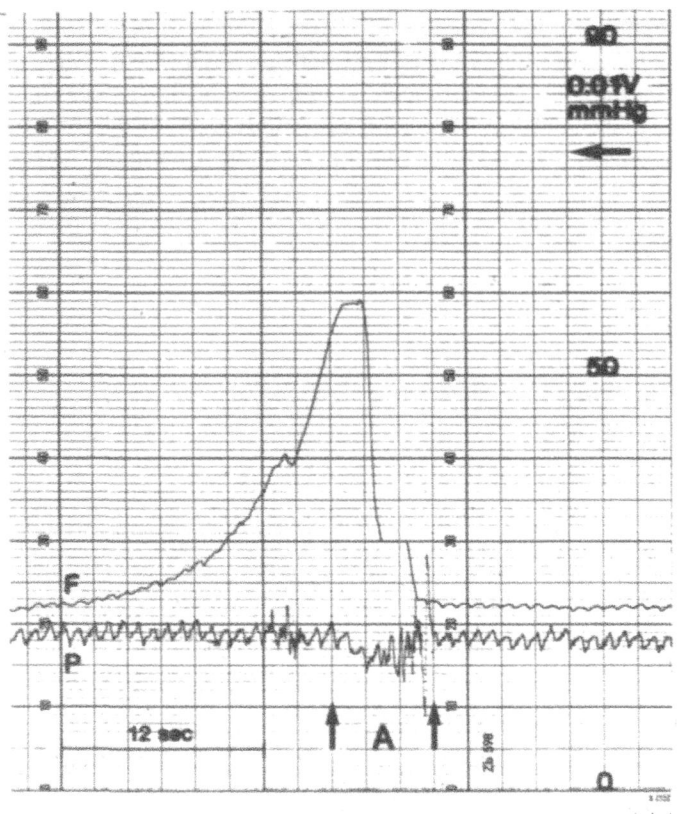

Figure 7. Pressure and flow chart from normal femoral head, before, during and after one forceful internal rotation of the femur (A).

In the living bone marrow of non-traumatic femoral head necrosis the tracings are also pulsatile, and the pressure is generally at a higher level than in osteoarthritis.
Intraosseous pressure during joint loading and movements was measured in patients with unilateral osteoarthritis of the hip joint or non-traumatic femoral head necrosis, lying supine. Pressure at rest was recorded, as well as pressure variations during simulated loading and passive joint movements. The point of measurement was the centre of the femoral head. These measurements were in some cases combined with measu-rement of intraosseous blood flow rate at the same level by means of the laser Doppler flowmetry method, using a Periflux PF 3 apparatus (Perimed, Stockholm, Sweden).
Figures 7 and 8 show pressure and flow tracings from a normal and an osteoarthritic femoral head before during and after one forceful movement from the neutral position into internal rotation. Figure 9 shows the effect of two movements into internal- and one into external rotation on intraosseous pressure and rate of blood flow. In the normal femoral head the flow increases, but the pressure only slightly. In the osteoarthritic femoral head the pressure variation is considerable, but the flow variation modest. In the osteonecrotic case, the initial pressure is high and it rises still higher during the manoeuvres, whereas the flow remains fairly constant at a very low level.

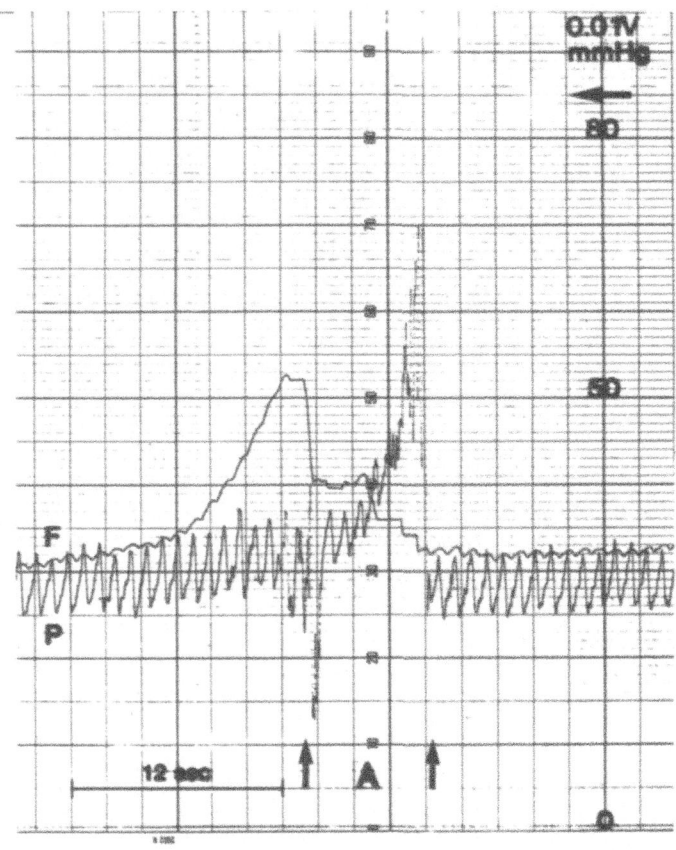

Figure 8. Pressure and flow chart from osteoarthritic femoral head, before, during and after one period of forceful internal rotation of the femur (A).

Scintigraphy

It has long been known that the uptake of ^{99m}Tc-phosphates is increased in painful osteoarthritis, in the intraosseous engorgement-pain syndromes and in the early stages of non-traumatic femoral head necrosis. Increased uptake is seen wherever the rate of build-up or break-down of bone is abnormally fast, the isotopes being bound to newly formed or recently denuded apatite crystals (Christensen, 1983, 1985). Thus, the increased uptake in these disorders indicates increased bone metabolism, anabolic and/or catabolic, and for both these activities a supply of blood is needed. "Cold-in-hot" spots are a late phenomenon in non-traumatic osteonecrosis.

Measurements of Gas Tensions in Subchondral Bone Marrow

Evidence of reduced oxygen tension in juxtachondral bone marrow of osteoarthritic joints was presented by Brookes and Helal (1968) and Pujol et al. (1973), and recently, mass spectrometry measurements have confirmed these observations in both osteoarthritis and non-traumatic femoral head necrosis (Kiær et al., 1986, 1990; Svalastoga, 1988; Pedersen et al., 1989).

207

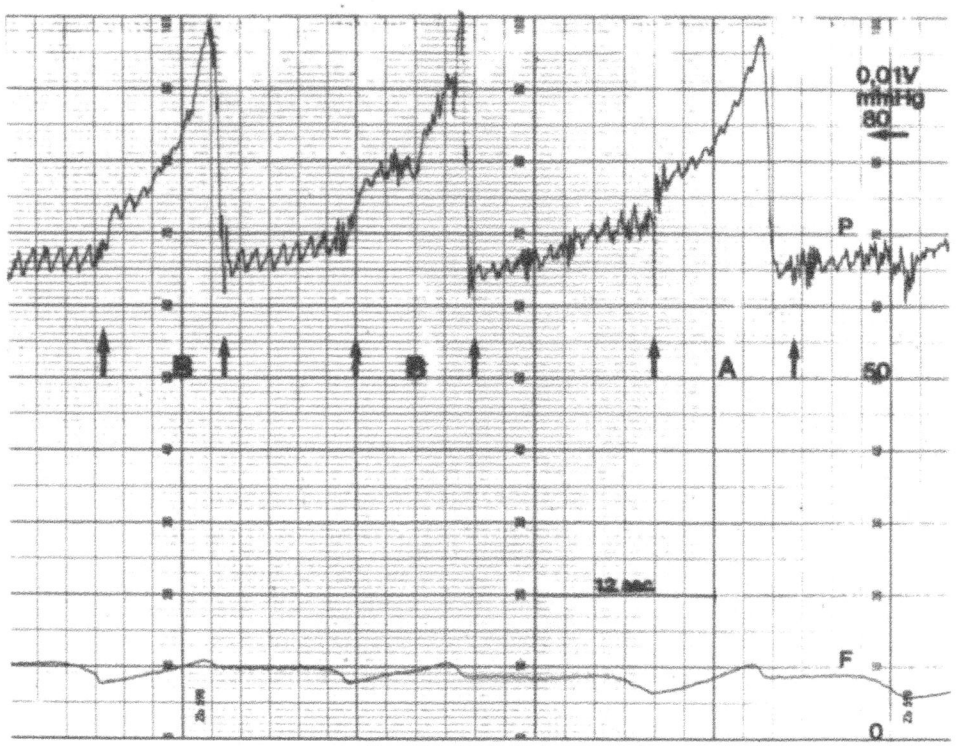

Figure 9. Intraosseous pressure and flow tracings from the center of the femoral head (outside the necrotic area). Steroid induced osteonecrosis. A = period of external rotation; B = periods of internal rotation.

Changes in the Synovial Fluid in Osteoarthritis

It is well established that the oxygen content in osteoarthritic synovial fluid is decreased (Lund-Olesen, 1970). This is also the case in early experimental osteoarthritis (Kofoed, 1986). Changes in the protein pattern of the synovial fluid have been observed by many authors. Reimann et al. (1980) found a relative increase of large-molecular proteins in osteoarthritic synovial fluid and ascribed this shift to increased capillary permeability in the synovium. It was especially pronounced in fluid from proliferative synovitis. Finally, Reimann et al. (1976) observed a reduction of boundary lubrication in osteoarthritic synovial fluid.

Histological Observations in Osteoarthritis

Seen with the naked eye the internal surface of normal synovium is smooth, moist, glistening and pink. Villi are few and small. Microscopic examination reveals an inner layer of synovial "lining cells" (synoviocytes), covering a loose fibrous stroma containing nerves, blood vessels and lymphatics. Figure 10 shows an example of normal synovium from the knee of an elderly patient. The vessels in the stroma are not a prominent feature. Proliferative synovitis is characterized by stromal oedema, enormously increased number of vessels, mostly dilated venules and veins, very often showing intraluminar aggregation and agglutination of erythrocytes and fibrin thrombi of various ages. The minute vessels show increased wall permeability, indicated by numerous extravascular erythrocytes in the oedematous stroma (Figures 11 and 12). Patches of collagen bundles increase as the stage of fibrous synovitis is approached, and the final stage is dominated by poorly vascularized fibrous tissue, containing areas of haemosiderin deposits (Figure 13).

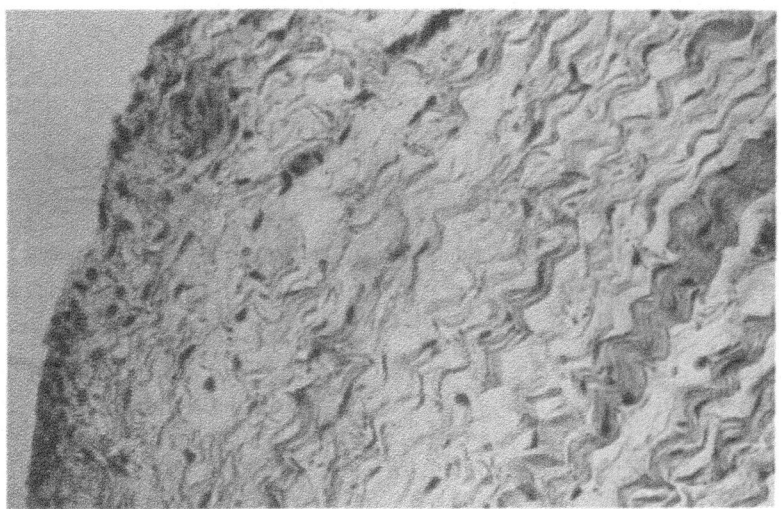

Figure 10. Normal synovium removed from knee of an elderly patient.

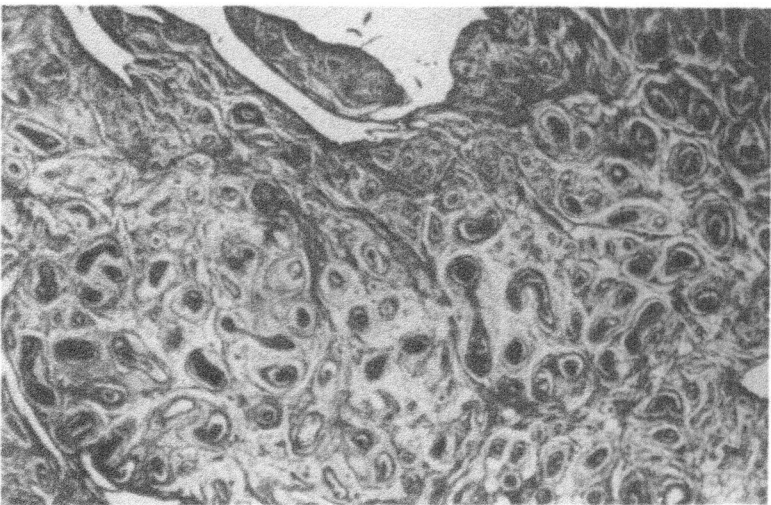

Figure 11. Light microscopy of histological section of the synovial membrane from a 70-year-old man with osteoarthritis of the hip joint, showing dilation of venules in the highly vascularized membrane and oedematous stroma (Martius scarlet blue staining, MSB).

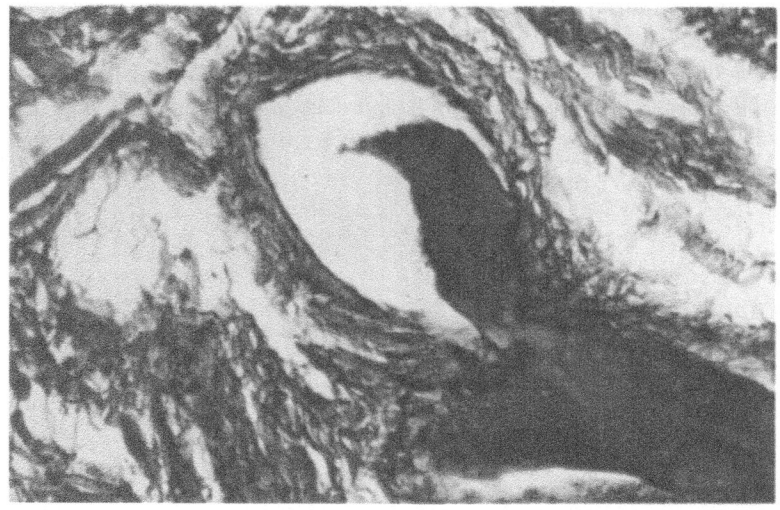

Figure 12. Large fibrin thrombus in vessel from osteoarthritic synovium (MSB).

The general histology of cartilage degeneration is well known from several textbooks. Only the vascular changes and changes probably or possibly caused by vascular derangement will be discussed here and later: Our findings varied from very moderate surface fibrillation to complete denudation of the osteo-chondral end-plate. Cloning of chondrocytes was seen in all stages (Figure 14). In most sections the tidemark was single, but duplication was sometimes observed. In such cases lacunae with small vessels, partly blocked by fibrin, were sometimes observed peripheral to the outer tidemark.

In the end-plate a common, almost typical, finding was areas of vascularized, more or less primitive tissue invading the end-plate from the bone marrow. Fibrin thrombi were common in the small vessels of these tissues and were seen occasionally in the Haversian canals. In the bone marrow small vessels (sinusoids and venules) very often showed erythrocyte aggregations and agglutinations. Free erythrocytes and haemosiderin deposits were common. Over considerable areas the normal bone marrow tissue was replaced by more primitive fibrous tissue. In the trabeculae the bone structure varied between laminar and woven. The vessels in the Haversian canals were usually open, but areas of agglutinations and fibrin thrombi were not uncommon.

Histological Observations in the Osteonecrotic Femoral Head

Where it was present, the necrosis proper consisted of non-vascularized debris. The rest of the bone marrow looked very much like the more heavily changed parts of osteoarthritic bone marrow, but all abnormal traits seemed enlarged. Thus, the invasions of primitive fibrous and fibrocartilagineous tissues had taken over greater areas. They were even more vascularized and the signs of stasis more pronounced. The frequency of fibrin thrombi of all ages in Haversian canals was notable (Figure 15). Large areas of cartilage lying either loose or attached to more or less necrotic parts of the femoral head looked almost normal. Cloning was rare and was in fact only seen where shreds of bone or cartilage had produced secondary synovitis.

210

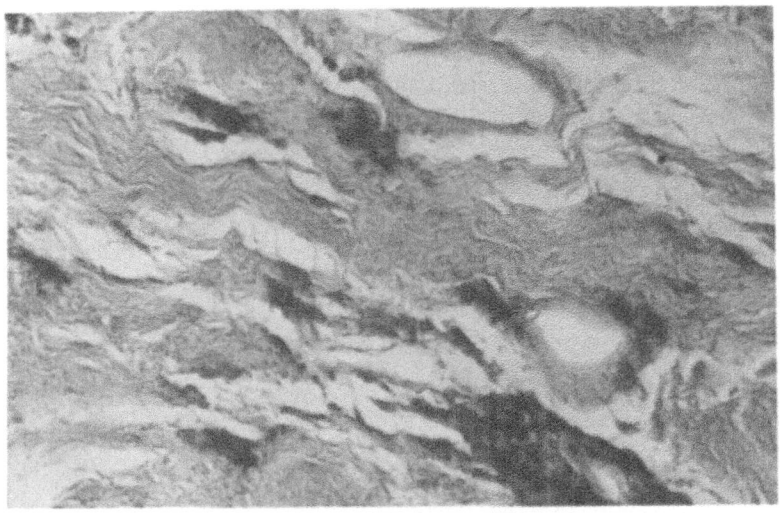

Figure 13. Haemosiderin deposits in osteoarthritic synovial tissue, in close juxtaposition to solid bands of collagen. Fibrous synovitis (Perl's prussian blue).

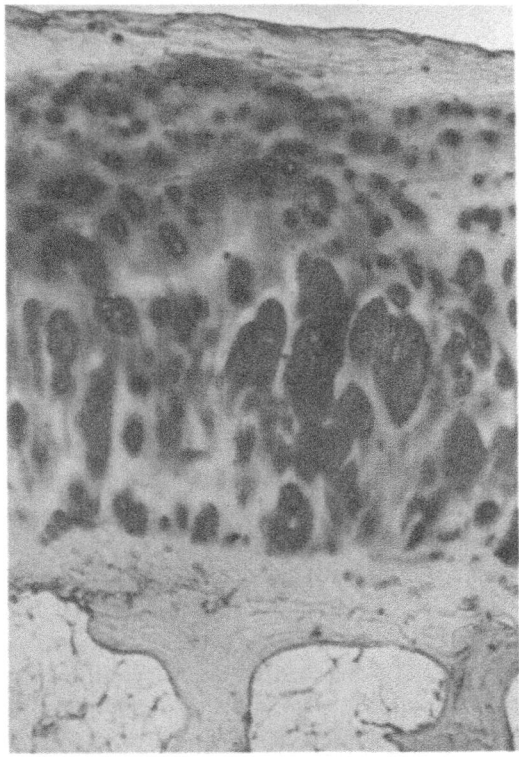

Figure 14. Cartilage from osteoarthritic hip joint. The surface almost normal to the naked eye, but cloning of chondrocytes has appeared (Safranin-0).

To sum up the histological observations: The overall impression of vascular conditions in the synovial membrane, as well as in the bone marrow, was a state of impaired venous drainage, the intraluminar blocking especially prominent in venules and small veins. Thrombi in Haversian canals were observed in the osteoarthritic femoral head, but were far more numerous in non-traumatic osteonecrosis.

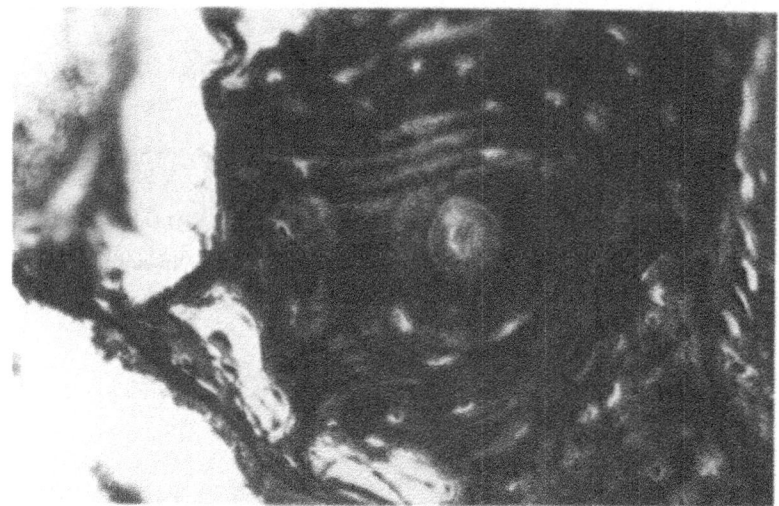

Figure 15. Old thrombus in Haversian canal. Empty osteocyte lacunae. Rim zone of steroid-induced femoral head necrosis (MSB).

Joint Innervation

Pain is one of the main causes of disability in patients with joint disorders. In fact, pain is the deciding factor for most surgical interventions. Pain is caused by irritation of nocireceptive nervous elements in the joint structures. Whereas normal hyaline cartilage is devoid of such pain receptors, investigations by means of conventional histological methods, e.g. silver impregnation, have demonstrated nerves in all other joint structures. In bone tissue the bone marrow trabeculae and the Haversian canals were reported to contain nerve fibres and in certain pathological states, such as degenerative osteoarthritis, the nerves in the bone marrow (femoral head) seem to increase in number (Reimann and Christensen, 1977). Grönblad et al. (1988 a, b) found nerve elements staining with neurofilament antiserum perivascularly in both normal synovium and synovium from patients with rheumatoid arthritis and osteoarthritis. Free nerve fibres were also present in these structures, and some of them stained with antisera to the neuropeptides Substance P (SP) and calcitonin gene-related peptide (CGRP). SP is a sensory transmitter and vasodilator, increases vascular permeability and thus protein extravasation, and stimulates fibroblast proliferation, and CGRP is also a potent vasodilator, as well as a sensory transmitter.

SYNTHESIS

The various investigations mentioned above indicate that in the degenerative joint disorders mentioned here the high intraosseous pressure is due to increased flow resistance in the veins draining the joint-bearing bone marrow. Whereas histological examinations of non-traumatic osteonecrosis of the femoral head point to intraosseous venous blockade as the primary cause of intraosseous stasis and hypertension, the findings in osteoarthritis indicate that the initial impairment of joint drainage is located in the synovium and that the changes in bone marrow and cartilage are secondary phenomena.

Observations from experimental osteoarthritis show two stages in the development of vascular "synovitis": An initial phase of arterial hyperemia with increased volume of the synovium, followed by a stage of venous stasis and increased production of synovial fluid. Osteophytes appear very early in the hyperemic stage, whereas changes in cartilage and subchondral bone leading to manifest osteoarthritis are later manifestations in the stage of stasis (Christensen, 1983).

That increased intraarticular pressure rises marrow pressure in adjacent bone has been shown by several authors (Arnoldi et al., 1979; Bünger et al., 1981, 1983).

In human osteoarthritis the very early stages are but imperfectly known. However, proliferative synovitis was a constant finding in early X-ray negative osteoarthritis of the hip joint (Arnoldi, to be published).

Recent observations on the innervation of the synovium have disclosed other aspects of synovial pathology, and the origin of joint pain. Thus, Konttinen et al. (1989) suggested that the neuropeptides SP and CGRP may be involved in both pain and synovial inflammation. The inflammatory effect they attribute to the well-known effects of both neuropeptides as vasodilators and on vascular permeability and protein extravasation. The illustrations given by these authors show the nerve fibres in close contact with the vessel walls. Considering the sometimes extreme vascularization of the synovial membrane in osteoarthritis and rheumatoid arthritis, where most of the vessels are dilated venules, their discoveries suggest a circulus viciosus, starting perhaps with traumatic hyperemia, proceeding to the stasis stages, and perpetuated by the influence of a rising content of vasodilatory neuropeptides. Many other observations, such as the high intraosseous pressure and venous distension in non-traumatic femoral head necrosis, in osteoarthritis and in engorgement-pain syndromes with rest pain and the sometimes extreme rise of intraosseous pressure in painful joint manoeuvres, strengthen the conception, that joint pain is connected with nocireceptors activated by venous distension.

In experimental osteoarthritis, intraosseous engorgement-pain syndromes (including X-ray negative osteoarthritis), osteoarthritis, rheumatoid arthritis and haemophilic arthropathy the group studied by the author), synovitis is a very early feature and probably the cause of cartilage degeneration. This group of disorders has the following charac-teristics in common: 1) the cartilage degeneration is "symmetrical", i.e. both opposing cartilage surfaces are involved, 2) the synovial membrane appears to pass through an initial "proliferative" and productive stage, before it ends up as fibrous and non-productive (the dry joints). Especially in the productive phase the chemical com-position of the synovial fluid is altered, presumably affecting cartilage components that are dependent on this fluid for normal metabolic activity. The observations suggest that the severity of changes in the composition of the synovial fluid is proportional to the extent of cartilage damage and to the rapidity of its progress; 3) the histological picture is dominated by degenerative changes, beginning at the joint surface, and early and conspicuous cloning of chondrocytes. In other degenerative joint disorders, such as non-traumatic femoral head necrosis and the patellar pain syndrome (chondromalacia), synovitis is not a primary feature. It only appears in the late stages when secondary osteoarthritis develops. In these disorders, where the primary pathological changes are found in subchondral bone or the basic layers of cartilage, cartilage involvement is asymmetrical in the early stages, and cloning of chondrocytes either absent of very modest (Arnoldi, 1991).

Whereas total lack of blood supply leads to death of all tissues involved, the hypoxia demonstrated in the bone marrow of osteoarthritis and non-traumatic femoral head necrosis by e.g. Kiær et al. (1986, 1990), Svalastoga (1988) and Pedersen et al. (1989) seems to be a stimulating factor for some mesenchymal tissue activity. Thus, Stern et al. (1966) studied the effect of various oxygen tensions on the synthesis and degradation of collagen in bone. With an oxygen content of 10-20 percent, collagen synthesis exceeded collagen degradation; at 30 percent, both processes increased and at 50 percent they were equal. Thus, both experimental and clinical observations confirm that hypoxia may stimulate growth of mesenchymal tissue. However, the histological evidence cited above suggests a considerable difference in the reaction of the various tissues of juxtachondral bone marrow and the other joint structures. Osteoarthritis, as well as non-traumatic osteonecrosis are characterized histologically by ingrowth of masses of primitive tissues into the bone marrow of the femoral head, and in osteoarthritis this was observed even in the early X-ray negative stage. There seems to be a relationship between the extent of this invasion and the degree of hypoxia. Thus, the substitution of normal marrow tissue and bone trabeculae by fibrous or fibro-cartilagineous tissues was more extensive in non-traumatic femoral head osteonecrosis, with a lower oxygen tension than in osteoarthritis.

CONCLUSION

The evidence cited indicates that in the disorders discussed here, reduced capillary flow and tissue oxygenation, due to impaired venous drainage from the synovium and from subchondral bone, are the primary factors in the pathogenesis of synovial and bone marrow changes, and, via the changes in the chemistry of synovial fluid are indirectly responsible for cartilage degeneration. The exceedingly slow flow in the bone marrow of the osteonecrotic femoral head is probably also due to arteriolar compression by high tissue tension (Figure 9) (Arnoldi, 1991).

REFERENCES

Arnoldi, C.C.: Patellar Pain. Acta Orthop. Scand. Suppl. 244, 1991. Munksgaard, Copenhagen 1991.

Arnoldi, C.C., R.K. Lemperg, H. Linderholm: Immediate effect of osteotomy on the intramedullary pressure of the femoral head and neck in patients with degenerative osteoarthritis. Acta Orthop. Scand. 42: 357-365 (1971).

Arnoldi, C.C., H. Linderholm, H. Müssbichler: Venous engorgement and intraosseous hypertension in osteoarthritis of the hip. J. Bone Joint Surg. (Br) 54: 409-421 (1972).

Arnoldi, C.C., I. Reimann, S.B. Christensen, S. Mortensen: The effect of increased intraarticular pressure on juxtachondral bone marrow pressure. IRCS Med. Sci. 7: 471 (1979).

Arnoldi, C.C., I. Reimann, P. Bretlau: The synovial membrane in human coxarthrosis. Light and electron microscopic studies. Clin. Orthop. 148: 213-220 (1980).

Brookes, M., B. Helal: Primary osteoarthritis, venous engorgement and osteogenesis. J. Bone Joint Surg. (Br) 50: 493-502 (1968).

Bünger, C., S. Harving, J. Hjermind, E.H. Bünger: Relationship between intraosseous pressure and intraarticular pressures in arthritis of the knee. Acta Orthop. Scand. 54: 188-201 (1983).

Bünger, C., J. Sorensen, J.C. Djurhuus, U. Lucht: Intraosseous pressures in the knee in relation to simulated joint effusion, joint position, and venous obstruction. Scand. J. Rheum. 10: 283-291 (1981).

Christensen, S.B.: Localisation of bone seeking agents in developing experimentally induced osteoarthritis in the knee joint of the rabbit. Scand. J. Rheum. 12: 243-249 (1983).

Christensen, S.B.: Changes of bone, cartilage and synovial membrane in relation to bone scintigraphy. Acta Orthop. Scand. 56: Suppl. 214. Munksgaard. Copenhagen (1985).

Grönblad, M., Y.T. Konttinen, O. Korkala: Neuroanatomical basis for pain reception in joint desease. XXII Scandinavian Congress of Rheumatology. Reykjavik, Iceland. Scand. J. Rheumatol. 15: Suppl. 72: 39 (1988).

Grönblad, M., Y.T. Konttinen, O. Korkala: Neuropeptides in the synovium of patients with rheumatoid arthritis and osteoarthritis. J. Rheumatol. 15: 1807-1810 (1988).

Harrison, M.H.M., F. Schajowicz, J. Trueta: Osteoarthritis of the hip: A study of the nature and evolution of the disease. J. Bone Joint Surg. (Br) 35: 598-626 (1953).

Kiær, T., K.K. Sorensen, J. Gronlund: Intraosseous pressures of oxygen and carbon dioxide in coxarthrosis. Acta Orthop. Scand. 57: 115-121 (1986).

Kiær, T., N.W. Pedersen, K.D. Kristensen, H. Starklint: Intraosseous pressure, oxygen tension in avascular necrosis and osteoarthritis of the hip. J. Bone Joint Surg. (Br) 72: 1023-1030 (1990).

Kofoed, H.: Synovitis causes hypoxia and acidity in synovial fluid and subchondral bone. Injury 17: 391-396 (1986).

Konttinen, Y.T., M. Grönblad, E. Hukkanen: Pain fibers in osteoarthritis: A review. Sem. Arthrit. Rheum. 18: Suppl. 2: 3540 (1989).

Lemperg, R.K., C.C. Arnoldi: The significance of intraosseous pressure in normal and diseased states, with special reference to intraosseous engorgement-pain syndrome. Clin. Orthop. 136: 143-156 (1978).

Lund-Olesen, K.: Oxygen tension in synovial fluids. Arthrit. Rheum. 13: 769-773 (1970).

Lynch, J.S.: Venous abnormalities and intraosseous hypertension associated with osteoarthritis of the knee. In: Ingwesen (ed): The Knee Joint. Proceedings of the international congress. Rotterdam Sept. 1973. American Elsevier Publication Comp. Inc. New York 1974.

Pedersen, N.W., T. Kiær, K.D. Kristensen, H. Starklint: Intraosseous pressure, oxygenation, and histology in arthrosis and osteonecrosis of the hip. Acta Orthop. Scand. 60:: 415-417 (1989).

Reimann, I., J. Stougaard, A. Northeved, S.J. Johnsen: Demonstration of boundary lubrication by synovial fluid. Acta Orthop. Scand. 46: 1-10 (1975).

Reimann, I., S.B. Christensen: A histological demonstration of nerves in subchondral bone. Acta Orthop. Scand. 48: 345-358 (1977).

Reimann, I., C.C. Arnoldi, O.S. Nielsen: Permeability of synovial membrane to plasma protein in human coxarthrosis. Clin. Orthop. 147: 296-300 (1980).

Stern, B., M.J. Glimcher, P. Goldhaber: The effect of various oxygen tensions on the synthesis and degradation of bone collagen in tissue culture. Proc. Soc. Exp. Biol. Med. 121: 869-872 (1966).

Svalastoga, E.: Local tissue hypoxia in the pathogenesis of osteoarthritis. The Royal Veterinary and Agricultural University Copenhagen. Thesis (1988).

CIRCULATORY ASPECTS OF REFLEX SYMPATHETIC DYSTROPHY

M. DRIESSENS

Fysische Geneeskunde en Revalidatie
Universitair Ziekenhuis Antwerpen
Antwerpen, Belgium

INTRODUCTION

Reflex sympathetic dystrophy (RSD) is a condition of the locomotor system characterized by clinical symptoms and vascular changes which vary in time. The wrist and hand or ankle and foot are affected most frequently. The disease has a spontaneous evolution over 3 stages, followed by stabilization (stage III).

In stage I there is an increase of the blood flow and of the blood volume in the bone which can be demonstrated. After a certain period of time (3 to 8 months) this condition evolves progressively into a hypovascularization which is typical of stage II. Bone scintigraphy is essential for the diagnosis of RSD and a negative bone scan excludes the condition.

Based on empirical findings and on the available literature, it can be stated that the treatment of RSD is not well defined and that our knowledge of the pathogenesis of the disease is limited and in practice, there is no agreement as to the best treatment.

Experience teaches us that many cases of RSD are not recognized as such, or are treated incorrectly. Reasons that can be put forward for this are its relative rarity, and the many different forms it takes. In addition, diagnosis requires specialized investigations (including bone scintigraphy) which may not be available to everybody. Moreover, although many kinds of treatment have been proposed over the years, each method has its own supporters. Some of the therapies have vascular effects and therefore it is important to know the vascular changes in each stage of RSD.

CLINICAL AND RADIOLOGICAL ASPECTS

The course of RSD can be divided into approximately three phases, each of which is characterized by different clinical features.

- Stage I: pain, oedema and swelling, excessive transpiration, hyperaemia, sometimes redness (this is also known as the hypertrophic stage, with inflammatory characteristics) (figure 1);

Bone Circulation and Vascularization in Normal and Pathological Conditions
Edited by A. Schoutens *et al.*, Plenum Press, New York, 1993

217

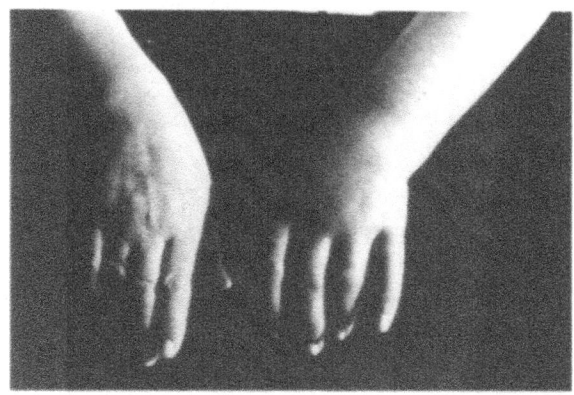

Figure 1. RSD stage I (Left forearm and hand, on the right).

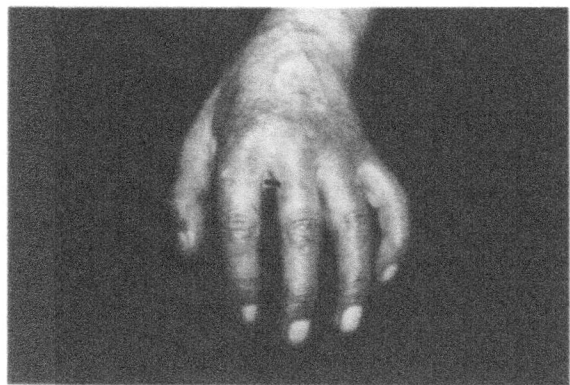

Figure 2. RSD stage II.

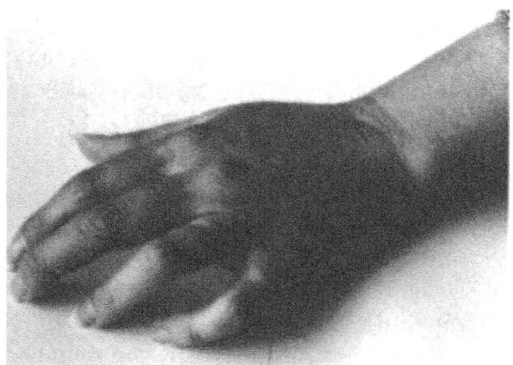

Figure 3. RSD stage III.

- Stage II: pain, cyanosis, skin atrophy, coldness and stiffening joints (this is also known as the atrophic stage) (Figure 2);
- Stage III: the atrophy of the skin may or may not continue, accompanied by coldness, continued stiffening of the joints, and reduction or disappearance of the pain (stabilized or recovery stage) (Figure 3).

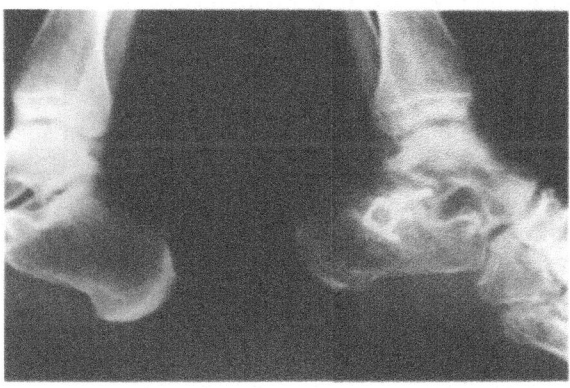

Figure 4. Typical radiological picture of RSD.

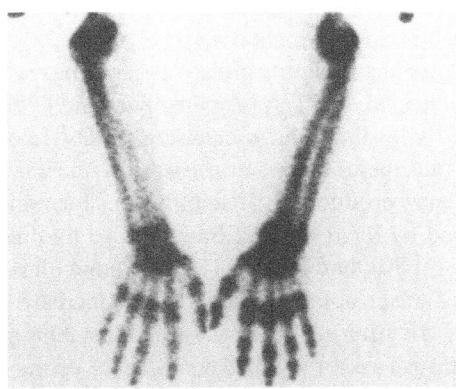

Figure 5. Hypercaptation of bone tracer on the affected left side.

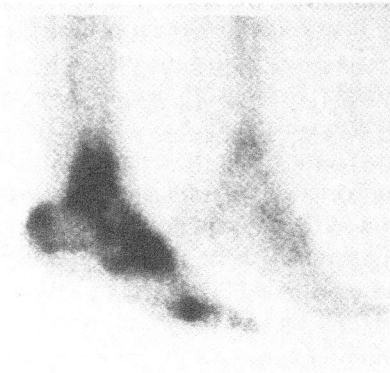

Figure 6. Bone scan in RSD of the ankle and foot.

During the course of the disease there is a fluent and continuous transition from one stage to the next. The transitional stage between stages I and II is known as the dystrophic stage. Clinically, the distinction is difficult to make at certain points in the course of the disease. Some of the symptoms can be present in the different stages. Also neurologic signs can be seen: paresis, hypertonia, discoordination, muscle tremor (Goris et al., 1988).

There are also considerable differences in terms of the topography of the disease. RSD of a more peripheral area, such as a foot, hand or wrist, generally shows more developed symptomatology than that found in a deep-lying joint, such as the hip.

There are also considerable differences in speed of evolution. Stage I generally lasts for 2 to 4 months, stage II takes up the next 3 to 6 months (if treatment is not initiated), and stage III usually begins after 6 to 12 months. There are however many variations, and very rapidly evolving cases are by no means rare.

The typically occurring abnormality on radiology is patchy osteoporosis in the affected parts of the skeleton (Darnault et al., 1987; Kienboeck, 1901). However, this type of osteoporosis is found in only two-thirds of cases (Figure 4), while in the remaining one-third, homogeneous osteoporosis occurs (Mortier et al., 1985).

Radiological investigations cannot be used for staging of RSD. In addition, osteoporosis is not pathognomic to RSD; simple disuse can lead to either patchy or homogeneous osteoporosis. In RSD, osteoporosis usually affects an entire region (for example, all bones of the hand and wrist, or all bones of the knee joint). However, more localized cases of RSD have been described: radial RSD, zonal RSD, parcellar RSD (Doury et al., 1979; Lequesne et al., 1979; Chester, 1990; Cuartero-Plaza et al., 1992).

BONE SCINTIGRAPHY

This method is of the uppermost importance for the diagnosis of RSD. A negative bone scan removes the possibility of diagnosing RSD (Figures 5 and 6).

A few children with RSD showing "hypofixation" on the bone scan have been published, but these findings will be discussed later (Pseudo-dystrophy).

It should however also be noted that the bone scan will be positive for a substantial number of other conditions (Paget's disease, metastasis, aseptic necrosis, etc).

The bone scan gives a direct picture of the affinity of the bone for phosphate complexes in the bone tissue. The most frequently used complexes are ^{99m}Tc pyrophosphate and ^{99m}Tc diphosphonate, which are incorporated very rapidly by the hydroxyapatite crystals. Local tracer hyperfixation points to increased local bone metabolism, consisting of increased absorption (osteoclasts) and increased secondary production (osteoblasts). Lavender (1979) showed that the tracer uptake is influenced by local blood flow, but also by other factors of increased bone formation. Charkes (1980) and Sagar (1979) found that a decrease of bone blood flow induces a decrease of tracer uptake. Otherwise, an increase of systemic blood flow has no effect on tracer fixation. But in cases of pathologic vasodilatation in bone (section of the sympathetic nerves) a significant increase of tracer uptake was found.

Normally one third of capillaries seems to be closed. In pathologic situations, the increase of tracer fixation must be due to opening of normally closed blood vessels, so that more bone contact surface is reachable for tracer molecules. Bone scans show very precisely where the disease is active, before any radiological changes are visible. This can be of great importance for diagnosis (Darnault et al., 1987). Even after clinical recovery or spontaneous remission (stage III), the bone scan will sometimes continue to be positive for a while. During stage II, a gradual reduction of excess tracer uptake follows, and usually after one year normalization or stabilization is reached. According to Gaucher (1982), bone scans do enable practitioners to follow the evolution of RSD, because the reduction in hyperfixation is a precursor of recovery. Nevertheless, it can not be used for staging. The use of MR imaging of RSD is still a matter of debate (Koch et al., 1991).

PHYSIOPATHOLOGY

The local reflex theory was put forward by Leriche and his co-workers (1924, 1926, 1927, 1928, 1932, 1941) on the basis of the following anatomical data. It is assumed that afferent sympathetic impulses from the traumatized tissue are send to the spinal cord, where they are controlled by higher central paths; the reflex arc to the efferent sympathetic axons is regulated here in the internuncial pool. The theory states that in cases of RSD, there are too much afferent sympathetic impulses, which are relayed through the internuncial pool to the efferent axons of the autonomic nervous system and are insufficiently suppressed. The phenomena of hyperexcitation in a closed circuit by intercalary neurons, and of after-discharge play a role in this process. Hyperexcitation also occurs in the nuclei of the spinothalamic tract via the internuncial pool. The increase in incoming pain impulses causes derangement of the higher centres which normally control the reflex arc in the medullary levels, thus contributing to the process of the pathological algodystrophic reflex.
These disturbances give rise to hyperexcitability of the efferent autonomic axons.

Experiments by Jänig and other authors support Leriche's theory. According to Jänig (1985), the initial trauma causes damage to the afferent neurons, resulting in abnormal activity in the primary afferent neurons (C-nociceptors), with myelinated and non-myelinated axons, as a consequence of abnormal chemosensitivity and mechano-sensitivity. This abnormal behaviour sets the reflex arc in action. The smooth muscle cells in the walls of the blood vessels are innerved by autonomic axons (Serratrice and Eisinger, 1967). This is one of the systems regulating the local blood flow by means of variable degrees of contraction in the different parts of the vessel. In cases of RSD it is believed that a functional disorder of the microcirculation occurs. Anatomical studies (Ravault et al., 1945) have demonstrated the existence of metarterioles with their precapillary sphincters from which blood is distributed around the capillary bed, before draining back into the venules. There is also a description of a direct link between arteries and veins in the form of Sucquet's canal, which is regulated by Masson's glomi. For some time now, physiological studies have demonstrated the role of various humoral factors in the regulation of the vascularization of tissues (noradrenaline, adrenaline, histamine, bradykinin, serotonin, etc.). The C-nociceptors in the skin, when mechanically stimulated, sensitize the wide dynamic range neurons (WDR) in the cornua of the spinal cord. The WDR neurons are to be held responsible for perception of pain. In the event of continous sensitization by the C-nociceptors, these neurons become more sensitive to both A- and C-neurons, activating pain perception and causing reflex excitation of the motoneurons and the efferent sympathetic neurons.

The efferent sympathetic neurons are believed to excite the afferent terminal in its turn, and to stimulate the reflex arc. This is done by means of vaso-motor disturbances, changes in vascular permeability, modification of tissue metabolites, a direct recruitment mechanism, etc. These effects are believed to be induced by various neuropeptides (tachykinins). (Blumberg et al., 1985; Bueno et al., 1986; Burnstock, 1985; Minvielle et al., 1986; Roberts, 1985; Sigrist et al., 1986; Van Zwieten, 1985; Drummond et al., 1991). Neuropeptide substance P (SP) is widespread in the central and peripheral nervous system, where it plays an essential part in a number of physiological processes, such as pain transmission and neurogenic inflammation. Various new tachykinins have been discovered with a structure analogous to that of SP: these include neurokinin A (NKA), neurokinin B (NKB) and neuropeptide K (NPK). A number of tachykinin receptors have also been found, which appear to be specific to SP, NKA and NKB. Primary C-axons contain a number of peptides which act as transmitters or mediators. In addition, some of these C-axons contain a calcitonin gene related peptide (CGRP) (Minvielle et al., 1986). It has been demonstrated that CGRP is the most important vasodilative peptide; SP, on the other hand, is the most important permeability mediator. However, the precise role of these tachykinins, or neuropeptides, in the nociceptive transmission of autonomic reflexes must be further investigated. Tachykinin antagonists, which were developed only recently, could well play an important role in the treatment of neurogenic inflammation or of RSD. Most authors agree today that hemodynamic abnormalities resulting from a disturbance of the autonomic innervation of the blood vessels are the cause of algodystrophic symptoms.

According to Belenger (1956) the RSD process is divided into two phases: a brief initial phase of vasodilatation, followed by a second phase in which the metarterioles and venules close, so that the main blood flow is through Sucquet's canal, and, as a result of the venous spasm, a reflex occurs, with filling of the capillaries, capillary overpressure, increased permeability, hemostasis, extravazation and formation of oedema. The third phase consists of a global atony of the smooth muscle cells. These hypotheses are supported by the observations of various authors: anatomo-pathological work has in fact shown that signs of capillary dilatation and hemostasis occur in cases of RSD, while other experiments (Bolliger, 1954) point to an increased concentration of oxygen in venous blood, which would appear to confirm the existence of arteriovenous bypasses (via Sucquet's canal) (Renier and Masson, 1987). Ficat and Arlet, who also performed oxygen measurements, found intraosteal hypersaturation of oxygen in RSD.

According to Ficat (1973) the key to the vicious circle of algodystrophy is to be found in "capillary stasis". This phenomenon leads to excess pressure and increased permeability, as well as to ischemia and localized acidosis, which in turn gives rise to a further increase in permeability. This results in oedema, progressive hyperaemia, and intra-osteal overpressure. The latter phenomenon has in fact been experimentally demonstrated.

However, these hypotheses are not consistent with the results of the dynamic isotopic studies (Gaucher et al., 1985; Kienboeck, 1901; Renier et al., 1979; Vattimo et al.) which suggest in some cases of RSD an increase of bloodflow and not a decrease.

VASCULAR SCINTIGRAPHY WITH HUMAN SERUM ALBUMIN

A very great number of treatment protocols have been used for RSD. Most of these include vasoactive substances. The stage of the disease is rarely used to determine the correct treatment. Therefore, the success rate of all these forms of therapy is variable and not very high. Clinical signs point to serious disorders in the vascularization of the affected part of the body and to the opposite finding in the hypertrophic and atrophic stages. A method has been developed to obtain vascular parameters in the form of a dynamic vascular scintigraphy with marked human serum albumin.

Originally, Renier (1979) published work on a threefold scintigraphy in a case of RSD. He used ^{99m}Tc-pyrophosphate for the conventional bone scan, ^{99m}Tc- blood corpuscles for detection of vascular volume, ^{99m}Tc-pertechnetate to study the interstitial compartment. Using this dynamic scintigraphy technique, he obtained an activity curve consisting of an entry phase and a plateau phase. The results of this method in 20 cases of RSD showed that there was no difference between the blood flow on the affected side and that on the normal side. The plateau curve however showed an average rise of 40 % on the affected side with respect to the normal side, pointing to increased vascular volume. In four of these cases, Renier (1979) also calculated the blood flowrate. He found a significant reduction in blood flowrate on the affected side with respect to the normal side. His conclusion was that "circulatory stasis" occurs in RSD, and that this circulatory stasis is characterized by increased vascular volume and reduced flowrate. These findings were found to be consistent with the observations of Arlet and Ficat (1982) in transosseous phlebographic intravenous pressure measurements.

Gaucher (1982) obtained the same findings with regard to the increased volume of blood (vascular scan with ^{99m}Tc-diphosphonate); however, he also found increased arterial activity during the first entry phase. His conclusion, that there was increased blood flow rate on the affected side though contradicts the findings of Renier. Some other authors reported on vascular changes but most of them used the early vascular inflow curves of bone scintigraphy (three phase bone scan). However, it was shown that this method is not a good choice for the staging and monitoring of RSD (Blockx et al., 1989; Driessens et al., 1990).

Therefore a more suitable vascular scan has been developed. In this method ^{99m}Tc-Human Serum Albumin (HSA) is injected intravenously into a non-affected part of the body (at the elbow in the case of RSD of a lower limb, at the foot in the case of RSD of an upper limb) (Blockx and Driessens, 1991; Driessens et al., 1984). The isotope radiations are measured using a gamma camera connected to a computer. These measurements are made simultaneously on the affected and non-affected limbs. The measurements at the healthy parts act as reference values or normal values. The regions of interest (ROIs) are delimited on a computer screen around the algodystrophic area and on the corresponding heterolateral area.

A time activity curve is calculated for both ROIs with corrections for differences in surface area. This curve, which is obtained after intravenous injection of the tracer, can be divided into three phases (Figure 7): 1° A vascular inflow phase with an average duration of three minutes; 2° An intermediate phase, with an average duration from the fourth to the tenth minute, in which tracer activity increases progressively towards equilibrium and plateau formation, as a result of the progressive homogeneous distribution of the tracer in the vascular compartment; 3° An equilibrium phase, reached after a maximum of ten minutes, in which tracer activity remains constant, as a result of the homogeneous distribution of the tracer, and which is measured for a duration of ten minutes.

In view of the fact that the tracer used is not metabolized or absorbed by bone tissue, but remains in the blood system for the duration of the investigation, this method yields measurable hemodynamic data, allowing comparison between the corresponding ROIs on the right or left. During the equilibrium phase the tracer is homogeneously mixed into the vascular compartment and a number of counts is directly proportional to blood volume. During the inflow phase, the homogeneous mix of tracer elements has not yet occured and the number of recorded counts in this phase shows the relationship between flowrate in the two areas investigated. The inflow phase is therefore a reflection of blood flowrate, hence of vascular resistance, while the equilibrium phase corresponds to the total quantity of blood.

It is clear that in healthy individuals the curves are more or less symmetrical. The blood flowrate in both limbs or ROIs is comparable, as is the quantity of blood present. The ratio between left and right never exceeds 112 % or falls below 88 %. In the early stage of RSD a very significant increase of the P/N ratio is observed, both in the inflow phase and the equilibrium phase (Figure 8). This means that both the local blood flow and the quantity of blood is higher on the affected side. The P/N ratio in the affected limb during the inflow phase increased by a significantly greater amount than during the

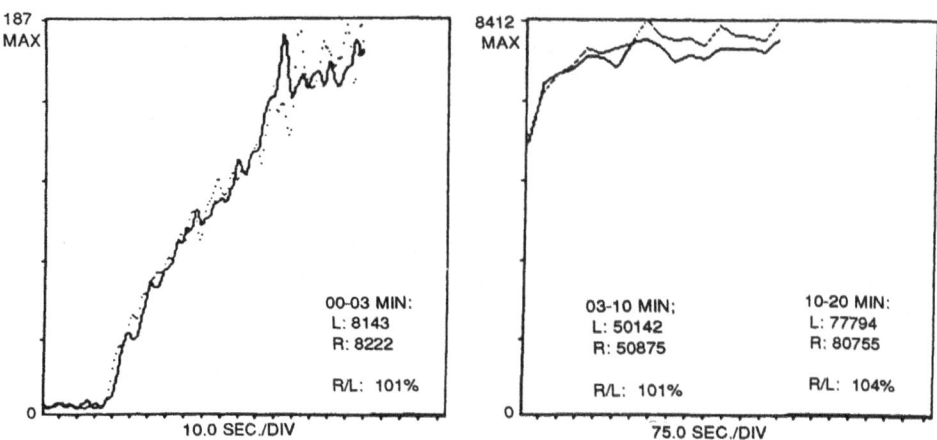

Figure 7. Normal vascular scintigraphy with human serum albumin.

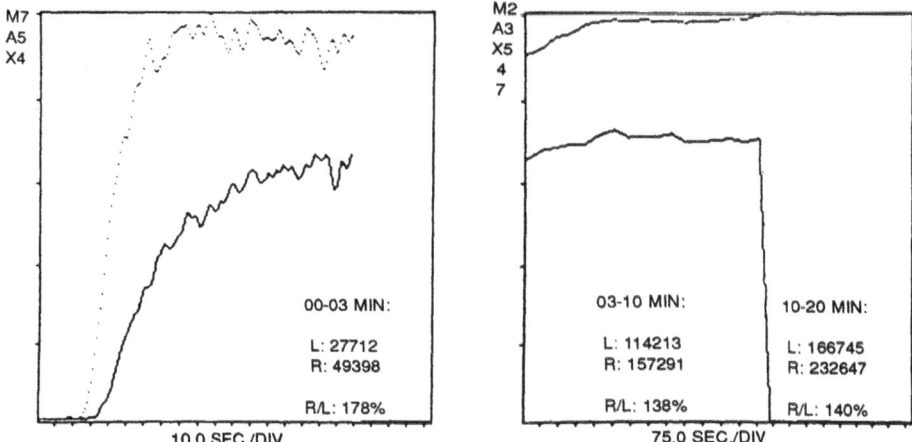

Figure 8. Vascular scintigraphy in RSD stage I.

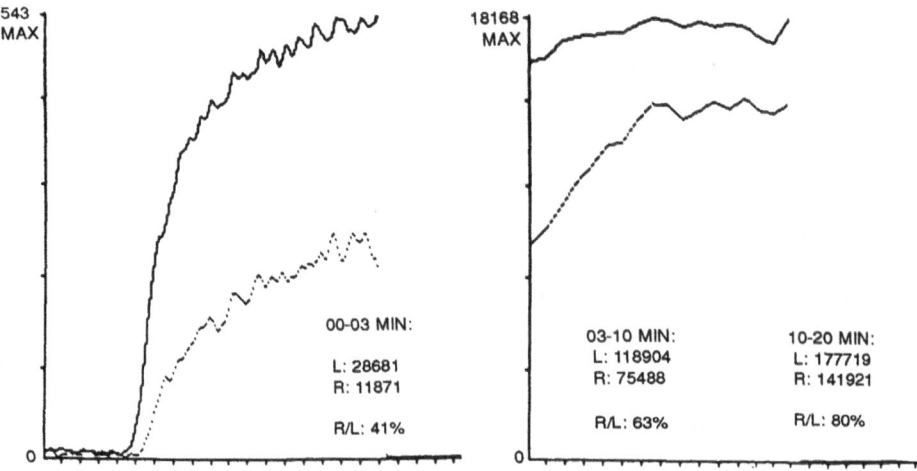

Figure 9. Vascular scintigraphy in RSD stage II.

equilibrium phase. In addition, the curve generally begins at an earlier stage in the areas affected by RSD than on the healthy side. The higher P/N ratio during the inflow phase therefore reflects not only an increase in blood flowrate as a result of dilatation - if that were the case the P/N ratios would only increase by the same amount as during the equilibrium phase - but also a higher blood flow. These observations therefore indicate that in stage I of RSD there is a condition of reduced peripheral resistance in the affected tissue, vaso-dilatation, increased blood volume and increased blood flowrate.

In stage II of RSD the findings are precisely the opposite to those of stage I. The P/N ratio is reduced both during the inflow phase and during the equilibrium phase (Figure 9). The decrease is more pronounced and more significant during the inflow phase. A late initiation of tracer activity is frequently observed, so that the curve in the affected area starts later. These findings indicate in stage II a reduced quantity of blood is present as compared to the healthy side and that the vascular resistance has increased, resulting in reduced blood flowrate.

The results of the vascular scan can be of great importance in the choice of treatment. In view of the opposite hemodynamic characteristics between the early and late stages of algodystrophy, the correct administration of vasoactive substances must be carefully considered and tested at all stages of the disease. Drugs with vasoconstrictive effects are not recommended for stage II and in fact their use should be discouraged. Conversely, vaso-dilative drugs, which cause a reduction in peripheral resistance, are not appropriate to the treatment system for stage I. In this stage there is pathological vaso-dilatation. The unjustified administration of these medications is therefore the most likely cause of the variable results in many studies. The HSA vascular scintigraphy provides good parameters for the evaluation of the current hemodynamic situation in the affected limb, and therefore for identification of the stage of the disease. The most appropriate treatment method can be deduced from these conclusions. The method is very simple to apply and involves no discomfort for the patient.

The findings of these investigations do not agree with the general assumptions concerning the pathogenesis of RSD, which may however not be totally eliminated. The classical hypothesis proposes the following reasoning: after an abnormal constriction of the metarterioles in the proximal bone blood vessels of the capillary bed through excitation of the sympathetic nervous system, a secondary by-pass occurs via Sucquet's canals, causing retrograde filling via the venous system to the capillaries. Stasis then occurs, resulting in hypoxia and development of the classic characteristics of Südeck's atrophia (1987). In these circumstances however, peripheral resistance would not be expected to fall or blood flow rate to rise as they have been shown to do on vascular scans. If the increased P/N ratio in the inflow phase is purely the result of increased filling of the capillaries (increased local plasma volume with possible stasis), a proportional increase should be found in the equilibrium phase. This is not however the case. The P/N ratio in the inflow phase is significantly higher than in the equilibrium phase. This points to increased local blood flow rate as does the premature start of the curve. It might still be argued that all this happened through the large-scale bloodpooling via Sucquet's canal, which overcompensates for the increased resistance of the metarterioles, resulting in a global fall in resistance. If this were the case, however, the only stasis that would occur in the capillary bed would be stasis of very slowly exchangeable blood. In this steady state the tracer molecules of the vascular scan would only be able to reach this area very late or in very small quantities as a result of progressive exchange in a latent equilibrium phase, but certainly not within the 20 minutes duration of the investigation. They would therefore disappear into the venous circulation without being able to cause the observed increase in plasma volume. The vascular scan would then show an increased inflow curve, but not an increased equilibrium phase (blood volume), unless of course the bloodflow to the soft bone tissues (pseudo-inflammatory characteristics) generates a surplus of counts with respect to the circulation in the bone, thereby influencing the vascular scan. This however would mean that the effect of calcitonin was not consistent with the observed results.

In the experimental animal, calcitonin caused prolonged vasoconstriction in the bone blood vessels (Driessens and Vanhoutte, 1981). Moreover it is well-known that this hormone causes flushing in man, in other words, brief vasodilatation of the blood-vessels of the skin. The findings with the vascular scan before and after treatment with calcitonin, which demonstrated a reduction of blood volume and vascular dilatation in the affected area, suggest that the situation in the soft tissues must be considered as an epiphenomenon while the physiophathological processes in the blood vessels must be accepted as decisive during the evolution of RSD and its treatment with calcitonin.

The recent developments involving tachykinines and neuropeptides in neurogenous inflammation do not fit in with the former hypothesis either; it has recently been shown that calcitonin gene related peptide (CGRP) is responsible for the pathological dilatation in the affected area which is consistent with the findings of the vascular scan but by no means consistent with the hypothetical premature constriction of the metarterioles which would in fact lead to increased peripheral resistance. This phenomenon was not demonstrated by the vascular scan in stage I; in fact the opposite was suggested.

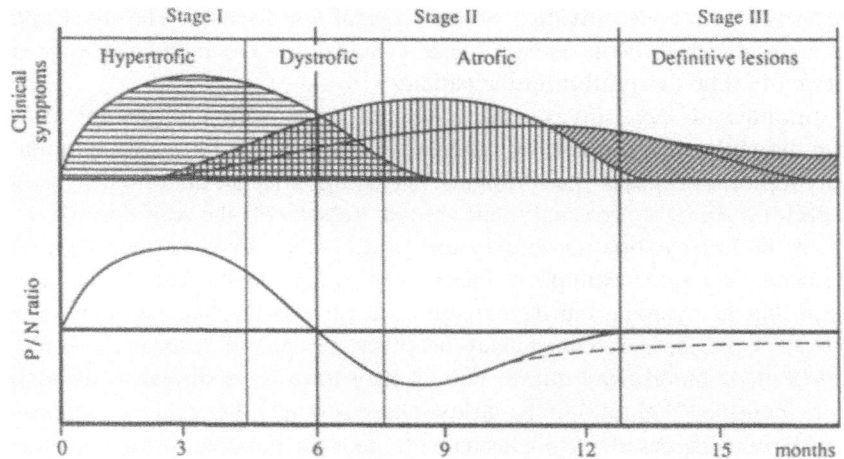

Figure 10. Clinical symptoms and vascular parameters (Pathologic/Normal ratio) during normal evolution of RSD (untreated).

The hemodynamic abnormalities observed in stage I of RSD do not therefore fit in with the generally accepted hypothesis of the disease. This earlier hypothesis is in closer agreement with the findings for stage II. In this phase the vascular scan showed up a circulatory situation which fits in with the current hypothesis; increased resistance (constriction of the metarterioles) and reduced blood volume associated with the classic symptoms of Südeck's atrophy (stages II or III). It is likely that the former hypothesis was expressed essentially on the basis of a disproportionately large number of stage II cases among the populations investigated. At that time diagnosis was generally made late on the basis of the typical clinical characteristics. Stage I was considered to be a posttraumatic arthritis and not recognized as stage I of RSD. It is not yet known with certainty how the transition from stage I to stage II takes place. There can be no question of capillary stasis initially. Previously there was no question of the occurrence of the rapidly exchangeable compartment, but this is in fact the case (as shown by the vascular scan).

Oedema formation does occur and can be attributed to the release of neurokinine A, resulting in an increased permeability of the capillaries. This may be what gives rise to the further evolution of the disease suggesting the disturbed diffusion of the necessary elements to the surrounding cells. This would eventually lead to a situation of hypoxia amongst other phenomena, and an accumulation of toxic metabolites in the interstices which would allow the typical symptoms of stage II to express themselves.

CALCITONIN AND RSD

In earlier studies an attempt was made to investigate the influence of calcitonin on vascular alterations in RSD. First of all the possible vascular effect of calcitonin on the bone circulation had to be investigated. The isolated tibia of the dog is an excellent in vitro preparation for evaluating the effects of drugs, particularly of calcitonin, on the tone of the bone blood vessels. It has been shown that calcitonin has a pronounced dose-dependent vasoconstrictive effect. The administration of calcitonin resulted in a global increase in vascular resistance, which, unlike other drugs, (e.g. noradrenaline) is not of limited duration, but is sustained (Driessens and Vanhoutte, 1981).

As a result of calcitonin treatment, relatively rapid improvement was obtained in the vascular alterations of stage I (Driessens et al., 1984). The pathologically high curves reached normal or, more frequently, limit normal values after an average of 6 weeks, coinciding with a clinical improvement or total cure. Slightly increased vascularization often continues for some time, but eventually normalizes after a few months. If appropriate treatment is administered, early evolution into stage II is seldom if ever observed. This is in contrast with the spontaneous evolution of RSD, which almost invariably progresses to stage II.

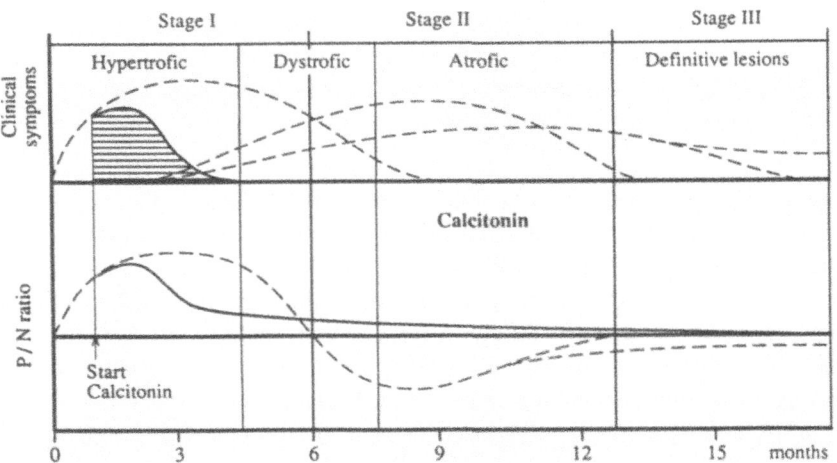

Figure 11. The effect of Calcitonin on clinical evolution and vascularization in RSD stage I.

227

The top part of figure 10 shows the spontaneous evolution of RSD in its most characteristic form, as regards clinical abnormalities. The lower part of figure 10 shows the spontaneous evolution of vascular parameters (blood flow-rate and volume). This shows the dynamic course of all these characteristics. In figure 11 the effect of calcitonin on clinical symptoms and vascular symptoms is shown. Clinical symptoms disappear soon, when treatment is started early. The vascularization evolves to normal values. Evolution into stage II does not occur in most cases (Driessens, 1984). It can therefore be stated that calcitonin is the suitable therapy of first choice in RSD stage I.

PSEUDODYSTROPHY

In some cases of so called reflex sympathetic dystrophy (RSD), the beginning of the disease appears to demonstrate in some patients typical signs of stage II (cyanosis, coldness, atrophy, pain and extreme loss of function) (Figure 12). French authors used to call it "les formes froides" of RSD (cold onset RSD) (Chester, 1990). However, the normal evolution of RSD usually shows an initial phase of pseudo-inflammation with warmness, hypersudation, oedema, redness and pain (stage I). The duration of this stage ranges between 2 and 8 months. On the contrary, typical symptoms of stage I of RSD cannot be found in the disease mentioned here. Any difference between this syndrome and RSD was never recognized (Wilder et al., 1992; Williame and Sand, 1991).

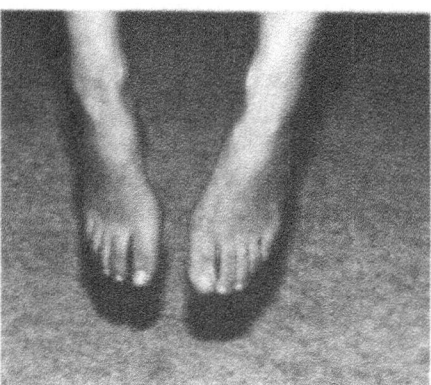

Figure 12. Clinical view of pseudodystrophy (Right foot, on the left).

Nevertheless, there are some important differences which lead to the exclusion of the diagnosis of RSD and suggest the existence of another disease. We would therefore suggest to call it pseudo-dystrophy. First of all, only children, adolescents and young adults seem to be affected (Dietz et al., 1990; Wilder et al., 1992) while the age distribution in RSD patients is totally different and include all ages of the adult population. Secondly, the diagnosis of RSD is made by a positive bone scintigraphy: typically a hyperfixation of bone tracer has to be found in the affected limb. This is not the case in pseudo-dystrophy: here normal tracer fixation or even a hypofixation is registered from the beginning in most patients. Hyperfixation is never found. This phenomenon reflects a situation of hypovascularization of the affected limb due to disuse, the tracer fixation on bone scan is indeed in relation to local blood flow.

The disorder seems to be always induced by serious psychosocial problems (of different kind), followed by psychosomatisation and loss of use of one limb. Most frequently one lower limb gets to be affected by active disuse. In some cases the upper limb is involved.

Patients fail to use the arm or leg and take care of it as if it was a foreign limb, resulting in total inactivity.

Radiological findings are nihil in most of the cases of early stages of the disease; later on some patchy or diffuse osteoporosis can develop.

The treatment of this disease must be totally different from RSD. It should include intensive physiotherapy, vasodilating agents, psychotherapy and social management.

The use of calcitonin is contra-indicated, because of its vasoconstrictive effect in bone.

REFERENCES

Arlet, J., P. Ficat.: Phlébographie transosseuse, pression intramédullaire et oxymétrie du sang osseux au cours des algodystrophies sympathiques réflexes. Revue du Rhumatisme 12: 883-885 (1982).

Belenger, M.: Aspects cliniques et méthodes thérapeutiques dans l'ostéoporose post-traumatiques des membres inférieurs. Xe Cong. Belg. Chir., 1956, Acta Chir. Belg., suppl.1, pp. 219 (1956).

Blockx, P., M. Driessens: The use of 99 Tcm -HSA dynamic vascular examination in the staging and therapy monitoring of reflex sympathetic dystrophy. Nucl.Med. 12: 725-731 (1991).

Blockx, P., M. Driessens, G. Mortier, A. Vrancken: Pitfalls of three phase bone scintigraphy in the staging of reflex sympathetic dystrophy. Trends and Possibilities. In: H.A.E. Schmidt & G.L. Buraggi (eds): Nuclear Medicine, pp. 463 - 466. Schattauer, Stuttgart (ISBN 3-7945-1305-3) 1989.

Blumberg, H., J. Habler., M. Hollerbach: Microneurographic studies in humans on sympathetic nervous system and pain. Séminaire Européen, Strasbourg, mai 1985.

Bolliger: Réactions vasculaires dans la genèse du syndrome de Südeck. Helv. Chir. Acta, 21: 61 (1954).

Bueno, L., M.J. Fargeas, P. Julie: Effects of calcitonin and CRGP alone or in combination on food intake and forestomach (reticulum) motility in sheep. Physiol. Behav. 36: 907-911 (1986).

Burnstock, G.: Sympathetic neurotransmission and Pain. Séminaire Européen, Strasbourg, 1985.

Charkes, J.: Skeletal blood flow: Implications for Bone-Scan Interpretation. Nuc.Med. 21, n°1: 91- 98 (1980).

Chester, M.H.:Segmental manifestation of reflex sympathetic dystrophy syndrome limited to one finger. Anesthesiology 73 (3): 558- 561 (1990).

Cuartero-Plaza, A., E. Martinez-Miralles, P. Benito-Ruiz, S. Martinez-Pardo, M.P. Sanz Marin: Abnormal bone scintigraphy and silent radiography in localized reflex sympathetic dystrophy syndrome. Eur. J. Nucl. Med. 19 (5): 330-333 (1992).

Darnault, A., G. Breton, J. Carzon, A. Courtillon, F. Daniel, J.N. Heuleu: La radiographie et la scintigraphie dans les algodystrophies récentes post-traumatiques ou post-opératoires du membre inférieur. In: "Les algo-dystrophies sympathiques réflexes", pp.42-46 (1987).

Dietz, F.R., K.D. Mathews, W.J. Montgomery: Reflex sympathetic dystrophy in children. Clin. Orthop. 258: 225-231 (1990).

Doury, P., R.P. Delahaye, R. Granier, S. Pattin, P.J. Metges, F. Fabresse: L'algodystrophie parcellaire. Revue du Rhumatisme 46: 37-44 (1979).

Driessens, M., P.M. Vanhoutte: Effect of calcitonin, parathormone and hydrocortisone on vascular reactivity of the isolated tibia of the dog. Am. J. Physiol. 241: H91-H94 (1981).

Driessens, M., P. Blockx, G. Mortier, J. De Bruyne: Evaluation of the vascular effect of calcitonin treatment in reflex sympathetic osteodystrophy. Bone Circulation, Chapter 60, 367-370 (1984).

Driessens, M., P. Blockx, G. Mortier, A. De Ridder, H. Dijs: Early bone tracer inflow curves cannot be used for staging of reflex sympathetic dystrophy. Bone Circulation and Bone Necrosis (ISBN 3-540-19220-4), Springer Verlag 1990.

Drummond, P.D., P.M. Finch, G.A. Smythe: Reflex sympathetic dystrophy: the significance of differing plasma catecholamine concentrations in affected and unaffected limbs. Brain 114: 2025-2036 (1991).

Ficat, F., J. Arlet, G. Lartique, M. Pujol, M.-A.: Algo-dystrophies réflexes post-traumatiques. Etude hémodyamique et anatomopathologique. Revue de Chirurgie Orthopédique 59: 401-141 (1973).

Gaucher, A., A. Bertrand, F. Tonnel, C. Hocquard, J. Adolphe, P. Pere.: L'angioscintigraphie osseuse. Revue du Rhumatisme 52: 701-705 (1985).

Gaucher, A., P. Raul, P. Wiederkehr, J. Pourel, A. Bertrand, J.N. Colomb, D.Ethgen: Etude scintigraphique des algodystrophies réflexes. Revue du Rhumatisme 11: 841-846 (1982).

Goris, A., A. Kolkman, H. Leenen, T. Van Bebber, M. Corstens, A. Heerschap: Symptomatologie van posttraumatische dystrofie. Chirurgie: 165-177 (1988).

Janig, W.: The generation of reflex sympathetic dystrophy: a neuro-biological hypothesis. Seminaire Européen, Strasbourg , 1985.

Kienboeck: Uber die acute Knochenatrophie bei Euntzundungsprocessen an den Extremitaten und ihre Diagnose nach der Röntgen Bilde. Wien Med. Wschr. 51: 1345 (1901).

Kim, C., S. Millin, J. Park: Scintigraphic patterns of the reflex sympathetic dystrophy syndrome of the lower extremities. Clin. Nucl. Med. 14: 657-661 (1989).

Koch, E., H.O. Hofer, G. Sialer, B. Marincek, G.K. Von Schultess: Failure of MR imaging tot detect reflex sympathetic dystrophy of the extremities. Am. J. Roentgenol. 156 (1): 113-115 (1991).

Lavender, J., R. Khan, S. Hughes: Blood flow and tracer uptake in normal and abnormal Canine bone: comparisons with Sr-85 microspheres, KR 81-m and Tc-99m MDP. J. Nucl. Med. 20: 413-418 (1979).

Leriche, R.: Des réflexes d'axone dans les traumatismes périphériques. Rev. Chir. 43: 379 (1924).

Leriche, R.: Sur quelques malaises osseux et articulaires d'origine vasomotrice et sur leur traitement. Bull. Soc. Nat. Chir. Paris 24: 1022 (1927).

Leriche, L. De l'importance en pathologie des réactions vasomotrices post-traumatiques. La Médecine (1928).

Leriche, R.. Oedème dur aigu post-traumatique de la main avec impotence fonctionelle complète. Transformation soudaine cinq heures après sympathectomie humérale. Lyon Chir. 20: 746 (1932).

Leriche, R.: Déséquilibres vasomoteurs post-traumatiques primitifs des extrémités. Lyon Chir. 20: 746 (1932).

Leriche, R.: Traitement de l'ostéoporose algique post-traumatique. Presse Méd. 49: 609 (1941).

Leriche, R., J. Policard: Les problèmes de la physiologie normale et pathologique de l'os. Masson Edit. Paris (1926).

Lequesne, M., M. Kerboull, M. Bensasson, C. Perez, M. Dreiser, A. Forest: L'algodystrophie décalcifiante partielle. Revue du Rhumatisme 46: 111-121 (1979).

Minvielle, S., M. Cressent, F. Lasmoles, A. Julienne, G. Milhaud, M.S. Moukhtar: Isolation and partial characterization of the calcitonin gene in a lower vertebrate. Predicted structure of avian calcitonin generelated peptide. FEBS Lett. 203: 7-10 (1986).

Mortier, G., M.F. Driessens, J.J. De Bruyne, A. De Ridder: Reflex algodystrofie - kliniek en medicamenteuze behandeling. Med. Phys. 2: 89-94 (1985).

Ravault, P.P., P. Guinet, P. Carrier: Une cause d'erreur dans le diagnostic des polyarthrites: le rhumatisme neurotrophique du membre supérieur. Arch. Rhumat. 6: 129-134 (1945).

Renier, J.C., CH. Masson: Physiopathogénie des algodystrophies. In: L. Simon, Ch. Hérisson (eds): "Les algodystrophies sympathiques réflexes", pp.15 (1987).

Renier, J.C., R. Moreau, M. Bernat, M.Basle, P. Jallet, J.F. Minier: Apport des explorations isotopiques dynamiques dans l'étude des algodystrophies. Revue du Rhumatisme 46: 235-241 (1979).

Roberts, W.J.: Neuronal Model for sympathetically maintained pain. Séminaire Européen, Strasbourg, 1985.

Sagar, V., J. Piccone, N. Charkes: Studies of Skeletal Tracer Kinetics. III. Tc-99m(Sn) methylenediphosponate Uptake in the Canine Tibia as a Function of Blood Flow. J. Nucl. Med. 20: 1257-1261 (1979).

Serratrice, G., J. Eisinger: Innervation et circulation osseuses diaphysaires. Rev. du Rhumatisme 9: 505-519 (1967).

Sigrist, S., A. Franco-Cereceda, R. Muff, H. Henke, J.M. Lundberg, J.A. Fischer.: Specific receptor and cardiovascular effects of calcitonin generelated peptide. Endocrinology 119: 381-389 (1986).

Van Zwieten, P.A.: The receptors of the adrenergic nervous system. Division of Pharmacotherapy. University of Amsterdam: 1 (1985).

Vattimo, A., R. Nuti, F. Lore, A. Caniggia: Human Calcitonin therapy of Südeck's atrophy: pathophysiological aspects. In: "Human Calcitonin", p.127-137.

Wilder, R.T., C.B. Berde, M. Wolohan, M.A. Vieyra, B.J. Masek, J. Micheli: Reflex sympathetic dystrophy in children. Clinical characteristics and follow-up of seventy patients. J. Bone Joint Surg. 74 (6): 910-919 (1992).

Williame, L.M., A. Sand: Hypofixation on bone scintigraphy in reflex sympathetic dystrophy syndrome. Clin. Rheumatol. 10 (1): 73-75 (1991).

OSTEONECROSIS

ATRAUMATIC NECROSIS OF THE FEMORAL HEAD: GENERAL REPORT

Jacques ARLET

Service de Rhumatologie
CHU de Rangueil
Toulouse, France

INTRODUCTION

The necrosis or osteonecrosis (ON) of the femoral head consists in the death of the cells of the proximal end of the femur.

A bone as a femur is an organ in itself, made of two tissues: a soft tissue, the marrow with lipocytes, hemopoietic cells and vessels; a hard tissue, the trabeculae with their attached cells, osteoblasts, osteoclasts and osteocytes. Both tissues are closely linked together: for exemple, the formation of trabecular architecture follows the pattern of the sprouting capillaries (Burckhardt et al., 1984). Both tissues are simultaneously involved in the process of osteonecrosis. If we exclude the cases of ON secondary to severe trauma, infections, malignant diseases and roentgentherapy, the atraumatic necrosis of the femoral head (ONFH) are, for most of the authors, secondary to a reduction of the bone blood supply, an ischemia. In most of the cases, ONFH presents as a chronic and potentially reversible lesion, because of the extraordinary capacity of repair of the bone tissue, as it was demonstrated by experimental studies.

On a practical point of view, the disease progress in two fundamental periods. The first one is more or less silent without any X-ray abnormalities and is able to cure. The second one is characterized by the fact that an area (usually antero-superior) of the femoral head is totally and definitively dead without any possibility of spontaneous recovery: it is called the sequestrum. So the diagnosis and the treatment of ONFH have to be made at the time of the first period what is becoming easier through the use of MRI and core decompression. The disease is more frequent in males (about three times) and in the middle age of adult life, but it can be observed in children and aged people. Its frequency seems to be increasing since thirty years: In Japon, about 3,000 new cases of ONFH were observed in the sole year 1988 (Ninomiya, 1989).

Bone Circulation and Vascularization in Normal and Pathological Conditions
Edited by A. Schoutens *et al.*, Plenum Press, New York, 1993

235

ETIOLOGICAL FACTORS

It is convenient to classify the etiological forms of ONFH in three groups.

1st group. ONFH recognized as definite complications of three specific diseases: Gaucher's disease, which is rare; Sickle cell anemia which is very frequent in groups of black people: both are inherited; Caisson's disease, an acquired and professional pathology in tunnel workers and divers, less frequent since preventive measures were carried out.

2nd group. Toxic ONFH: Two toxics are considered by the majority of the experts as playing an important role in the frequency and the spreading out of the disease: Corticosteroid treatment and alcoholism. Corticosteroid treatment was part of the clinical picture in 25 to 50 % of the cases according to the series: toxicity is related to the dose, the majority of the patients are females, bilaterality and multiple sites of ON are often noted in the same patient. Alcoholism would be responsible for 10 to 40 % of ONFH; a part of these patients have a involvement of the liver.

3nd group. Finally in most of the other cases, history and investigations reveals one or several of the following risk factors: obliterative arteritis, past history of severe phlebitis, pregnancy, micro-trauma to the hip (Post-contusive form), osteoporosis, hyperlipidemia, hyperuricemia (with or without Gout), hyperviscosity of the blood, clotting abnormalities, propensity to vascular thrombosis (eventually due to an anticardiolipid antibody).

CLINICO-RADIOLOGICAL EVOLUTION

For reason of simplicity we will describe it in four stages since the first symptom.

Stage I or preradiological stage. In many cases, if we do not neglect symptomatic patients because of normal X-rays, we can observe the first clinical and preradiological stage. The patient complains of moderate but often nocturnal pain in the groin, the buttock and/or the thigh. Minutious physical examination reveals slight but definite painful stiffness of the hip. In our experience (Ficat and Arlet, 1980) those two signs, the one functional and the other physical, if they are associated with normal X-rays (frontal and profile), make probable the diagnosis of ONFH in the absence of history of metastatic carcinoma or tendinitis, and still more in the presence of risk factors of ON. The diagnosis is almost certain if the other hip has a stage III necrosis.

Stage II. The patient has the same symptoms and signs as in the stage I, but there are some X-rays abnormalities inside the femoral head: areas of hyperclarity (sometimes organized in a pseudocyst), areas of hyperdensity (often concave and in the center of the head) more or less associated in a mottled image. However X-rays show perfect preservation of the outline of the head and the joint space is normal or slightly narrowed.

Stage III. That is the "no return point". X-rays show either in frontal or in profile wiews or both, a subchondral fissure or crack, actually a fracture, usually situated in the weight-bearing area, with or without collapse. The earlier image is called the crescent sign: it is definitively an image of fracture without deformation.

These images with a good joint space are characteristic of an advanced, we would say complicated ONFH. Unfortunately it often happens that the diagnosis is only made at that stage, either because the patient did not complain earlier or because the doctor was falsely reassured by normal X-rays.

Stage IV. At that stage ON is complicated with secondary osteoarthritis, i.e. narrowing of the joint space, osteophytes and involvement of the acetabulum.

The time elapsed between the diagnosis and a severe deterioration leading to major surgery was about three years in 50 % of the cases. On conservative treatment 81 % of the stages I and II cases are worsening (Steinberg et al., 1984). At the time of the first examination both hips are involved in 30 to 70 % of the cases.

DIAGNOSTIC PROCEDURES

In order to make the diagnosis as early as possible, one can propose the following rules:

1. To suspect a painful but radiologically normal hip.
2. Take into account the risk factors, the patient at risk.
3. Ask for an MRI. In more than 90 % of the stages I and II cases, MRI give significant images, i.e. an area of low intensity signal on T1 weighted images, more or less large in the head of the femur, either oval and subchondral, or band like through the head (Mitchell et al., 1987), more rarely diffuse. The same area give a low or intermediate intensity signal on T2 weighted images, contrary to what we observed in the cases of transient osteoporosis of the hip where the low intensity signal area on T1 has a normal or even high intensity signal on T2. That is a very important point in matter of differential diagnosis.
4. If MRI is normal or not significant, it is justified to ask for a computed tomography (CT) and a bone marrow pressure measurement (IMP). In some early cases, CT shows abnormal architecture of the trabeculae inside the head (asterisk sign described by Dihlmann, 1982). IMP is abnormally high (more than 30 mmHg) specially after injection of saline (stress test) in most of the early cases (Ficat and Arlet, 1980; Zizic et al., 1986). IMP determination has to be done with a special needle introduced through the cortex of the great trochanter after local anesthesia.
Bone scintigraphy with 99m technetium MDP is definitively less sensitive than MRI in precollapse stages.
5. Finally most of the authors think that biopsy can assure the diagnosis in the absence of others proofs. It is made with a special needle, either of little calibre (3mm) (Hauzeur et al., 1986) or bigger, as the Ficat's trephine (8 to 10 mm) (Ficat and Arlet, 1980). The latter give better specimens and provokes a therapeutic decompression (Ficat et al., 1972).
6. In stages III and IV, decisions concerning the choice and the modalities of a major surgery have to be based on an appreciation of the precise site and size of the lesion. It will be made easily according to the method of examination and staging proposed and controlled by Steinberg et al. (1984). Their method of staging in six stages has been adopted by the ARCO association. It would permit not only better surgical decisions but also to establish a prognosis.

HISTOPATHOLOGY

1. Advanced cases. They were perfectly described by many authors. In these cases the femoral head presents, in the weight-bearing zone, a conic or oval shaped area of totally dead bone, without any living cell and with desorganized trabeculae and fatty tissue, eventually replaced by granular, reticular or amorphous necrosis. It is the sequestrum, in which there is a crack, an osteochondral fracture usually situated into the subchondral plate. The osteo-chondral proximal fragment is more or less displaced to the center of the

head, realizing a collapse, a flattening of the femoral head contour. The above cartilage is (Stage IV) or is not (Stage III) deteriorated. Underneath the sequestrum, there is a transitional and repairing zone, made of thickened trabeculae and of marrow spaces filled with fibrosis and new vessels.

It is worth to add that in the areas of the inferior part of the head and in the neck, underneath the transitional zone, we observed in many cases scattered and spotted abnormalities in the marrow spaces: fat necrosis, hemorrhages, oedema, fibrosis. Our observations were confirmed by Hauzeur and Pasteels (1990). They probably are early and more or less repaired ischemic lesions.

2. Early lesions. They can only be observed in specimens extracted by biopsy. We (Arlet et al., 1984), Belgian (Hauzeur et al., 1986) and Japanese (Saito et al., 1987) authors described them. They specially involved the tissues of the bone marrow which are experimentally the first to suffer from the ischemia (Rutishauser and Taillard, 1966) and where the lesions are easier to observe than the lesions of the hard tissue. We classified our specimens in four types: Type 1. Spots of fibrosis, necrosis, stasis and hemorrhages surrounded by areas of apparently normal fatty marrow. Type 2. Extensive necrosis of the hemopoietic and fatty tissues of the bone marrow, occupying most of the spaces. It often presents as a fine eosinophilic network with the H.E. Staining. Type 3. The above marrow lesions are accompanied by trabeculae deprived from their osteocytes. We considered, by comparison with controls, that deprivation is abnormal when at least 50 % of the osteocytic lacunae are empty. Emptying of the lacunae is the first clear sign of osteocytic necrosis that we can observe on light microscope. Type 4. Signs of repair similar to those observed in the transitional zone underneath the sequestrum in advanced cases. We estimated that the types 2 and 3 are consistant with the diagnostic of ONFH.

PHYSIOPATHOLOGY AND PATHOGENESIS

As we said, most of the authors think that ONFH is secondary to ischemia. However the experimental and clinical studies by Kawai et al. (1983), present an alternative physiopathology based on the primary fat loading and fat necrosis of the osteocytes.

We still think that in the first etiological group of ONFH the cause of the necrosis is a primary restriction of the blood supply in the capillary and/or sinusoidal circulations, either by extrinsic compression (Gaucher's and Sickle cell diseases) or by intrinsic obstruction with nitrogen bubbles (Caisson's disease). However it is very probable that in the latter example, the phenomenons are more complex with the implication of fat emboli, release of vasoactive amines, platelet aggregation and finally intravascular coagulation (Jones, 1993). This sequence of events intervenes also probably in the others etiological forms, but in those, what is the primary site of ischemia? Arterial obliteration, venous obstruction, capillary circulation disturbance? Arguments in favour of the arterial restriction do exist: Arteriosclerosis of the intraosseous arteries were frequently observed by Ramseier (1962), in a series of post mortem examinations; Arteriographic studies of ONFH, either in living patients, firstly presented by Theron (1977), or in surgically extracted femoral heads, show thinning and absence of arterial branches into the femoral head; ONFH can be due to obliterative arteritis of iliac artery; lesions of intraosseous arteries, specially ruptures of elastic membrane can be observed in cases of ONFH (Saito et al., 1987); In steroid treated collagenosis, specially in SLE, the frequency of ONFH increases with the severity of the vasculitis; Biologic abnormalities, as those presented by diabetic and arteriosclerotic patients, are often noted in ONFH patients (See the list of risk factors).

It is more difficult to argue that venous obstruction can be the primary phenomenon, even if perosseous phlebography of the proximal femur shows a definite slowness of the venous drainage in early cases of ONFH.

Capillary obliteration by fat emboli was observed in patients specimens, specially in the subchondral plate of the femoral head, by Jones (1965), and since by others authors and it really can be the first link of the chain. Capillary restriction can be due to a hypertrophy of the lipocytes, as demonstrated by Wang et al. (1977) in rabbits treated with high doses of cortisone. Finally compression of the vessels in the marrow spaces following an important and durable increase of the intramedullary pressure, could realize a sort of "compartment syndrome" with subsequent ischemia and necrosis (Ficat and Arlet, 1980).

I would say, in summary, that it is unprobable that there is an unique mechanism to explain all the different etiological forms of ONFH, but chronic ischemia is very probably the common denominator of most of them, whereas intravascular coagulation and thrombosis would be the final event. In many patients, several risk factors of vasculitis are found together. On another hand, abnormalities in the lipid metabolism and in the marrow fat tissue, can play an important role in the progress of the lesions.

REFERENCES

Arlet, J., R. Durroux, C. Fauchier, M. Thiechard: Histopathology of the nontraumatic necrosis of the femoral head: topographic and evolutive aspects. In: J. Arlet, P. Ficat, D. Hungerford (eds): Bone Circulation. Williams and Wilkins, Baltimore 1984.

Burkhardt, R., R. Bartl, B. Frish: The structural relationship of bone forming and endothelial cells of the bone marrow. In: J. Arlet, P. Ficat, D. Hungerford (eds): Bone Circulation. Williams and Wilkins, Baltimore 1984.

Dilhmann, W.: CT analysis of the upper end of the femur. The asterik sign and ischaemia bone necrosis of the femoral head, Skel. Radiol. 8: 251-258 (1982).

Ficat, P., J. Arlet, R. Vidal, A. Ricci, J.C. Fournial: Résultats thérapeutiques du forage-biopsie dans les ostéonécroses fémoro-capitales primitives (100 cas). Rev. Rhum. Mal. Ostéoart. 38: 269-276 (1972).

Ficat, P., J. Arlet: Ischemia and Necrosis of Bone, Williams and Wilkins, Baltimore 1980.

Hauzeur, J.P., S. Orloff, J. Taverne-Verbank, J.L. Pasteels: Diagnosis of aseptic necrosis of the femoral head by percutaneous transtrochanteric needle biopsy. Clin. Rhumatol. 5: 346 -358 (1986).

Hauzeur, J.P., J.L. Pasteels: Pathology of bone marrow distant from the sequestrum in nontraumatic aseptic necrosis of the femoral head. In: J. Arlet and B. Mazières (eds): Bone Circulation and Bone Necrosis. Springer-Verlag, Berlin 1990.

Jones, J.P.: Fat embolism as a possible mechanism producing avascular necrosis. Arthritis Rheum. 8: 449 (1965).

Jones, J.P.: Intravascular coagulation and osteonecrosis. Clin. Orthop. 277: 41 (1992).

Kawai, K., H. Maruno, K. Hirohata: Fat necrosis of osteocytes as a causative factor of idiopathic necrosis of the femoral head in man, ORS 29th annual meeting, Anaheim, California, Mar. 8-10, p. 263 (1983).

Mitchell, D.C., V.M. Rao, M.K. Dalinka: Avascular necrosis of the femoral head: Magnetic resonance imaging appearance correlated with radiographic staging and radionuclide, histologic and clinical findings. Radiology, 162: 709-715 (1987).

Ninomiya, S.: An epidemiological survey of idiopathic avascular necrosis of the femoral head in Japan. Annual report of Japanese Investigation Committee for intractable diseases. Ministry of Health and Welfare, Tokyo 1989.

Ramseier, E.: Untersuchungen über arteriosklerotische Veränderungen der Knochenarterien, Virchows Arch. Pathol. Anatom. 336: 77-86 (1962).

Rutishauser, E., W. Taillard: L'ischémie articulaire en pathologie humaine et expérimentale. Rev. Chir. Orthop. 52: 197-223 (1966).

Saito, S., A. Inoue, and K. Ono: Intramedullary hemorrhages as a possible cause of necrosis of the femoral head. The histology of 16 femoral heads at the silent stage. J. Bone Joint Surg. 69B: 346-351(1987).

Steinberg, M.E., G.D. Hayken, D.R. Steinberg: The conservative management of avascular necrosis of the femoral head. In: J. Arlet, P. Ficat and D. Hungerford (eds): Bone Circulation. Williams and Wilkins, Baltimore 1984.

Steinberg, M.E., G.D. Hayken, D.R. Steinberg: A new method for evaluation and staging of avascular necrosis of the femoral head. In : J. Arlet, P. Ficat and D. Hungerford (eds): Bone Circulation. Williams and Wilkins, Baltimore 1984.

Theron, J.: Superselective arteriography of the hip: technique, normal features and early results in idiopathic necrosis of the femoral head. Radiology. 124: 649-655 (1977).

Wang, G.J., D.E. Sweet, S.I. Reger, R.C. Thompson: Fat-cells changes as a mechanism of avascular necrosis of the femoral head in cortisone treated rabbits. J. Bone Joint Surg. 59A: 729-735 (1977).

Zizic, T.M., D.S. Marcoux, D.S. Hungerford, M.B. Stevens: The early diagnosis of ischemic necrosis of bone. Arthritis Rheum. 29: 1177-1186 (1986).

OSTEONECROSIS
1. General Aspects of Osteonecrosis

EPIDEMIOLOGY AND RISK FACTORS IN AVASCULAR OSTEONECROSIS OF THE FEMORAL HEAD

K. ONO* AND Y. SUGIOKA**

Osaka University Medical School *
Kyushu University **
Faculty of Medicine
Japan

OVERVIEW

Number of victims of nontraumatic avascular osteonecrosis of the femoral head (ONFH) is believed to have increased world widely. This is partly owing to an improved awareness of the condition and development of diagnostic imaging techniques sensitive to the condition. Yet a real increase in the disease burden is undeniable as seen in the figures from ONFH in Osaka University Hospital for the last 20 years (Figure 1). The recent increase in number of patients is remarkable, particularly in ONFH diagnosed as "steroid induced".

The Investigation Committee for ONFH (founded in 1974 under the auspices of the Ministry of Health and Welfare) have been collecting the data relating to ONFH epidemiology in order to promote etiological study and to develop preventive and therapeutic programmes for ONFH. Estimated number of the new patients in 1987 was 2,500 to 3,300 in Japan with 95 % confidence interval. This survey was carried out nation-widely with enthusiastic cooperation from 1,721 hospitals which reported 2,364 patients in 1987 including traumatic cases. Demographic data are obtained from this series and summarized in table 1.

In 31.4 % of the reported cases bilateral femoral heads were involved according to plain roentgen examination. Bilateral affliction were seen far often in steroid induced osteonecrosis. There was no obvious sex preponderance in the condition (ratio of males to females: 1.4). Female preponderance in collagen disease or autoimmune disease resulted in the same trend in steroid induced osteonecrosis but this seemed to be balanced, with a male preponderance in alcohol associated and in idiopathic osteonecrosis. As to the etiology of osteonecrosis, the majority of patients were documented as "of idiopathic nature". In this study, however, a nearly equal number of steroid induced osteonecrosis was discovered. We believe this is not due to a particular circumstance in Japan but to a common trend in developed countries. In the 1980s the majority of patients

Bone Circulation and Vascularization in Normal and Pathological Conditions
Edited by A. Schoutens *et al.*, Plenum Press, New York, 1993

243

Table 1. Investigation Committee for ONFH (Japan)

	Year	Patients	
Reported new patients with ONFH in each year	1984	1131	
	1885	1206	
	1986	1361	
	1987	2364	
Etiologies	Idiopathic	39.5 %	
	Steroid induced	37.1 %	
	Alcohol associated	23.3 %	
Surgical Series	Femoral Head Replacement	57.9 %	
1 403 joints, 1 230 patients	Total Joint Replacement	15.9 %	
(1987)	Rotation osteotomy (Sugioka)	10.0 %	
	Bone Grafting	7.5 %	
	Trepanation	2.0 %	
	Other Osteotomy	1.5 %	
	Miscellaneous procedures	5.2 %	

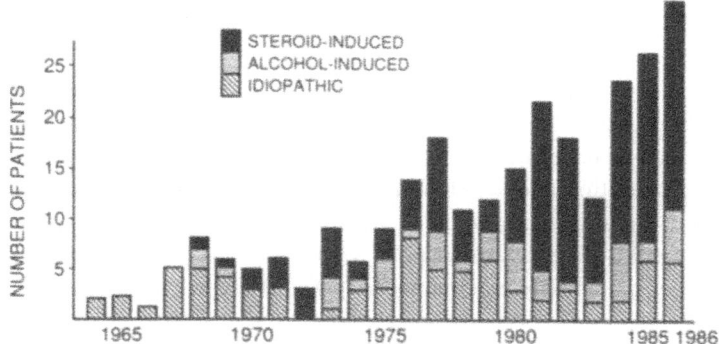

Figure 1. ONFH patients in Osaka University (284 cases).

were treated by femoral arthroplasty but now either this procedure or rotation osteotomy or bone grafting is selected in view of disease stage, type or bilateral involvement.

RISK FACTOR ANALYSIS

The Investigation Committee has undertaken two epidemiological studies for ONFH risk analysis, one in patients with systemic lupus erythematosus under high-dose corticosteroid therapy and the other in those patients with a regular alcohol intake and/or cigarette smoking (No steroid therapy). The former was carried out from 1984 to 1988 on prospective study basis and the latter from 1988 to 1990 on control study basis. We have identified a few risk factors related to steroid therapy and alcohol intake respectively. A similar study is being continued in the patients undergoing renal transplantation with steroid coverage.

Systemic Lupus Erythematosus (SLE)

Patients with SLE who received more than 30 mg/day prednisolone, for at least 30 days, were included in the study. Roentgenography, bone scintigraphy and laboratory examinations were carried out periodically before, during and after hypercorticosteroid therapy. Two hundred twelve patients entered the study and 62 patients completed this maximum five year study. In nine, ONFH developed within 640 days, on an average, since the introduction of high-dose corticosteroid therapy. To examine interdependence between ONFH affliction and quantitative or categorical data obtained from the subjects, various statistical methods were used. Thirty one patients receiving less than 29 mg/day prednisolone or nonsteroid antiinflammatory drugs or immune suppressor drugs, or both, were used as a control group.

Patients presenting with one or some clinical manifestations of alopecia, papulo-macular eruption, drug-induced lupus, lupus erythematosus cell positive rheumatoid arthritis, interstitial pneumonitis, and thrombocytopenic purpura, proved to be potentially at risk of ONFH.

A characteristic patterns of laboratory data abnormalities around the presumed onset of ONFH was regarded as increasing the risk of ONFH: Increased total cholesterol, glutamic oxaloacetic transaminase, glutamic pyruvic transaminase, alkaline phosphatase; increased red blood cell, hemoglobin, albumin/globulin; advanced renal failure, as reflected by increased blood urea nitrogen and creatinine.

A rash introduction of high-dose corticosteroid therapy (30 mg/day prednisolone or more) without corticosteroid preloading proved to be a risky mode of administration of the drug.

Regular Intake of Alcohol and/or Cigarette Smoking

Because alcohol intake or cigarette smoking or both constitute the necessaries of human life itself worldwidely, it has long been considered worthless to subject these habits separately to risk analysis. However, the reported enhanced ONFH risk of alcohol intake with cigarette smoking (Matsuo, 1988), urged our Committee to examine his results in a larger scale. The Committee has collected 118 patients with ONFH and well matched 236 controls for a case control study, 1988-1990, and elucidated a dose and duration dependent ONFH risk in alcohol intake. Patients suffering from orthopaedic diseases other than ONFH were collected as controls. Significant statistical differences between both groups were observed for: weekly alcohol consumption, number of cigarette smoked daily, occupation and its demanding energy consumption, and incidence of liver dysfunction (table 2). Relative risk (RR) and the 95 % confidence interval (CI) of RR were calculated for the factors under study. The RR was the ratio of the incidence rate in the group exposed to those not exposed. A conditional logistic regression model for matched sets was used in order to obtain adjusted RR while considering potentially confounding effects of other factors that showed apparent associations with ONFH. Increased relative risk was noticed in regular alcohol drinkers, current cigarette smokers and the subjects having liver dysfunction in their history (table 3). Increase in weekly alcohol consumption entailed a marked enhancement of relative risk of ONFH, particularly in those regular drinkers of more than 400 ml/week alcohol (table 4). A cumulative ill effect of both alcohol and cigarette were estimated by using drink-year and Brinkman and Coates Index respectively. A marked enhancement of relative risk of ONFH was noticed in more than 4,000 drink-years individuals, i.e. in drinkers of more than 400 ml/week for ten years. A cumulative ill effect was not remarkable in cigarette

Table 2. Demographic features of ONFH patients and their controls

Characteristic	Cases	Controls
No. of subjects	118	236
Male/Female	83/35	166/70
Mean age at onset of symptoms	46.5	45.7
Mean age at initial examination	48.4	48.3
Mean age at last school completion	18.1	18.5
Mean weekly alcohol consumption (ml/week)**	62.3	225.8
Mean daily no. of cigarettes smoked*	17.5	13.5
Mean body mass index (kg/m^2)*	22.0	22.7
Mean daily occupational energy consumption (Kcal/day)*	2355.6	2196.8
Blue-collar workers (%)#	52.6	40.5
History of liver dysfunction (%)##	23.7	7.3

*p<0.05, **p<0.01. t-test
#p<0.05, ##p<0.01.X^2-test

Table 3. Adjusted RR of Selected Factors for ONFH

Characteristic	Level	RR	95 % CI
Alcohol drinking	Never	1.0	
	Former	1.0	0.2-6.2
	Occasional	3.2	1.1-9.2
	Regular	13.1	4.1-42.5
			(Trend p<0.001)
Cigarette smoking	Never	1.0	
	Former	3.3	0.9-11.6
	Current	4.7	1.5-14.5
			(Trend p<0.001)
Daily occupational	< 1900	1.0	
energy	1900-2500	1.0	0.4-2.4
consumption	≥ 2500	2.0	0.9-4.6
(Kcal/day)			(Trend p<0.05)
Obesity	No	1.0	
	Yes	0.8	0.4-1.7
History of liver	No	1.0	
dysfunction	Yes	2.2	1.0-5.2

Table 4. Distribution of Cases, Controls and RR for ONFH According to Weekly Alcohol Consumption

Alcohol Consumption (ml/week)	Cases	Controls	RR	95 %CI
Nondrinker	23	87	1.0	
<400	25	93	2.8	1.0-7.8
400-10.000	51	48	9.4	3.0-29.0
≥10.000	19	8	14.8	3.8-57.2
				(Trend p<0.001)
Total	118	236		

246

Table 5. Adjusted RR of ONFH According to the Classification of Drink-Years and Brinkman and Coates Index

Characteristic	Level	RR	95 % CI
Alcohol drinking	Never	1.0	
(drink-year)	<4 000	2.2	0.7-6.9
	4 000-10 000	9.7	2.6-36.1
	≥ 10 000	12.9	3.8-43.4
			(Trend p<0.001)
Cigarette smoking	Never	1.0	
(Brinkman and Coates Index)	<200	1.6	0.4-6.3
	200-400	6.6	1.7-25.7
	≥ 400	6.5	1.9-21.9
			(Trend p<0.005)

smoking as compared to alcohol drinking (table 5). Whether or not cigarette smoking or history of hepatic dysfunction enhance synergetically ONFH risk of Alcohol intake was tested negatively in both Alcohol-cigarette and Alcohol-hepatic dysfunction pairs.

SUMMARY

Current increase of the disease burden is remarkable, particularly in steroid induced osteonecrosis of the femoral head. Estimated number of the new patients in 1987 was 2,500-3,300 in Japan (including traumatic cases).

Keeping an increasing number of steroid induced ONFH in mind, epidemiologic risk analysis had been undertaken in a prospective manner among systemic lupus erythematosus (SLE) patients under hypercorticosteroid therapy. This study clarified three categories of risk factors related to SLE types, laboratory data abnormalities and mode of corticosteroid administration. The following clinical manifestations of SLE identified patients at risk of ONFH: alopecia, papulo-macular eruption, drug-induced lupus, lupus erythematosus cell positive rheumatoid arthritis, interstitial pneumonitis, and thrombocytopenic purpura. A characteristic pattern of laboratory data abnormalities around the presumed onset of ONFH and a high dose corticosteroid therapy (30 mg/day prednisolone or more) without preloading of corticosteroid increased the risk of ONFH.

More than 400 ml/week regular alcohol intake proved particularly risky for ONFH. Both dose-related and cumulative effect were statistically clarified in alcohol associated ONFH. Synergetic risk enhancements by coupling alcohol intake with cigarette smoking or alcohol intake with hepatic dysfunction were not demonstrated.

REFERENCES AND FURTHER READING

Matsuo, K., T. Hirohata, Y. Sugioka, M. Ikeda and A. Fukuda: Influence of alcohol intake, cigarette smoking, and occupational status on idiopathic osteonecrosis of the femoral head. Clin. Orthop. 234: 115-123, 1988.

Nishio, A.: Annual report of the investigation committee for adult idiopathic avascular necrosis of the femoral head, Tokyo, Ministry of Health and Welfare, 1975-1980.

Ono, K.: Annual report of the investigation committee for adult idiopathic avascular necrosis of the femoral head, Tokyo, Ministry of Health and Welfare, 1984-1989.

Ono, K., T. Tohjima, and T. Komazawa: Risk factors of avascular necrosis of the femoral head in patients with systemic lupus erythematosus under high-dose corticosteroid therapy . Clin . Orthop . 277: 89-97, 1992 .

Sugioka, Y.: Annual report of the investigation committee for adult idiopathic avascular necrosis of the femoral head, Tokyo Ministry of Health and Welfare, 1990-1991.

Tohjima, T., and Y. Shiokawa: Survey of aseptic bone necrosis of the femoral head in collagen diseases by retrospective analysis. Report of investigation committee for adult idiopathic avascular necrosis of the femoral head in Japan, Tokyo, Ministry of Health and Welfare, 1984, 87-93.

PATHOPHYSIOLOGY OF OSTEONECROSIS

John Paul JONES, Jr.

Diagnostic Osteonecrosis Center and Research Foundation
Kelseyville, USA

Death of all the cellular elements of bone indicates osteonecrosis (ON). Irreversible osteocytic necrosis requires a minimum of two hours of complete ischemia with total anoxia (James and Steijn-Myagkaya, 1986). ON without repair occurs with physiologic death of the human skeleton, and it can be seen in paleopathologic specimens that were dehydrated and not permineralized (Figure 1). Unfortunately, the pathophysiology of ON still remains controversial, even though Martin and Rothschild (1989) have found that this disease occurred in the bones of deep-diving reptiles who were living 64 to 100 million years ago.

TRAUMATIC OSTEONECROSIS

Disruption of the arterial blood supply to bone results in post-traumatic ON, which usually involves those vulnerable bones (femoral and humeral heads, body of the talus, carpal scaphoid and lunate, and others) covered extensively by cartilage, with few vascular foramina and limited collateral circulation. Since oxygenated blood to the superolateral two-thirds of the femoral head comes almost exclusively from the lateral epiphyseal branches (Figure 1A) of the medial femoral circumflex artery, this area is especially susceptible to ON.

Four-part proximal humeral fractures or intracapsular femoral neck fractures interrupt most blood flow through the subsynovial retinacular vessels. Therefore, 78 % of post-fracture femoral head specimens are found to be partially or totally avascular, as shown by Calandruccio and Anderson (1980). However, late segmental collapse of the articular surface occurs in about 30 % of displaced subcapital fractures and 10 % of hip dislocations without fractures, since dislocations usually rupture the vessels of the ligamentum teres. Intracapsular tamponade of arteries can cause ON in immature animals. Likewise, hemarthrosis caused by undisplaced fractures may cause reversible femoral head ischemia, as demonstrated by Wingstrand et al. (1986).

Bone Circulation and Vascularization in Normal and Pathological Conditions
Edited by A. Schoutens *et al.*, Plenum Press, New York, 1993

249

NONTRAUMATIC OSTEONECROSIS

Although chemotherapy, thermal injuries, and radiation usually produce ON by cytotoxicity and obliterative endarteritis, L-asparaginase chemotherapy causes ON by a thrombotic coagulopathy (Hanada et al., 1989). ON also results from occlusive vascular disease with thrombosis. For instance, ON in Fabry-Anderson disease is probably caused by thrombotic obliteration of vascular lumens by progressive lipid accumulation within endothelial cells. Presumably, the painful crises in acute leukemia, Gaucher disease, sickle cell disease, and decompression sickness (DCS) all result from acute marrow thrombosis, ischemia and infarction, since bone scintigraphy shows decreased uptake ("cold" marrow lesions) shortly after symptoms appear (Katz et al., 1991; Macleod et al., 1982; Sebes, 1989; Sissons and Hayward, 1989).

Intravascular coagulation (IC), either focal or disseminated (DIC), now appears to be the most likely final common pathway in the early pathogenesis (Jones, 1991, 1992). Thrombotic complications and bone infarctions probably result from a consumptive coagulopathy in pancreatitis, Leriche syndrome, hemophilia B (treated with factor IX concentrates), various malignancies, and several other conditions (Harigaya et al., 1977). DIC can also occur during rapid organ rejection. However, IC is not the actual cause of ON but is an intermediary event, which is always triggered by some other underlying etiology (Bick, 1988; Halleraker, 1970).

Fat Embolism

In 1965, the association of ON with intraosseous fat embolism was first demonstrated clinically, and it was confirmed experimentally in 1966 (Jones, 1985). Jacobs (1978) found that 89 % of 269 patients with nontraumatic ON had disorders known to be complicated by disturbed fat metabolism or fat embolism. Fat emboli may arise from a fatty liver (Mechanism A), destabilization and coalescence of plasma lipoproteins (Mechanism B), and/or disruption of fatty bone marrow or other adipose tissue depots (Mechanism C).

Absolute Fat Overload. An absolute overload (Haber et al., 1988; Shapiro et al., 1985) of intraosseous fat emboli, exceeding the ischemic threshold, with impaired clearance and IC, appears to be the most likely cause of ON in alcoholism and hypercortisonism, which now account for about two thirds of the cases. Either exogenous corticoid use or endogenous hypercortisonism (Alexakis and Wallack, 1989) can cause ON. Elevated serum and urine cortisol can also occur in alcoholics with ON (Rico et al., 1985). The corticoid or alcohol-induced fatty liver is probably the source of the emboli, which may also account for delirium tremens, as suggested by Horowitz et al. (1977) in an alcoholic patient with bilateral femoral head lesions.

Thompson et al. (1969) performed in vivo fat infusion studies and directly visualized platelet aggregation occurring over the surface of intravascular fat globules. Two hours later, fibrin-platelet microthrombi had formed around the fat globules. Intravascular fat can also migrate through canaliculi to become deposited within osteocytic lacunae. Lipid accumulates in subchondral osteocytes that subsequently become necrotic, as shown both experimentally by Kawai et al., (1985) and clinically by Muratsu et al. (1990), in both alcoholism and hypercortisonism.

There appears to be a cumulative alcohol-corticoid dose-related ON response. In the author's experience, the exposure threshold for alcohol-associated ON (at a consumption of

400 ml or more of absolute ethanol a week) is about 150 liters of 100 % ethanol, and the corticoid exposure threshold is about 2,000 mg of prednisone, for ON to appear in adults (Jones, 1992). However, several prothrombotic factors may decrease this threshold, including septic shock (O'Brien and Mack, 1992), cerebral trauma, cerebral and spinal surgery, exogenous thrombin administration, and others.

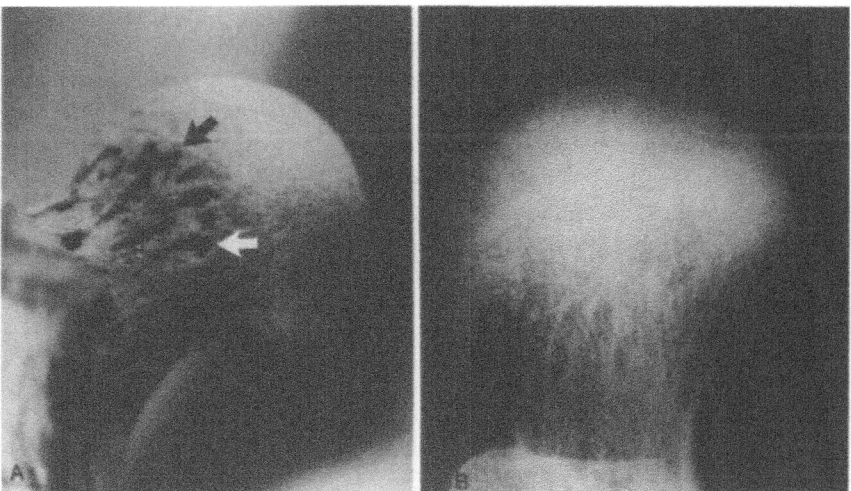

Figure 1. 3,000-year-old proximal femur from an American Indian, preserved by natural dehydration without permineralization, reveals that dead bone without repair does not spontaneously collapse. There are several foramina (arrows) for the lateral epiphyseal (superior capital) arteries (left, A). High-resolution radiograph of this femoral head with physiologic ON shows a remarkably unaltered trabecular pattern without articular incongruity (Specimen courtesy of John W. Parker, Ph.D.).

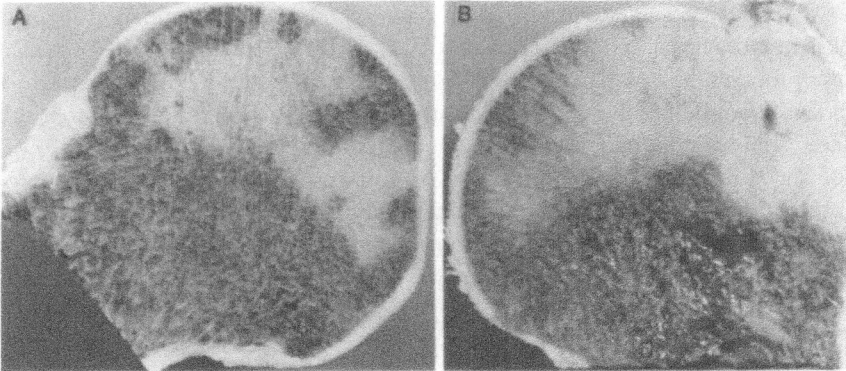

Figure 2. Coronal section of right femoral head (left, A) and left humeral head (right, B) in a patient who expired 18 hours after developing anaphylactic shock. Disseminated intravascular coagulation had resulted in fibrin thromboses of subchondral vessels, with necrosis of hematopoietic and adipocytic tissue (light areas are probably non-viable) and fibrinolysis and interadipocytic hemorrhage of adjacent marrow (dark areas). This classic distribution of ON also appeared in the left femoral and right humeral heads.

Relative Fat Overload. A relative overload of intraosseous fat emboli, which is below the ischemic threshold and insufficient to trigger IC, may be a cause of decreased osteoblastic bone formation in glucocorticoid- or alcohol-induced osteoporosis. Arlot et al. (1983) found co-existent osteoporosis in histomorphometric studies from 68 of 77 (89 %) ON patients. Prolonged heparin therapy with increased lipolytic activity and hyperlipemia prevents IC but causes osteoporosis. Also, Wang et al. (1989) showed that lipid-clearing agents can prevent corticoid-induced osteoporosis, perhaps by diminishing the steroid-induced accumulation of lipids within osteocytes, as demonstrated by Maruno et al.(1991). Perhaps this is the mechanism by which tamoxifen, an anti-estrogen compound with lipid-clearing effects, prevents osteoporosis.

Corticosteroid-treated rabbits were studied in nine laboratories (Jones, 1991). All instances involved hyperlipemia, fatty liver, pulmonary and systemic fat embolism, subchondral fat embolism of the femoral (and humeral) heads and osteoporosis. Focal osteocyte death in the femoral heads occurred in six of the eight series studied. This sequence of intraosseous fat embolism, endothelial cell necrosis, extravascular lipid migration, reduced osteoblastic activity, and fatty osteocytic necrosis has recently been confirmed in osteopenic space-flight rats with possible endogenous hypercortisonism (Doty et al., 1990).

Endotoxic Reactions

In the generalized Shwartzman phenomenon, rabbits given endotoxins not only develop DIC, but also hyperlipemia, fatty liver, systemic fat embolism and fibrin thrombosis. Duncan and Ramsay (1984) found ON complicating meningococcemia and DIC. Fifty-one ON lesions, including the humeral and femoral heads, complicated DIC in 8 children (Jones, 1992). Grogan et al. (1989) performed histologic studies revealing ischemia and fibrin thrombi within the microvasculature adjacent to the necrotic bone.

Hypersensitivity Reactions

A very early case of nontraumatic ON, affecting both humeral and femoral heads (Figure 2), with subchondral fat embolism (Figure 3A) and DIC, occurred 18 hours after anaphylactic shock (Jones, 1993). There was both subchondral fibrin thrombosis (Figure 3B) and peripheral inter-adipocytic hemorrhage, resulting from secondary fibrinolysis and reperfusion of necrotic vessels.
Two animal models (Type III hypersensitivity, immune-complex mediated) also suggest IC is the final common pathway. Two to three hours following antigen injection into the rabbit knee, Mahowald et al. (1988) showed that focal areas of subchondral bone were devoid of secondary and tertiary branches of epiphyseal arteries. At one to two weeks there were occlusions of subchondral vessels adjacent to avascular subchondral bone. Matsui et al. (1992) demonstrated that serum sickness and corticoids produced arteriolar interruptions and necrotic lesions in 14 of 20 (70 %) femoral metaphyses and diaphyses.

Dysbaric Phenomena

Dysbaric ON can occur in humans and sheep (Lehner et al., 1992) after a single hyperbaric air exposure with inadequate decompression. Although there is no evidence that ON directly results from venous and/or arterial gas embolism alone, intravascular fat and tissue factor (thromboplastin, factor III) accelerate DIC after DCS. Philp (1974) showed that fibrinogen, lipid, and platelet aggregation at the blood-bubble interface are associated with post-dive thrombocytopenia, accelerated platelet and fibrinogen turnover, decreased antithrombin III activity, prolongation of the prothrombin time and increased fibrin degradation products.

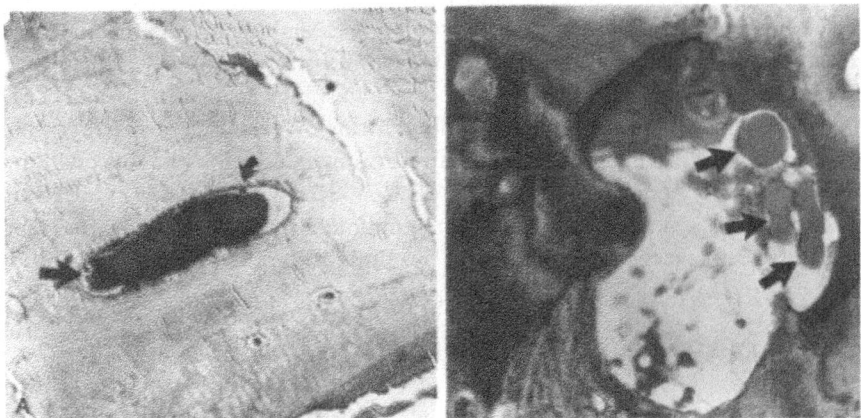

Figure 3. Photomicrograph of subchondral bone from humeral head showing pyknotic-appearing endothelial cell nuclei (arrows) within a Haversian canal which is completely occluded by a deformed fat embolus, Osmium-potassium dichromate (left, A); photomicrograph of humeral head at the tidemark revealing subchondral capillaries occluded with multiple fibrin thrombi (arrows), Masson trichrome (right, B).

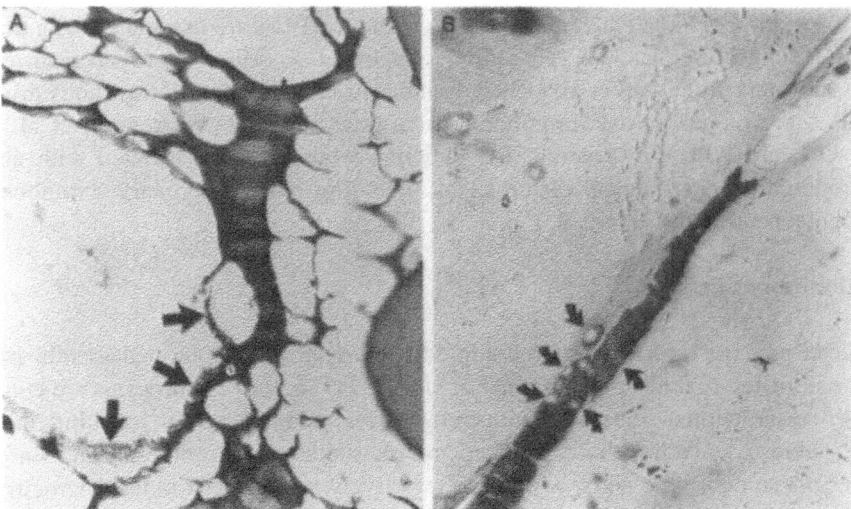

Figure 4. Photomicrograph of fatty bone marrow from the right femoral head of a scallop diver who had inadequate decompression, resulting in DCS and death 70 minutes later. Two gas bubbles are surrounded by disrupted adipocytes, with platelet aggregation (arrows) and fibrin thrombosis of sinusoids, Hematoxylin and eosin (left, A); subchondral capillary of the right femoral head, immediately beneath the tidemark, which is completely occluded by multiple deformed fat emboli, including a terminal saddle embolus. Small voids within the fat may represent minute intravascular gas bubbles (arrows), Osmium-potassium dichromate (right, B).

A commercial scallop diver who remained at 92 feet of sea water for 4.5 hours on surface-supplied compressed air was recently studied by Jones et al.(1992). DCS occurred after a no-stop ascent to the surface, and he died 70 minutes later. Autopsy revealed multiple gas bubbles, not only within the great vessels, but in the fatty marrow of the femoral and humeral heads. Lipid and platelet aggregates were found on the surface of marrow bubbles. Fibrin-platelet thrombi were present within dilated sinusoids adjacent to the bubbles (Figure 4A), and were also detected in marrow veins, arteries, and capillaries. Pulmonary and renal fat embolism was observed. Fat embolism (Figure 4B) and fibrin thrombosis of subchondral capillaries were also demonstrated.

Our observations are consistent with those of Kawashima et al. (1978) and Kitano and Hayashi (1981) and suggest that expanding nitrogen gas injures marrow adipocytes which release liquid fat. This lipid, along with fibrinogen, envelops bubbles within venous sinusoids and locally induces platelet aggregation, fibrin-platelet thrombosis and fatty marrow necrosis. Subchondral fat emboli, toxic unbound free fatty acids, tissue factor, and other vasoactive substances probably play a systemic role in triggering DIC and dysbaric ON.

Antiphospholipid Antibody Syndrome

Asherson et al. (1989) discovered that antiphospholipid antibodies (APLA) can cause venous and arterial thromboses, as well as ON of the femoral heads. Protein C deficiency with lupus anticoagulants can result in occlusion of the femoral and profundus arteries (Resnick, et al., 1985; Harrison and Alperin, 1992), and probably their branches supplying the femoral head (Jones, 1985). Systemic lupus erythematosus (SLE) is the most common underlying disease causing DIC. APLA with ON has been confirmed in patients with SLE (Nagasawa et al., 1989), and without SLE, who have never received corticoids (Seleznick et al. (1991); Vela et al. (1991). Although APLA are associated with thrombo-cytopenia, suggesting platelet aggregation, Stinson et al. (1990) detected recurrent DIC in this syndrome. Belmonte et al. (1991) found two HIV-infected intravenous drug users with APL, prolonged partial thromboplastin times, and ON. Nassonova et al. (1991) studied ON and APLA in several SLE patients, which were associated with abortion, livedo reticularis, neurological disturbances including stroke, coronary thrombosis, and homozygous beta-thalassemia.

Hemoglobinopathies

Those patients with the hemoglobin SS genotype and alpha-thalassemia have the highest risk of femoral head ON (Milner et al., 1991). Sickled erythrocytes are considered to cause stasis within sinusoids, with progressive deoxygenation resulting in further sickling, acidosis, endothelial damage, microcirculatory thrombosis and infarction. However, these dense red cells, which are selectively sequestered in the microcirculation during vaso-occlusive crises, probably do not cause ON by themselves, since Billett et al. (1988) have found that they produce stasis without thrombosis. However, a hypercoagu-lable state with increased platelet activation, thrombin generation, fibrin deposition and impaired fibrinolysis occur at the same time as erythrocyte sequestration. (Beurling-Harbury and Schade, 1989). Thrombosis with increased serum fibrinogen degradation products most likely results from activation of the coagulation system in the sickle hemoglobinopathies by a combination of fat and marrow embolism, thrombocytosis, hyperfibrinogenemia, hyperviscosity, increased factor VIII, plasma B-thromboglobulin, and thromboxane B2, and decreased factor V, plasminogen, proteins C and S, and anti-thrombin III.

Obstetrical Complications

During pregnancy there is already a hypercoagulable state with hyperlipemia, depression of the fibrinolytic system, and occasionally, decreased antithrombin III. ON occurs in late pregnancy or in the early post-partem period. It is related to fatty liver of pregnancy, (Jones, 1985) retained fetus in utero, (Jones, 1992) and several other obstetric problems, virtually all of which are capable of triggering DIC, (Bick, 1988) since intravascular placental tissue is rich in thromboplastin.

Prethrombotic Conditions

Malignancies, especially metastatic carcinoma and promyelocytic leukemias, may have DIC and ON. Femoral head ischemia in Perthes' disease may be due to an intravascular decrease in fibrinolytic activity (Gregosiewicz et al., 1989). Thrombosis is also potentiated by hypofibrinolysis resulting from increased plasminogen activator inhibitor and/or increased alpha-2 plasmin inhibitor in the nephrotic syndrome, hemolytic-uremic syndrome, thrombotic thrombocytopenic purpura, and hypertriglyceridemic syndromes (types IIB and IV). Hyperlipemia is associated with increased plasma factor VII activity. The hyperlipemia of the nephrotic syndrome may be exacerbated by corticoid therapy, further increasing ON susceptibility. Moreover, co-existent protein C and S deficiencies, antithrombin III deficiency, hyperviscosity, hyperfibrinogemia, poly-cythemia, thrombocytosis, and other unknown factors would produce a decreased threshold resistance to thrombogenesis and ON.

INTRAVASCULAR COAGULATION

There is direct histological evidence in humans of intraosseous fibrin-platelet thromboses within prenecrotic femoral and humeral head segments, both 70 minutes (Jones et al., 1992), and 18 hours (Jones, 1993) after a known ischemic event, and prior to complete autolytic reduction of the avascular zone. Intracapital arterial and arteriolar thromboses have also been observed histologically within later lesions (Bonfiglio, 1976; Ficat and Arlet, 1980; Saito et al., 1992; Spencer and Brookes, 1988). A combination of three factors produces microcirculatory thrombosis of these susceptible intraosseous end-organs: (1) stasis, (2) hypercoagulability, and (3) endothelial damage.

Stasis

The initial juxta-articular lesion is localized to the subchondral bone, which has a vulnerable microanatomy which facilitates vascular stasis. Terminal arteries with few collaterals supply subchondral areas, which favors embolic occlusion and thrombosis, especially with localized vasoconstriction, or possibly vasospasm (Raynaud's phenomenon). The fatty subchondral bone of epiphyseal regions has reduced intramedullary pressure and blood flow with long, narrow arcades of end-capillaries. Also, Wang et al. (1984) showed in rabbits that after 10 weeks of corticoid treatment there was steroid-induced marrow adipocytic hypertrophy and a 32 % reduction in blood flow.

Hypercoagulability

An increased amount and potency of procoagulants, decreased anticoagulants, vasoconstriction of the subchondral arteriolar bed, and probably the most important, decreased endogenous fibrinolysis (Muller-Berghaus, 1989) result in a hypercoagulable state. The subchondral capillary bed has a high surface/volume ratio which results in a marked increase of endothelial cells in direct contact with blood. In the perturbed state, subchondral endothelial cells can be thrombogenic through the synthesis of von Willebrand factor antigen, plasminogen activator inhibitor, and especially tissue thromboplastin. For example, hypercoagulability, hyperlipemia and platelet hyperaggregability can occur in hemodialysis patients with ON.

Endothelial Damage

The most likely event triggering platelet aggregation and fibrin thrombosis, with progressive involvement of venules, veins, arterioles, and extraosseous arteries is endothelial damage to marrow capillaries and venous sinusoids, exposing procoagulant subendothelial collagen. Subchondral thrombosis is followed by some degree of secondary, endogenous fibrinolysis. During the initial phase of reflow there is probably intracellular and intercellular swelling. Ischemia-reperfusion injuries cause lipid peroxidation of endothelial cell membranes by toxic oxygen free radicals and could potentiate microfocal bleeding with extravasation of erythrocytes between marrow adipocytes with hemosiderin deposition.

Although early reperfusion may salvage viable but ischemic bone, late reflow (reperfusion after six hours of occlusion) of necrotic arterioles may cause peripheral hemorrhage into marrow margins which are already necrotic. Saito et al. (1992) studied core biopsies of 24 precollapse femoral heads in steroid-treated patients which showed damaged vessels with multifocal marrow hemorrhages. Within about 12 hours of arterial and/or venous marrow ischemia there is probable sinusoidal distension and rupture, with additional interadipocytic hemorrhage. After a few days the thrombosed and necrotic vessels in the avascular zone apparently disappear altogether, since they could not be visualized on subsequent microangiographic or histological studies by Ohzono et al. (1992). With thrombosis of the marrow vasculature there is deficient venous drainage with increased hydraulic outflow resistance, intraosseous hypertension, and further impaired marrow perfusion (Ficat and Arlet, 1980). In contrast, transient osteoporosis of the hip could potentially be caused by a reversible circulatory disturbance (Atsumi and Kuroki, 1992), i.e. DIC with thrombosis, immediate and complete fibrinolysis, hyperemia, and edema.

REPAIR WITHOUT COLLAPSE

A lag period occurs before the repair response is activated. Angiogenesis may be signaled by platelet-derived growth factor (PDGF), insulin-like growth factor (IGF-1), and endothelial cell growth factor (ECGF). Osteoclast migration may be stimulated by necrotic collagen degradation products. Complete healing may occur depending upon the size and location of the necrotic lesion. Spontaneous revascularization of segmental lesions of the femoral head from the cancellous region is possible, within certain limits (15-20 mm). Repair is initiated along the outer perimeter of the avascular segment (Figure

5A), at the junction of the ischemic zone surrounding the dead area and the viable area with intact circulation. Focal hemorrhages may occur in the reparative margin of the infarction (Figure 5B). Dead trabeculae are partially or completely resorbed by osteoclasts and are replaced or covered with new appositional bone, resulting in thickened, reinforced trabeculae.

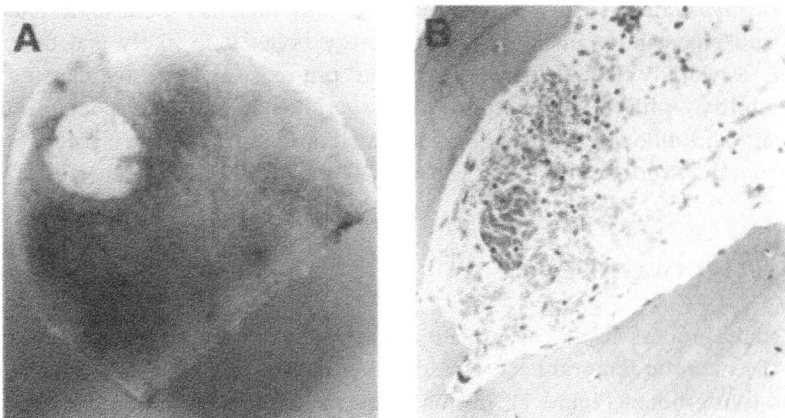

Figure 5. A) Gross coronal section of left femoral head from an alcoholic patient showing a necrotic lesion with revascularization (dark areas). B) Photomicrograph within the viable perinecrotic marrow showing extravasation of erythrocytes as a focal intertrabecular hemorrhage, Hematoxylin and eosin.

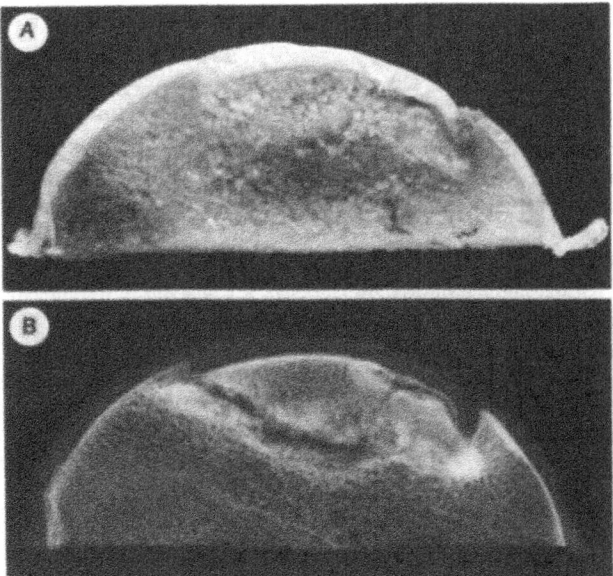

Figure 6. Gross section of left humeral head from an alcoholic patient which shows a localized anemic infarction with late segmental collapse of the central subchondral bone and a darkened peripheral zone of hyperemia, (top, A); slab radiograph of this specimen showing the depressed segment of dead bone, with a subchondral fracture (radiolucent crescent), and a deeper fracture which probably extends within the reparative front.

In rabbit studies, Gold et al. (1978) and Surat (1984) both detected intraosseous fat embolism with increased free fatty acids and prostaglandins in steroid-induced ON. Similarly, Tsai and Liu (1992) recently found intraosseous fat embolism in 9 of 16 (56%) ON patients, as well as increased lymphocytes, plasma cells, and eicosanoids in the reparative front, including prostaglandin E2 (PGE2), thromboxane B4 (TXB4), and leukotrienes LTC4, and especially LTB4. Oleic acid, the principal fatty acid of neutral trioleum, causes a stripping of capillary endothelium, passive congestion, and edema formation. Intravascular hydrolysis of fat emboli may increase these toxic unbound free fatty acids and eicosanoids in the repaired spongy bone. It is likely that these elevated eicosanoids, especially leukotrienes, aggravate pre-existing ON by causing vascular hyperpermeability with inflammatory marrow edema and additional intraosseous hypertension. Thromboxane-induced vasoconstriction and platelet aggregation at the reparative front may further promote IC and ON (Jones, 1992).

REPAIR WITH COLLAPSE

The repair process itself causes late segmental collapse. If repair were prevented altogether, the necrotic femoral head would not collapse because dead bone is essentially as strong as living bone (Figure 1). In this stage the revascularization front has resorbed and weakened subchondral bone at the supero-lateral chondro-osseous junction. Shear-induced microfractures of the previously necrotic trabeculae result from impulse loading, beginning at stress risers, which propagate into the femoral head. Finally, there is irreversible collapse and depression of the antero-superior portion of the femoral head or the central portion of the humeral head (Figure 6), with progressive articular incongruity and secondary osteoarthritis.

REFERENCES

Alexakis, P.G., M. Wallack : Idiopathic osteonecrosis of the femoral head associated with a pituitary tumor. J. Bone Joint Surg. 71A: 1412-1414 (1989).

Arlot, M.E., M. Bonjean, P.M. Chavassieux, P.J. Meunier: Bone histology in adults with aseptic necrosis. Histomorphometric evaluation of iliac biopsies in seventy-seven patients. J. Bone Joint Surg. 65A: 1319-1327 (1983).

Asherson, R.A., M.A. Khamashta, J. Ordi-Ros, R. Derksen, S.J. Machin, J. Barguinero, H.H. Outt, E.N. Harris, M. Vilardell-Torres, G.R.V. Hughes: The "primary" antiphospholipid syndrome: major clinical and serological features. Medicine 68: 366-374 (1989).

Atsumi, T., Y. Kuroki Role of impairment of blood supply of the femoral head in the pathogenesis of idiopathic osteonecrosis. Clin. Orthop. 277: 22-30 (1992).

Belmonte, M.A., R. Garcia-Portales, I. Domenech, I. Galvez, A. Fernandez-Nebro, M.T. Camps, E. Ramon: Human immunodeficiency virus (HIV) infection and avascular osteonecrosis (AON). XIIth European Congress of Rheumatoloy, Budapest, Hungary (1991).

Beurling-Harbury, C., S.G. Schade: Platelet activation during pain crisis in sickle cell anemia patients. Am. J. Hematol. 31: 237-241 (1989).

Bick, R.L.: Disseminated intravascular coagulation and related syndromes: A clinical review. Sem. Thromb. Hem. 14: 299-338 (1988).

Billett, H.H., R.L. Nagel, M.E. Fabry: Evolution of laboratory parameters during sickle cell painful crisis: Evidence compatible with dense red cell sequestration without thrombosis. Am. J. Med. Sci. 296: 293-298 (1988).

Bonfiglio, M.: Development of bone necrosis lesions. In: C.J. Lambertsen (ed.): Underwater Physiology V. Proc. Fifth Symposium Underwater Physiology, pp. 117-132. Fed. Am. Soc. Exp. Biol., Bethesda 1976.

Calandruccio, R.A., W.E. Anderson: Post-fracture avascular necrosis of the femoral head. Correlation of experimental and clinical studies. Clin. Orthop. 152: 49-84 (1980).

Doty, S.B., E.R. Morey-Holton, G.N. Durnova, A.S. Kaplansky: Cosmos 1887: morphology, histochemistry and vasculature of the growing rat tibia. FASEB 4: 16-23 (1990).

Duncan, J.S., L.E. Ramsay: Widespread bone infarction complicating meningococcal septicaemia and disseminated intravascular coagulation. Brit. Med. J. 288: 111-112 (1984).

Ficat, R.P., J. Arlet: Ischemia and Bone Necrosis. Williams and Wilkins, Baltimore 1980.

Gold, E.W., O.D. Fox, S. Weissfeld, P.H. Curtiss: Corticosteroid-induced avascular necrosis : An experimental study in rabbits. Clin. Orthop. 135 : 272-280 (1978).

Gregosiewicz, A., M. Okonski, D. Stolecka, G. Kandzierski, M. Szponar: Ischemia of the femoral head in Perthes' disease : Is the cause intra- or extravascular. J. Ped. Orthop. 9: 160-162 (1989).

Grogan, D.P., S.M. Love, J.A. Ogden, E.A. Millar, L.O. Johnson: Chondro-osseous growth abnormalities after meningococcemia. A clinical and histopathological study. J. Bone Joint Surg. 71A: 920-928 (1989).

Haber, L.M., E.P. Hawkins, D.K. Seilheimer, A. Slaeem: Fat overload syndrome. An autopsy study with evaluation of the coagulopathy. Amer. J. Clin. Path. 90:223-227 (1988).

Halleraker, B.: Fat embolism and intravascular coagulation. Acta Pathol. Microbiol. Immunol. Scand. 78: 432-436 (1970).

Hanada, T., Y. Horigome, M. Inudoh, H. Takita: Osteonecrosis of vertebrae in a child with acute lymphocytic leukaemia during L-asparaginase therapy. Eur. J. Pediatr. 149: 162-163 (1989).

Harigaya, K., S. Watanabe, Y. Watanabe, K. Kageyama, K. Nakazawa: Multiple bone marrow necrosis and disseminated intravascular coagulation. Arch. Path. Lab. Med. 101: 652-654 (1977).

Harrison, R.L., J.B. Alperin: Concurrent protein C deficiency and lupus anticoagulants. Am. J. Hematol. 40: 33-37 (1992).

Horowitz, I., R.J. Klingenstein, R. Levy, M.J. Zimmerman: Fat embolism syndrome in delirum tremens. Am. J. Gastroent. 68: 476-480 (1977).

Jacobs, B.: Epidemiology of traumatic and non-traumatic osteonecrosis. Clin. Orthop. 130: 51-67 (1978).

James, J., G.L. Steijn-Myagkaya: Death of osteocytes. Electron microscopy after in vitro ischaemia. J. Bone Joint Surg. 68B: 620-624 (1986).

Jones, J.P., Jr.: Fat embolism and osteonecrosis. Orthop. Clin. North Am. 16: 595-633 (1985).

Jones, J.P., Jr.: Etiology and pathogenesis of osteonecrosis. Seminars Arthroplasty 2: 160-168 (1991).

Jones, J.P., Jr.: Intravascular coagulation and osteonecrosis. Clin. Orthop. 277: 41-53 (1992).

Jones, J.P., Jr.: Fat embolism, intravascular coagulation and osteonecrosis. Clin. Orthop. in press (1993).

Jones, J.P., Jr., S. Ramirez, S.B. Doty: The procoagulant role of fat in dysbaric osteonecrosis, unpublished data (1992).

Katz, K., S. Mechlis-Frish, I.J. Cohen, G. Horev, R. Zaizov, E. Lubin: Bone scans in the diagnosis ot bone crisis in patients who have Gaucher disease. J. Bone Joint Surg. 73A : 513-517 (1991).

Kawai, K., A. Tamaki, K. Hirohata: Steroid-induced accumulation of lipid in the osteocytes of the rabbit femoral head. A histochemical and electron microscopic study. J. Bone Joint Surg. 67A: 755-763 (1985).

Kawashima, M., T. Torisu, K. Hayashi, M. Kitano: Pathological review of osteonecrosis in divers. Clin. Orthop. 130: 107-117 (1978).

Kitano, M., K. Hayashi: Acute decompression sickness-Report of an autopsy case with widespread fat embolism. Acta Pathol. Jpn. 31: 269-276 (1981).

Lehner, C.E., T.F. Lin, E.H. Lanphier, M.A. Wilson, A.A. DeSmet, M.D. Markel, R. Dueland: Early pathogenesis and detection of dysbaric osteonecrosis induced in sheep. Undersea Biomed. Res. 19 (Suppl.): 52-53 (1992).

Macleod, M.A., A.J.B. McEwan, R.R. Pearson, A.S. Houston: Functional imaging in the early diagnosis of dysbaric osteonecrosis. Brit. J. Radiol. 55: 497-500 (1982).

Mahowald, M.L., P.J. Majeski, S.R. Ytterberg: Microvascular pathology of antigen induced arthritis. Arth. Rheum. (Suppl.) 31: S91 (1988).

Martin, L.D., B.M. Rothschild: Paleopathology and diving mosasaurs. Amer. Scientist 77: 460-467 (1989).

Maruno, H., T. Shimizu, K. Kawai, K. Hirohata: The response of osteocytes to a lipid clearing agent in steriod-treated rabbits. J. Bone Joint Surg. 73B: 911-915 (1991).

Matsui, M., S. Saito, K. Ohzono, N. Sugano, M. Saito, K. Takaoka, Ono, K.: Experimental steroid-induced osteonecrosis in adult rabbits with hypersensitivity vasculitis. Clin. Orthop. 277: 61-72 (1992).

Milner, P.F., A.P. Kraus, J.I. Sebes, L.A. Sleeper, K.A. Kukes, S.H. Embury, R. Bellevue, M. Koshy, J.W. Moohr, J. Smith: Sickle cell disease as a cause of osteonecrosis of the femoral head. New Eng. J. Med. 325: 1476-1481 (1991).

Muller-Berghaus, G.: Pathophysiologic and biochemical events in disseminated intravascular coagulation: dysregulation of procoagulant and anticoagulant pathways. Sem. Thromb. Hemost. 15: 58-87 (1989).

Muratsu, H., T. Shimizu, K. Kawai, K. Hirohata: Alcohol-induced accumulations of lipids in the osteocytes of the rabbit femoral head. Trans. Ortho. Res. Soc. 15: 402 (1990).

Nagasawa, K., Y. Ishii, T. Mayumi, Y. Tada, A. Ueda, Y. Yamauchi, T. Kusaba, Y. Niho: Avascular necrosis of bone in systemic lupus erythematosus: possible role of haemostatic abnormalities. Ann. Rheum. Dis.48: 672-676 (1989).

Nassonova, V.A., Z. Alekberova, M. Folomeyev, N. Kamova, E.L. Nassonova: Antiphospholipid antibodies and avascular bone necrosis. Third Meeting Assoc. Res. Circ. Osseous, Basel, Switzerland, December 7, 1991.

O'Brien, T.J., G.R. Mack: Multifocal osteonecrosis after short-term high-dose corticosteroid therapy. A case report. Clin. Orthop. 279: 176-179 (1992).

Ohzono, K., K. Takaoka, S. Saito, M. Saito, M. Matsui, K. Ono: Intraosseous arterial architecture in non-traumatic avascular necrosis of the femoral head. Microangiographic and histologic study. Clin. Orthop. 277: 79-88 (1992).

Philp, R.B.: A review of blood changes associated with compression-decompression: Relationship to de-compression sickness. Undersea Biomed. Res. 1: 117-140 (1974).

Resnick, D., C. Pineda, D. Trudell: Widespread osteonecrosis of the foot in systemic lupus erythematosus: Radiographic and gross pathologic correlation. Skel. Radiol. 13: 33-38 (1985).

Rico, H., F. Gomez-Castresana, J.A. Cabranes, I. Almoguera, L.L. Duran, J.A. Matute: Increased blood cortisol in alcoholic patients with aseptic necrosis of the femoral head. Calcif. Tiss. Int. 37: 585-587 (1985).

Saito, S., K. Ohzono, K. Ono: Early arteriopathy and postulated pathogenesis of osteonecrosis of the femoral head. The intracapital arterioles. Clin. Orthop. 277: 98-110 (1992).

Sebes, A.J.: Diagnostic imaging of bone and joint abnormalities associated with sickle cell hemoglobinopathies. AJR 152: 1153-1159 (1989).

Seleznick, M.J., L.H. Silveira, L.R. Espinoza: Avascular necrosis associated with anticardiolipin antibodies. J. Rheumatol. 18: 1416-1417 (1991).

Shapiro, S.C., F.C. Rothstein, A.J. Newman, B. Fletcher, T.C. Halpin, Jr.: Multifocal osteonecrosis in ado-lescents with Crohn's disease: A complication of therapy ? J. Ped. Gastro. Nutr. 4: 502-510 (1985).

Sissons, G.R.J., M.W.J. Hayward: The hemipelvis: An unusual site of osteonecrosis. A report ot two cases. Clin. Radiol. 40: 494-497 (1989).

Spencer, J.D., M. Brookes: Avascular necrosis and the blood supply of the femoral head. Clin. Orthop. 235: 127-140 (1988).

Stinson, J., G. Tomkin, G. McDonald, F. Jackson, M. Harrison, A. Murray, C. Feighery, J. Jackson: Recurrent disseminated intravascular coagulation and fulminant intrahepatic thrombosis in a patient with the anti-phospholipid syndrome. Am. J. Hemat. 35: 281-282 (1990).

Surat, A.: Isolation of prostaglandin E2-like material from osteonecrosis induced by steroids and its prevention by kallikrein Inhibitor, aprotinin. Prostaglandins Leukotrienes Med. 13: 159-167 (1984).

Thompson, P.L., K.E. Williams, M.N. Walters: Fat embolism in the microcirculation: An in-vivo study. J. Path. 97: 23-28 (1969).

Tsai, C.-L., T.-K. Liu: Evidence for eicosanoids within the reparative front in avascular necrosis of human femoral head. Clin. Orthop. 281: 305-312 (1992).

Vela, P., E. Salas, E. Batlle, P. Marco: Primary antiphospholipid syndrome and osteonecrosis. Clin. Exper. Rheum. 9: 545-546 (1991).

Wang, G.-J., S.L. Hubbard, S.I. Reger, E.D. Miller, W.G. Stamp: Femoral head blood flow in long-term steroid treatment (study of rabbit model). In: J. Arlet, R.P. Ficat, D.S. Hungerford (eds.): Bone Circulation, pp. 35-37. Williams and Wilkins, Baltimore 1984.

Wang, G.-J., K.C. Chung, W.J. Shen, W.G. Stamp: Preventing steroid induced osteoporosis in the femoral head using lipid clearing agents. Trans. ORS 35th Meeting, Las Vegas, February 6-9, 1989.

Wingstrand, H., B. Stromqvist, N. Egund, T. Gustafson, L.T. Nilsson, K.-G. Thorngren: Hemarthrosis in undisplaced cervical fractures. Tamponade may cause reversible femoral head ischemia. Acta Orthop. Scand 57: 305-308 (1986).

OSTEONECROSIS
2. Methods of Diagnosis

DIAGNOSIS OF OSTEONECROSIS OF THE FEMORAL HEAD

David S. HUNGERFORD and Lynne C. JONES

Good Samaritan Hospital
Orthopaedic Surgery
Baltimore, USA

The proper diagnosis of ischemia and necrosis of bone is of primary practical importance because of the primacy of early diagnosis and the outcome of treatment, whatever it may be. Many authors, supporting core decompression, electrical stimulation, bone grafting or osteotomy have linked success to the stage at which the diagnosis is made (Steinberg et al., 1984; Hungerford et al., 1990). The purpose of this paper is to provide an overview of the issues and diagnosis, some of the obstacles and disputes, and finally to present our diagnostic algorithm as currently practiced in trying to arrive at the proper diagnosis for the patient presenting with a painful hip for evaluation.

WHAT CONSTITUTES A DIAGNOSIS OF OSTEONECROSIS OF BONE?

While everyone agrees that a positive biopsy showing dead trabeculae devoid of lacunae remains "the gold standard" for the diagnosis of osteonecrosis, to insist that it be the "sine qua non" for the diagnosis simply impedes progress and condemns the physician to never making an early diagnosis. It also ignores the possibility of sampling error if one has anything less than the whole femoral head available for sectioning. Moreover, such insistence ignores the marrow element of bone as an integral part of the organ system and denies the fundamental landmark work of Rutishauer, Rohner and Held (1960) who showed that bone trabecular death was the last item on a well defined pathophysiological cascade of bone ischemia and necrosis. The initial stages of ischemia are evidenced only in morphological changes in the marrow.

The error of insisting that only the finding of dead trabeculae in the biopsy justified the diagnosis of osteonecrosis is demonstrated in the publication of Camp and Colwell (1986). These authors divided their 42 cases of core decompression into histologically positive and histologically negative cases. Following the "sine qua non" line of reasoning, the former were considered to have definite osteonecrosis and the latter definitely not have osteonecrosis. This was in spite of the fact that 6 of 15 in the biopsy negative group had X-rays that were characteristic of osteonecrosis and during the follow-up period an

Bone Circulation and Vascularization in Normal and Pathological Conditions
Edited by A. Schoutens *et al.*, Plenum Press, New York, 1993

265

additional 3 progressed to characteristic changes. Nonetheless, the bone scan, bone marrow pressure tests, and venography were judged to be falsely positive in this group of patients rather than consider that the biopsy may have been subject to sampling error.

Many other vascular disorders of organ systems, e.g. Leriche Syndrome, cerebral and cardiac ischemia, etc., do not demand biopsy proof of dead muscle, brain, or heart, to establish the diagnosis and the insistence on the part of some for this criteria for bone is a serious impediment to early treatment and particularly noninvasive treatment. The situation for bone ischemia and necrosis is not dissimilar to the situation with rheumatoid arthritis before the American Rheumatism Association adoption of major and minor criteria, the combination of which would be sufficient to establish diagnosis (Ropes et al., 1959). While it is certainly not the place of this chapter or these authors to establish such criteria, Table 1 suggests at least some major and minor criteria which could be considered in such ischemia.

Table 1. Diagnostic Criteria for Diagnosis of Osteonecrosis

Major Criteria

1. Femoral head collapse
2. Subchondral radiolucency
3. Anterolateral Sequestrum
4. "Cold in hot" bone scan
5. Double band on T-2 MRI scan
6. Positive bone biopsy

Minor Criteria

1. Collapse with joint space narrowing
2. Mottled cyst/sclerosis pattern in femoral head
3. Increased uptake on bone scan
4. MRI changes of marrow edema/fibrosis
5. Hip painful range of motion with normal radiography
6. History of alcohol abuse/steroid use

Major criteria could be considered sufficient to establish diagnosis if certain other factors could be definitely excluded. For example, marginal subchondral collapse is virtually pathognomonic for osteonecrosis if major trauma, the only reasonable alternative etiology, can be excluded. It would require a committee of international experts to establish a comprehensive set of criteria to establish diagnosis, but it would be a worthwhile and enormously useful exercise. In the absence of such standards, researchers will continue to argue what is and what is not osteonecrosis.

RADIOGRAPHY

The first step in the diagnostic algorithm of osteonecrosis is good quality AP and frogleg laterals of the hip. The AP radiographs show the principal area of interest through the anterior and posterior margins of the acetabulum. Figure 1 shows an AP radiograph in which the acetabular margins have been outlined with a dotted line. Because the anterior and posterior acetabular margins overlap the superior portion of the femoral head, subtle abnormalities in the subchondral region may be missed. It is imperative that good quality lateral X-rays of the femoral head be obtained.

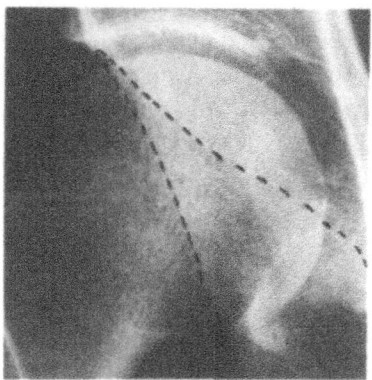

Figure 1. Anterior-Posterior radiograph of the hip showing the outline of the rim of the acetabulum which covers the area of interest for detecting early osteonecrosis.

A cross-table lateral X-ray is less satisfactory than a frogleg lateral or Lowenstein lateral to show the architectural details of the femoral head, since the former must penetrate more soft tissue, blurring bony outlines. The lateral is also important for staging purposes since it is often the anterior segment of the femoral head which first undergoes collapse or exhibits subchondral radiolucency (Figure 2). If the X-ray is positive and shows pathognomonic changes of avascular necrosis, no further diagnostic tests are necessary and treatment options can be chosen from the information available.

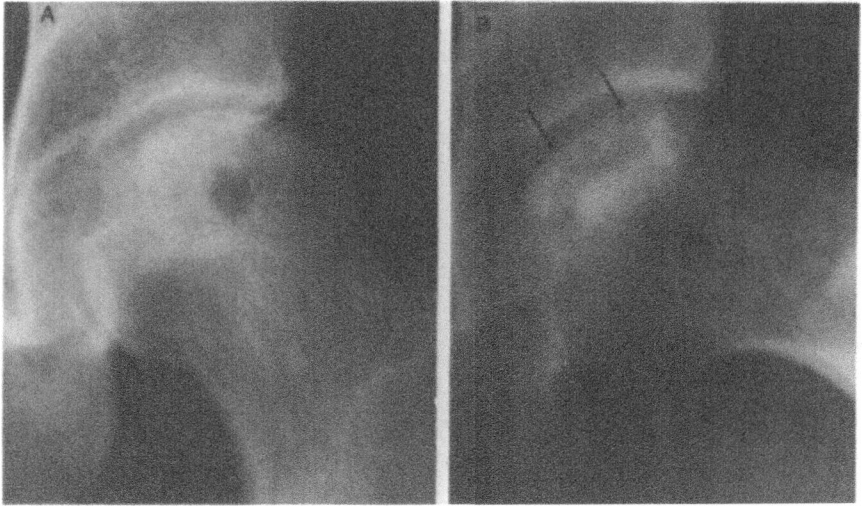

Figure 2. (A) The AP radiograph shows a completely spherical head while the frog-leg lateral (B) shows subtle flattening (arrows) indicative of early collapse.

COMPUTERIZED TOMOGRAPHY

Computerized tomography (CT) scanning is expensive, exposes the patient to considerable radiation and is usually unnecessary for establishing the diagnosis of osteonecrosis. However, it may be useful in separating late precollapse stages of osteonecrosis from the early collapse stage. Minimal collapse will only be shown if the area of collapse is perfectly aligned tangential to the X-ray beam, and, therefore, will often escape detection by routine radiography but would be detected on CT scan. Therefore, when considering more conservative treatment for pre-collapse disease, that one would not consider for post-collapse disease, CT scanning may be helpful in separating the two.

RADIONUCLIDE IMAGING

^{99m}Tc-diphosphonate imaging is a useful technique for detecting osteonecrosis. In general, the reactivity of bone around the infarcted segment shows increased uptake on the delayed image. This represents accumulation of the radionuclide in the area of increased bone turnover at the junction between dead and reactive bone. The image of increased uptake alone, however, is not restricted to osteonecrosis but would be seen in reflex sympathetic dystrophy, transient migratory osteoporosis, osteoarthritis, rheumatoid arthritis, infection, and tumor to cite a few. However, rarely one will see a "cold in hot" image which is virtually pathognomonic for osteonecrosis. This is a very early finding and is only seen when the area of infarction is relatively large and the reaction to it not yet maximal. Otherwise, the area of reaction will obscure the area of decreased uptake. Although single photon emission computerized tomography (SPECT) can overcome this deficiency, it is not as widely available as other imaging forms - specifically MRI.

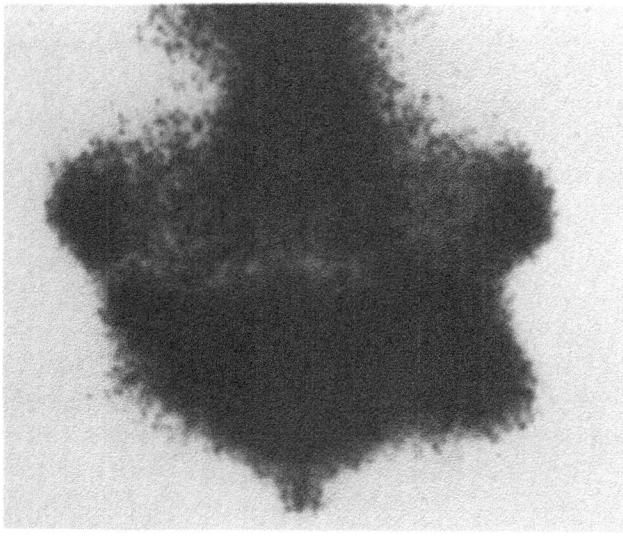

Figure 3. ^{99m}Tc-diphosphonate bone scan in unilateral osteonecrosis is highly reliable. As shown here the affected left side shows diagnostic increased uptake.

^{99m}Tc-diphosphonate bone scanning is best in evaluation of unilateral disease, since its highest sensitivity is reached when asymmetry of uptake is detected (Figure 3). If both sides are involved, there is no reference for the detection of subtle changes even though more dramatic changes could be detected (Figure 4). Even when the X-ray of the pelvis shows two normal hips, bilateral preradiologic disease in the earlier stages could be missed on bone scan because of the lack of asymmetry to detect subtle uptake. From the diagnostic algorithm at the end of this chapter, it will be seen that, for these reasons, bone scanning is a diagnostic technique of secondary importance .

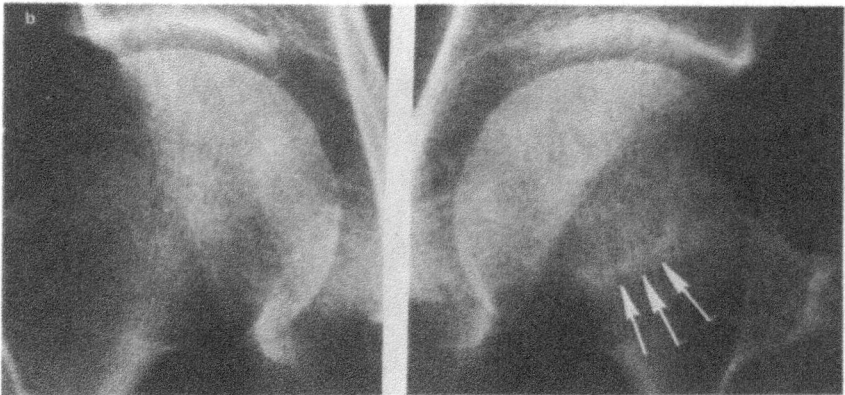

Figure 4. (A) bone scan; (B) AP radiograph both hips. The right side in this patient is also affected, but the changes are obscured by the more advanced side (arrows). The right side was falsely interpreted as normal.

MAGNETIC RESONANCE IMAGING

Magnetic resonance imaging is opening up a whole new perspective and a new set of questions in the diagnosis of osteonecrosis. A full explanation of the mechanism of detection of bone necrosis by MRI is beyond the scope for this chapter, but the interested reader is referred to the work of Mitchell and coworkers (Mitchell et al., 1987; Mitchell et al., 1986; Mitchell and Rao et al., 1987). The generation of signal from the MRI is dependent upon the tissue which is present in the femoral head. Therefore, the initial death of any segment of the femoral head would not be immediately detectable by MRI. MRI changes only are detectable seven to ten days after the event. Several authors have reported MRI to be the most accurate of all imaging modalities (Steinberg et al., 1984; Kokubo et al., 1992), although Kulkarni et al. (1987) reported 2 cases which were biopsy positive but MRI negative. Also the original MRI units were not as powerful and the programming not as sensitive as units in use today. Assuming that a 1.2 or 1.5 tesla unit is available, MRI detection of avascular necrosis should approach but not equal 100 %. The double line signal on T2 weighted image is virtually pathognomonic for osteonecrosis. This double density has been shown to represent the thickened reactive trabeculae at the margin of the osteonecrotic lesion, representing the area of decreased signal; and the increased water content in the blood of the hyperemic reaction adjacent to the increased trabecular density, representing the area of increased signal. Also, the single density line which is so often seen outlining the necrotic lesion on the T1 weighted image is thought to be highly specific for osteonecrosis (Figure 5). MRI also has two additional values. Firstly, the MRI effectively outlines the area of involvement. Steinberg and his colleagues (1992) have shown a closer correlation to the outcome of core decompression to the size of the area involved than to the stage of the disease. Confirmation of this work is required, but it would be enormously useful even in the preradiologic stages of the disease to be able to delineate the subgroup that will or will not benefit from core decompression. MRI will also allow the sequential evaluation of asymptomatic lesions that are being followed and cannot be done by conventional radiography and can also show the revascularization front and give objective evidence of tissue changes in response to treatment. Thickman et al. (1986) demonstrated that success in core decompression was characterized by either normalization of the MRI image or at least stabilization of the MRI image whereas failed cases showed evidence of disease progression on the MRI.

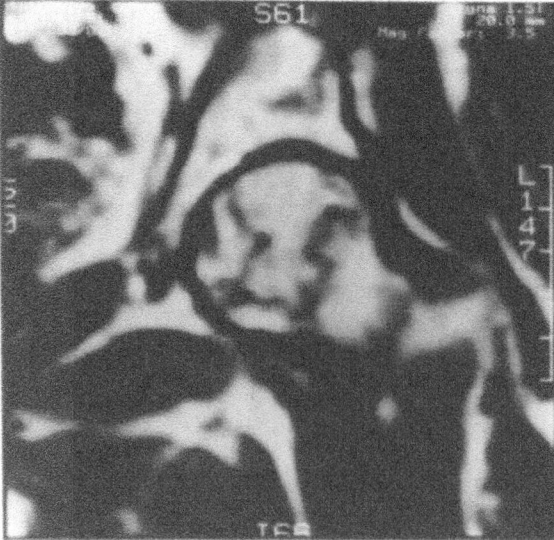

Figure 5. The single line of decreased signal intensity on this T-1 weighted image is outlining the area of necrosis.

In many instances the MRI shows extensive marrow changes involving the whole proximal end of the femur. Again, Mitchell and coworkers have shown high incidence of marrow abnormalities in the contralateral side in patients with unilateral disease (Mitchell et al., 1986). The relationship of these marrow changes to the eventual detection of full fledged necrotic lesions is not yet well understood, but there is some supporting evidence that bone infarction may be preceded by a period of bone ischemia. A great deal of additional follow-up work on the evolution of MRI changes in susceptible populations and in the asymptomatic side of patients presenting with unilateral disease should be helpful in elucidating these issues. Furthermore, much more work is urgently needed to correlate the histologic changes with the MRI findings.

Magnetic resonance imaging is a rapidly evolving discipline. Advances in the hardware, software, and signal capabilities are being exploited to enhance the capability of this diagnostic tool. New MRI angiographic techniques which enable noninvasive visualization of the vascular bed in combination with faster imaging methods may offer the capability of measurement of blood flow to assess osteonecrosis status (Cova et al., 1991; Tsukamoto et al., 1992).

BONE MARROW PRESSURE AND VENOGRAPHY

Bone marrow pressure (BMP) and venography were the mainstay of the preradiologic diagnosis of osteonecrosis prior to the advent of MRI. Unpublished data from our unit on 101 cases of biopsy-proven avascular necrosis seen between 1974, when we started to do the functional investigation of bone, and 1980 showed that bone marrow pressure was elevated as a baseline measurement in 90 of the cases. The stress test was positive in 88 of the cases; the venogram was abnormal in 90 of the cases. In fact, there was only 1 case in which all three diagnostic procedures were normal, but because of a high index of suspicion, core biopsy was obtained which showed histologic evidence of osteonecrosis. There were two cases of suspected osteonecrosis in which all three parameters were normal and the biopsy was also normal. Subsequently, other sources were found to account for their symptoms and no further evidence of osteonecrosis has surfaced. Therefore, bone marrow pressure, stress test, and venography taken as a composite diagnostic test have a high degree of sensitivity and specificity, if one considers any of the tests being abnormal as being indicative of osteonecrosis. Because BMP and venography are invasive and painful under local anesthesia, with the advent and widespread availability of MRI, bone marrow pressures have been reserved for a secondary diagnostic measure when the MRI and X-rays are both normal and a patient in whom one has a high degree of suspicion of osteonecrosis. Nonetheless, bone marrow pressure measurements showing a high percentage of abnormality in the intertrochanteric region suggests that the vascular abnormality of osteonecrosis affects a more widespread area than that which is delineated by the area of absolute necrosis and it may well be that the bone marrow pressure, itself, plays an important role in the pathogenesis of osteonecrosis. This consideration is further strengthened by our experience in measuring bone marrow pressures on the asymptomatic radiologically negative contralateral side in patients presenting with unilateral osteonecrosis. Between 1980 and 1985 we measured the bone marrow pressure and carried out venography in 48 bones in 42 patients (Zizic et al., 1989). Thirty-six cases had abnormality of baseline bone marrow pressure and/or positive stress test. In twelve joints, the bone marrow pressure and stress tests were normal. All patients were followed a minimum of four years. Fifteen out of the 36 cases

(42 %) with abnormal diagnostic tests developed signs and symptoms of avascular necrosis. All had eventual histologic proof of dead bone marrow and trabeculae. None of the patients with complete normalcy of the bone marrow pressure tests (n = 12) subsequently developed avascular necrosis or radiologic changes (p < 0.005). Moreover, those patients who eventually developed avascular necrosis did so within two years and no patient who remained asymptomatic for two years subsequently developed signs and symptoms of avascular necrosis within the followup period (maximum 8 years). This allowed us to separate unilateral patients into a subgroup which was at a 42 % risk for developing disease and another subset which, apparently, had no risk for developing the disease. Unfortunately, this was before the era of MRI and comparable information is not available in a patient population that has had an MRI evaluation.

CORE BIOPSY

Core biopsy, as a diagnostic procedure, will be indicated in those patients in whom there remains a high index of suspicion for osteonecrosis, but for whom pathognomonic changes are not evident on the X-ray, MRI, or bone scan. The indication for core biopsy, as a diagnostic procedure, is separate from core biopsy as a therapeutic procedure and, in the authors' opinion, is only necessary for the establishment of diagnosis if the image modalities have not established the diagnosis with overwhelming but probably not absolute certainty. Nonetheless, as intimated earlier in this chapter, the biopsy is subject to several errors which are reviewed in detail in the chapter in this book by Bauer and Stulberg. To summarize, the biopsy is subject to sampling error in which the biopsy tract does not actually enter the area of necrosis, but passes tangential to it and to heat artifact due to thermal changes induced by the insertion of the biopsy trocar. This would be most likely to occur with small biopsy and power inserted biopsy devices and to processing error of overdecalcification which can cause loss of trabecular osteocytes. For this reason, a biopsy which fails to show osteonecrotic trabeculae in the face of pathognomonic MRI bone scan, SPECT scan, or X-ray changes should not constitute the absence of the diagnosis, but should be interpreted as a sampling error. Moreover, the bone marrow in the biopsy specimen must be as carefully evaluated as the trabeculae, and evidence of bone marrow necrosis should be reported as such and considered as the earliest stage of osteonecrosis. It may be reversible spontaneously or under treatment but to consider that it is not part of the disease process is to ignore perhaps the most treatable phase of the disease.

THE SUSCEPTIBLE HIP

In our experience, the majority of patients in whom the diagnosis of osteonecrosis of the hip is eventually established have some medical condition with which an increased incidence is associated. In fact, only a small percentage of patients who develop avascular necrosis are otherwise completely healthy, with both normal history and laboratory findings. This has not been a universal finding with some authors reporting an incidence of idiopathic osteonecrosis of the femoral head as high as 30 or 40 %. However, Matsuo and coworkers (1988) in a careful epidemiologic study have identified an increased risk of avascular necrosis with as little as 400 ml. of ethanol consumption per week which is approached by the per capita alcohol consumption in some parts of Germany and France.

Therefore, with closer questioning, it is felt that many of the cases reported as idiopathic could actually be considered alcohol-associated. Table 2 shows the etiologic associations of the first 90 patients that we treated for avascular necrosis. The most important information from this figure is that only 6 % had no detectable etiologic association. Therefore, a high index of suspicion of osteonecrosis of the femoral head must be entertained for anyone presenting with hip pain, negative X-ray, and particularly with a predisposing etiology.

Table 2. Clinical conditions associated with ON

Condition	N. Patients	N. Hips
Trauma	10	11
Steroids	39	65
SLE	23	39
RA	2	3
Transplant	6	9
Misc.	8	14
Gaucher	1	2
Alcohol	32	54
Sickle Cell	3	5
None	5	6
Total	90	143

In the report of Merle D'Aubigne et al. (1965) only 13 % of patients had evidence of disease at the time of first visit, increasing to 50 % bilaterality during the period of follow-up. Incidence of bilaterality as reported in the literature varied from 50 to 80 % (D'Aubigne et al., 1965; Jacobs, 1978; Kerboul et al., 1974). Therefore, particular attention must be paid to the asymptomatic side and the patient should be alerted to report for immediate evaluation should symptoms develop on that side while they are under treatment for the index joint.

SUMMARY: A PROPOSED DIAGNOSED ALGORITHM

Figure 6 shows the diagnostic algorithm that we are currently practicing for all patients who present with hip complaint. In addition to the obvious clinical history and physical exam, all patients receive AP and frogleg pelvis X-rays, so that both hips can be evaluated radiographically simultaneously. If the hip is positive for avascular necrosis, then treatment choices can be based upon the findings. If it is negative, then the patient is referred for MRI of both hips. If the MRI is positive, treatment choices can be based upon that finding and those treatment choices are beyond the scope of this article. If the MRI is negative, the patient is referred for a technetium 99 m bone scan. If the bone scan is positive, the patient is scheduled for a core biopsy to establish diagnosis. If the bone scan is negative, the patient is scheduled for bone marrow pressure study and venogram and, if any of the three tests - bone marrow stress test, bone marrow pressure stress test, or venogram - are positive, a core biopsy is carried out.

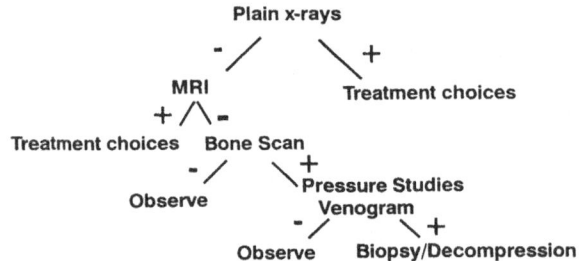

Figure 6. Diagnostic Algorithms.

The biopsy is for diagnostic purposes, although we also believe that it has therapeutic effect if the patient is suffering from early osteonecrosis of the femoral head. If the bone marrow pressure stress test and venogram are all negative, the patient is observed for change in symptoms, physical findings, and imaging findings. Under normal circumstances, if the symptoms persisted in equal or greater intensity, X-ray and MRI would be repeated at three months. We believe that it is important not to miss the early diagnosis of osteonecrosis and that the outlined algorithm is a responsible way to establish a diagnosis while keeping down medical expenses.

REFERENCES

Camp, J.F., C.W. Colwell: Core decompression of the femoral head for osteonecrosis. JBJS 68-A: 1313-1319 (1986).

Cova, M., Y.S. Kang, H. Tsukomoto, L.C. Jones, E. McVeigh, B.L. Neff, C.J. Herold, W.W. Scott, D.S. Hungerford, E.A. Zerhouni : Bone marrow perfusion evaluated with gadolinium-enhanced dynamic fact MR imaging in a dog model. Radiology 179: 535-539 (1991).

D'Aubigne', R.M., M. Postel, A. Mazabraud, P. Massias J. Gueguen: Idiopathic necrosis of the femoral head in adults. JBJS 47B: 612-633 (1965).

Hungerford, D.S., D.W. Lennox: Diagnosis and treatment of ischemic necrosis of the femoral head. In: C. McCollister Evarts (ed): Surgery of the Musculoskeletal System. 2nd ed. Churchill Livingstone Publ., New York 1990.

Jacobs, B. : Epidemiology of traumatic and non-traumatic osteonecrosis. Clin. Orthop. 130: 51-67 (1978).

Kerboul, M., J. Thomine, M. Postel, R.M. D'Aubigne': The conservative surgical treatment of idiopathic aseptic necrosis of the femoral head. JBJS 56B: 291-296 (1974).

Kokubo, T., Y. Takatori, S. Ninomiya, T. Nakamura, M. Kamogawa: Magnetic resonance imaging and scintigraphy of avascular necrosis of the femoral head. Prediction of subsequent segmental collapse. Clin. Orthop. 277: 54-60 (1992).

Kulkarni, M.V., R.R. Tarr, E.E. Kim, C.B. McArdle, C.L. Partain: Potential pitfalls of magnetic resonance imaging in the diagnosis of avascular necrosis. J. Nucl. Med. 28: 1052-1054 (1987) .

Matsuo, K., T. Hirohata, Y. Sugioka, M. Ikada, A. Fukuda: Influence of alcohol intake, cigarette smoking, and occupational status on idiopathic osteonecrosis of the femoral head. Clin. Orthop. 234: 115-123 (1988).

Mitchell, D.G., P.M. Joseph, M. Fallon, W. Hickey, H.Y. Kressel, V.M. Rao, M.E. Steinberg, M.K. Dalinka: Chemical shift MR imaging of the femoral head: An in vitro study of normal hips and hips with avascular necrosis. AJR 148: 1159-1164 (1987).

Mitchell, D.G., H.Y. Kressel, P.H. Arger, M.K. Dalinka, C.E. Spritzer, M.E. Steinberg: Avascular necrosis of the femoral head: morphologic assessment by MR imaging, with Cl correlation. Radiology 161: 739-742 (1986).

Mitchell, D.G., V.M. Rao, M.K. Dalinka, C.E. Spritzer, A. Alavi, M.E. Steinberg, M. Fallon, H.Y. Kressel: Femoral head avascular necrosis: correlation of MR imaging, radiographic staging, radionuclide imaging, and clinical findings. Radiology 162: 709-715 (1987).

Mitchell, D.G., V.M. Rao, M. Dalinka, C.E. Spritzer, L. Axel, W. Gefter, M. Kricum, M.E. Steinberg, H.Y. Kressel: Hematopoietic and fatty bone marrow distribution in the normal and ischemic hip: new observations with 1.5-T MR imaging. Radiology 161: 199-202 (1986).

Ropes, M.W., G.A. Bennett, S. Cobbs, R. Jacox, R.A. Jessar: Diagnostic criteria for rheumatoid arthritis, 958 revision. Ann. Rheum. Dis. 18: 49-53 (1959).

Rutishauser, E., A. Rohner, D. Held: Experimentelle Untersuchungen über die Wirkung der Ischämie auf den Knochen und das Mark. Virchows Arch. Path. Anat. 333: 101-118 (1960).

Steinberg, M.E., R.E. Bands, S. Parry, E. Hoffman, T. Chan, K.M. Hartman: Does lesion size affect outcome in avascular necrosis. 59th Annual Meeting of The Academy of Orthopaedic Surgeons. Final Program, p. 176 (1992).

Steinberg, M.E., C.T. Brighton, D.R. Steinberg, S.E. Tooze, G.D. Heyken: Treatment of avascular necrosis of the femoral head by a combination of bone grafting, decompression, and electrical stimulation. Clin. Orthop. 186: 137-153 (1984).

Thickman, D. , L. Axel, H .Y. Kressel, M. Steinberg, H. Chen, M. Velehick, M. Fallon, M.K. Dalinka: Magnetic resonance imaging of avascular necrosis of the femoral head. Skeletal Radiol. 15: 133-140 (1986) .

Tsukamoto, H., Y.S. Kang, L.C. Jones, M. Cova, C.J. Herold, E. McVeigh, D.S. Hungerford, E.A. Zerhouni: Evaluation of marrow perfusion in the femoral head by dynamic magnetic resonance imaging: effect of venous occlusion in a dog model. Investigative Radiology 27: 265-281 (1992).

Zizic, T.M., C.G. Lew, C. Marcoux, D.S. Hungerford: The predictive valve of hemodynamic studies in preclinical ischemic necrosis of bone. J. Rheumatol. 16: 1559-1564 (1989).

BONE BIOPSY AS DIAGNOSTIC CRITERIA FOR ASEPTIC NECROSIS OF THE FEMORAL HEAD

J-PH. HAUZEUR, and J.L. PASTEELS

Services de Rhumatologie et d'Histologie
Université Libre de Bruxelles
Bruxelles, Belgique

INTRODUCTION

Aseptic osteonecrosis of the femoral head, post-traumatic or not, is a disease that can lead to destruction of the hip. The principal diagnostic criteria are classically the radiological findings of a crescent sign (Norman et al., 1963) or collapse of the femoral head (Griffith, 1986; Jacobs, 1978; Jones, 1985). In the radiologically silent stages of the disease, other diagnostic procedures were proposed such as radionuclide uptake study (Alavi, 1977), intraosseous phlebography (Hulth, 1970), intramedullary pressure (Arlet et al., 1968) and more recently magnetic resonance imaging (MRI) (Moon et al., 1983). However the reference criteria were always the histological findings obtained after resection of the head or by biopsy.

BIOPSY

Bone biopsy of the femoral head can be obtained by two methods.
The first and the oldest was developed by Ficat et al. (1968) following Phemister (1949) and Bonfiglio et al. (1968) who developed this method aiming at a therapeutic effect by core decompression and grafting. This method is a surgical procedure using an 8 mm diameter drill. Some complications were described by Ficat (1977): bad direction of the core, too hard bone, hematoma, thrombophlebitis, skin necrosis, acute or subacute arthritis.
The second method was described by Hauzeur et al. (1986): the percutaneous transtrochanterian needle biopsy. Under fluoroscopic control, a special 3 mm diameter drill is introduced by hand through the greater trochanter and a core of bone is obtained from the intertrochanteric area to the subchondral bone or the cartilage of the femoral head. After 225 biopsies, only 3 aseptic trochanteritis and one asymptomatic exostosis were reported. Hauzeur's method seems easier and safer than Ficat's method, but it has

Bone Circulation and Vascularization in Normal and Pathological Conditions
Edited by A. Schoutens *et al.*, Plenum Press, New York, 1993

277

not yet been tested whether such a percutaneous needle procedure could have some therapeutic effects.

Since a bone biopsy cannot assess the whole femoral head, some sampling errors might be suspected. When the anterosuperior part of the femoral head is investigated, there have been no false negative biopsies except in two cases reported by Arlet (1981) and Stulberg (1991). The wide Ficat drill, to avoid the cartilage, did not reach limited

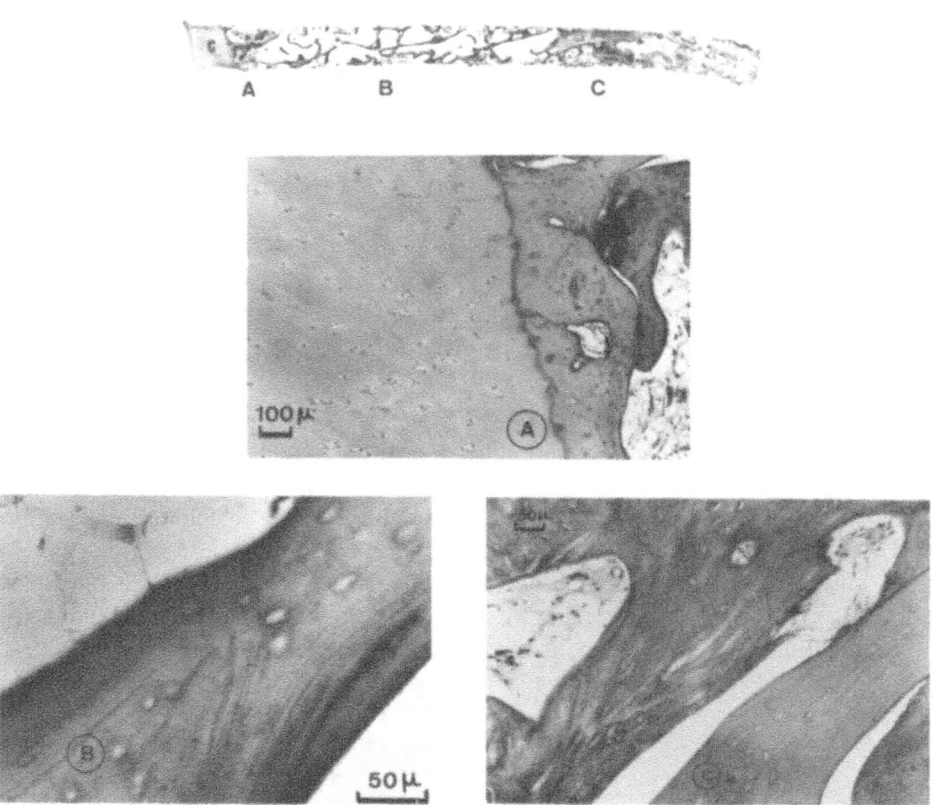

Figure 1. General view of a core biopsy specimen, 3 mm thick. The cartilage (c) is well preserved. From cartilage to deeper locations, the following aspects are found. A) Necrotic subchondral bone with apposition of viable bone by the mechanism of "creeping substitution", inducing characteristic thickening of the trabeculae. Incipient fibrosis in the neighbouring bone marrow; B) A large necrotic area where the osteocyte lacunae are completely empty and fat cells appear "mummified", as can been deduced from absence of nuclei and discontinuous outlines, but their general shape is still present; C) At the deeper border of the necrotic area, a zone of creeping substitution with thickening of the trabeculae is found again.

subchondral osteonecrotic lesions. In one case, Hauzeur's drill did not detect the necrotic lesion because of an unusual location in the anteroinferomedial part of the head. Fortunately the MRI study detected the error and a second biopsy in the right place was obtained (Hauzeur et al., 1989; Hauzeur et al., 1992). Such a MRI control of the biopsy trail can be used every time a sampling error is suspected.

HISTOLOGICAL CRITERIA FOR OSTEONECROSIS

In advanced stages of osteonecrosis, the lesions found in resected heads include bone marrow and trabecular bone modifications. They result from segmental necrosis of the subchondral bone followed by marginal demarcation of the infarcted areas by marrow fibrosis and reactive bone formation, the so-called creeping substitution (Phemister, 1949). Often a fracture could be found at the margin of the necrotic area with a collapse of this part of the femoral head. Unlike osteoarthritis, the articular cartilage is more or less preserved until the late stage, when an osteoarthritic reaction appears. Although no clear histological diagnostic criteria exist, the diagnosis for aseptic osteonecrosis seems to be histologically confirmed when the subchondral space contains a necrotic area surrounded by fibrosis of the bone marrow with enlarged bone trabeculae (Sweet et al., 1988).

In the earlier stages, the lesions may be less developed. The trabecular bone necrosis may be only partial or doubtful and the bone marrow abnormalities may be only hemopoietic, with fat cell necrosis replaced by a protein-rich exudate, an early fibroblastic reaction or recent or older hemorrhages (Arlet, 1973; Hauzeur, 1989).

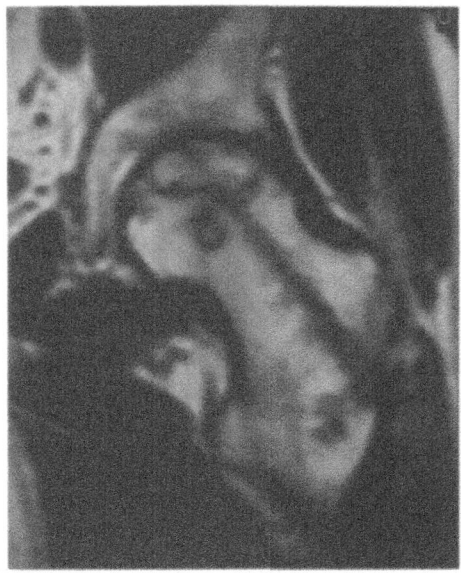

Figure 2. T1 weighted Magnetic Resonance image of a left hip, four weeks after biopsy. The trail of the biopsy is well seen and makes it possible to avoid a sampling error.

Such incomplete findings may be considered as diagnostic if three rules are respected. Firstly, at least partial trabecular bone necrosis must be found. Such partial trabecular bone necrosis is revealed by more than 50 % empty osteocyte lacunae. Secondly, abnormal bone marrow must be present around such partial necrotic trabeculae, because empty osteocyte lacunae with normal bone marrow could be due to aging (Wong, 1985). Lastly, no local septic or neoplastic process can be present. The relationship with osteoarthritis is less clear. Focal trabecular bone necrosis without extensive bone marrow necrosis under the eburnated surface or around cystic lesions is a feature of advanced osteoarthritis (Franchi et al., 1992). In some instances, significant lesions of aseptic necrosis extending to a deep location within the femoral head were found in cases of osteoarthritis of the hip (Franchi et al., 1992; Sissons et al., 1992). As there was no evidence of collapse and necrotic subchondral trabecular bone was markedly thickened indicating that the bone sclerosis preceded the necrosis, it is suspected that necrosis was secondary. Ficat et al. (1972) elsewhere described an association of osteonecrosis and early chondrolysis under the term of "ischemic coxopathy".

Finally, a differential diagnosis must be done with transient osteoporosis of the hip (TOH), also called algodystrophy of the hip. The pathology of this condition is not well known. Arlet et al. (1979) compared the pathological findings in 9 TOH with findings in osteonecrosis. In TOH, the bone marrow lesions were principally edema without significant fat necrosis and the trabecular bone was found to be fairly normal with some increase in the osteoblastic activity. Potter et al. (1992) reported more recently in 4 biopsies of TOH some fat cell necrosis, edema and fibroblastic reaction, with some increase in the osteoblastic and osteoclastic activity.

CORRELATION WITH THE OTHER DIAGNOSTIC PRODEDURES

In cases with normal or doubtful radiological findings (no crescent sign or collapse) we have compared bone biopsy with the other diagnostic procedures. The radionuclide study is positive in 51 %, the intramedullary pressure study in 80 % and MRI in 100 % when bone biopsy is positive (Hauzeur et al., 1989).

CONCLUSION

The best diagnostic criteria for aseptic osteonecrosis are still anatomo-pathologic confirmation. Bone biopsy can provide the material for it. The diagnostic criteria defined in advanced stages of osteonecrosis have sometimes an imcomplete presentation in early stages, partial trabecular bone and partial bone marrow necrosis. The diagnostic implications of such lesions are discussed by the Association for Research on Osseous Circulation (ARCO). Bone biopsy is an invasive method and, although tolerance is excellent, its use as a diagnostic test should be reserved for cases where non-invasive techniques like MRI give doubtful results.

REFERENCES

Arlet, J., R. Durroux: Diagnostic histologique précoce de l'ostéonécrose aseptique de la tête fémorale par le forage-biopsie. In: La circulation osseuse, compte rendu du 1er symposium international sur la circulation osseuse, pp. 293-302. INSERM, Paris 1973.

Arlet, J.: Histopathologie comparée de la nécrose de la tête fémorale et de l'algodystrophie sympathique réflexe de la hanche. Rhumatologie 10: 377-378 (1979).

Arlet, J., R. Durroux, C. Fauchier, C. Thiechart: Histopathology of the non traumatic necrosis of the femoral head: topographic and evolutive aspects. In: J. Arlet, P. Ficat, D.S. Hungerford (eds): Bone circulation, pp. 296-305. William and Wilkins, Baltimore 1981.

Bonfiglio, M., E.M. Voke: Aseptic necrosis of the femoral head and non union of the femoral neck. Effect of treatment by drilling and bone grafting (Phemister technique). J. Bone Joint Surg. 50-A: 48-66 (1968).

Ficat, P., G. Utheza: Le forage-biopsie de la hanche. Rev. Med. Toulouse 4: 223-230 (1968).

Ficat, P., J. Arlet : Coxopathise ischémiques. Rev. Chir. Orthop. 58: 543-561 (1972).

Ficat, P., J. Arlet: Méthodes histologiques, forage-biopsie. In: P. Ficat, J. Arlet (eds): Ischémie et nécrose osseuses. L'exploration fonctionnelle de la circulation intra-osseuse et ses applications, pp. 46-53. Masson, Paris 1977.

Franchi, A., P.G. Bullough: Secondary avascular necrosis in coxarthrosis: A morphologic study. J. Rheumatol. 19: 1263-1268 (1992).

Hauzeur, J.PH., S. Orloff, J. Taverne-Verbanck, J.L. Pasteels: Diagnosis of aseptic osteonecrosis of the femoral head by percutaneous transtrochanterian needle biopsy. Clin. Rheumatol. 5: 346-358 (1986).

Hauzeur, J.PH., J.L. Pasteels, S. Orloff: Bilateral non-traumatic aseptic osteonecrosis in the femoral head. An experimental study of incidence. J. Bone Joint Surg. 69 A: 1221-1223 (1987).

Hauzeur, J.PH., J.L. Pasteels, A. Schoutens, M. Hinsenkamp, T. Appelboom, I. Chochrad, N. Perlmutter: The diagnostic value of magnetic resonance imaging in non-traumatic osteonecrosis of the femoral head. J. Bone Joint Surg. 71-A: 641-649 (1989).

Hauzeur, J.PH., JR.S. Sintzoff, T. Appelboom, V. De Martelaer, J. Bentin, J.L. Pasteels: Relationship between magnetic resonance imaging and histologic findings by bone biopsy in non-traumatic osteonecrosis of the femoral head. J. Rheumatol. 19: 385-392 (1992).

Norman, A., P. Bullough: The radioluscent crescent line - An early diagnostic sign of avascular necrosis of the femoral head. Bull Hosp. J. Dis. 24: 99 (1963).

Phemister, D.B.: Treatment of the necrotic head of the femur in adults. J. Bone Joint Surg. 31-A: 55-66 (1949).

Potter, H., M. Moran, R. Schneider, R. Bansal, C. Sherman, T. Markisz: Magnetic resonance imaging in diagnosis of transcient osteoporosis of the hip. Clin. Orthop. 280: 223-229 (1992).

Sissons, H.A., M.A. Nuovo, G.L. Steiner: Pathology of osteonecrosis of the femoral head. Skeletal Radiol. 21: 229-238 (1992).

Stulberg, B.N., A.W. Davis, T.W. Bauer, M. Levine, K. Easley: Osteonecrosis of the femoral head. A prospective randomized treatment protocol. Clin. Orthop. 268: 140-151 (1991).

Sweet, D.E., J.E. Madewell: Pathogenesis of osteonecrosis. In: D. Resnick, G. Niwayama (eds): Diagnosis of Bone and Joint disorders, pp. 3188-3237. Saunders, Philadelphia 1988.

Wong, S.Y.P., J. Kariks, R.A. Evans, C.R. Dunstan, E. Hills: The effect of age on bone composition and viability in the femoral head. J. Bone Joint Surg. 67-A: 274-283 (1985).

THE HISTOLOGY OF OSTEONECROSIS AND ITS DISTINCTION FROM HISTOLOGIC ARTIFACTS

T.W. BAUER[*], and B.N. STULBERG[**]

[*] Departments of Pathology and Orthopaedic Surgery
The Cleveland Clinic Foundation
[**] The Center for Joint Reconstruction
Cleveland, Ohio, USA

INTRODUCTION

Current theories of the pathogenesis of osteonecrosis have led surgeons to try to diagnose and treat osteonecrosis very early in its course; before radiographic changes become evident and sometimes even before pain develops (Ficat, 1980; Stulberg, 1989, 1990, 1991). It follows that surgeons expect pathologists to be able to recognize very early osteonecrosis. The purpose of this discussion is to emphasize the difficulty of this histologic recognition, and to illustrate selected sampling and processing artifacts that, if misinterpreted could lead to false positive and false negative diagnoses. Theories of pathogenesis will be reviewed, followed by the histologic findings in osteonecrosis and a brief discussion concerning the recognition of selected histologic artifacts.

Pathogenesis of Osteonecrosis. We now recognize that there are a number of patient populations that show an increased incidence of osteonecrosis, especially of the femoral head. These groups include patients with traumatic disruption of the blood supply, hemoglobinopathies, patients with endogenous or exogenous corticosteroid excess, patients with alcoholism, Gaucher's disease, collagen vascular disorders or rheumatoid arthritis, and transplant recipients (Ficat, 1980). It was recognized early on that no single theory of pathogenesis would easily explain this epidemiology. Over the years a number of different mechanisms have been proposed, and some have been shown experimentally to result in identical end stage lesions of osteonecrosis. One mechanism is by disruption of the arterial blood flow, and the most common pathogenesis of osteonecrosis is post-traumatic necrosis involving disruption of the epiphyseal arteries. Atherosclerotic occlusion of the blood supply and even Buerger's disease have also been proposed. Extraosseous venous compression has also been shown experimentally to produce identical lesions, but is probably less common clinically. The theory of fat embolism has been promoted, and red cell emboli are a likely mechanism for necrosis in

Bone Circulation and Vascularization in Normal and Pathological Conditions
Edited by A. Schoutens *et al.*, Plenum Press, New York, 1993

283

patients with sickle cell disease. A direct toxic effect on the osteocytes themselves has also been proposed (Spencer, 1986), and lipid accumulation has been demonstrated in the osteocytes of animals and patients who have received corticosteroids (Kawai, 1985).

There are probably another group of patients, however, for whom the most likely explanation is extracellular, intraosseous compression of vessels. The femoral head might be thought of as a closed, rigid compartment. Within the compartment are trabeculae of bone, marrow elements, adipocytes and blood vessels. It has been hypothesized that an expansion of any one of the elements in this space can only occur at the expense of another (Ficat, 1985; Ficat, 1980). If there is an increase in marrow elements, histiocytes or adipocytes, for example, then there must be an increase in the extracellular pressure, at least transiently. This increased pressure might result in collapse of the least rigid compartment, capillaries and small veins. This vascular collapse will result in ischemia, edema, a further increase in pressure and an enlarging zone of ischemia.

There is now little question that patients with avascular necrosis of the femoral head have increased intraosseous pressures (Downey, 1988; Ficat, 1985; Zizic, 1986). At the time of reconstruction, many surgeons have measured the pressure in the femoral head and documented pressures far exceeding those of the venous blood system. However, this observation does not prove cause and effect, and intraosseous pressures have also been elevated in patients with osteoarthritis. Several authors have attempted to quantitate change in size and/or number of adipocytes in the femoral heads of animals given large doses of steroids. In several studies, Wang and co-workers (Wang, 1977) have shown an increase in the mean diameter of adipocytes in rabbits given pharmacologic doses of steroids. Other authors, however, have not been able to duplicate those observations in a different animal system (Bauer, 1987).

Core Decompression Biopsy. With this potential pathogenesis in mind, orthopedic surgeons have developed a treatment mechanism designed to alleviate intraosseous pressure elevations before necrosis progresses. This procedure is called the "core decompression biopsy", and consists of essentially drilling a large hole in the femoral head. This hole is intended to "decompress" the compartment, and allow new blood vessels to penetrate the femoral head. Interestingly, most patients report an immediate relief of hip pain when this procedure is performed, and there is evidence that it works well if applied early in the course of the disease. In fact, when a patient presents with osteonecrosis of the femoral head, some surgeons promote drilling the contralateral head because it probably represents Stage 0, i.e. pre-symptomatic AVN.

HISTOLOGY

Just as we cannot immediately recognize a myocardial infarct histologically, it is also not possible to recognize the earliest phases of osteonecrosis. The changes in bone are more variable than in the heart, probably because of less uniform vascularity. Although cytologic changes can be identified by electron microscopy as early as 4 hours (James, 1986), experimental studies suggest that at least 24 - 72 hours of anoxia are required before autolysis is sufficient for recognition at the light microscopic level. Because viable osteocytes sometimes appear shrunken and pyknotic, we cannot use nuclear pyknosis as a reliable sign of osteocyte death. Even with complete anoxia, experimental studies have shown that it may take 48 hours to 4 weeks before the cell nuclei disappear. Therefore, the presence or absence of nuclei cannot be the sole criterion

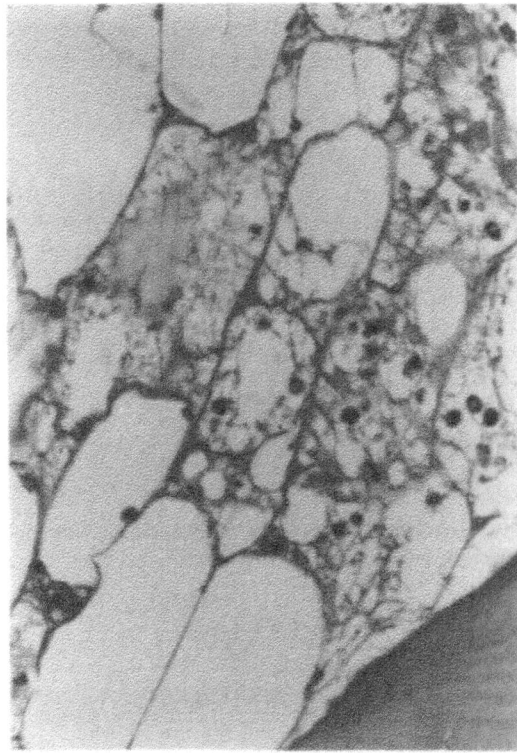

Figure 1. Photomicrograph of early ischemic osteonecrosis. "Microvescicular" fatty change of marrow adipocytes is present, and there is a histiocytic reaction to necrotic debris.

of viability.

The first reliable histologic signs of osteonecrosis are hemorrhage, loss of hematopoietic elements, loss of adipocyte nuclei and a microvescicular fatty change of the marrow adipocytes (Saito, 1987). This is followed by rupture of the adipocytes and histiocytic phagocytosis of debris (Figure 1). The ruptured adipocytes produce round cavities that may be surrounded by histiocytes ("liquefaction necrosis").

The histologic picture of the next phase depends largely on geographic factors. Just like a large bone allograft, if the infarct is not revascularized, then a central zone will remain acellular for many years (Figure 2). Tissue on the edge of the zone of necrosis, however, reflects dynamic changes between repair and ischemic necrosis (Sissons, 1992; Spencer, 1986). There may be convincing, complete loss of osteocyte nuclei and loose marrow fibrosis. Chronic inflammation may be prominent, and lipoid granulomas are often present. The trabecular bone adjacent to the infarct shows somewhat increased vascularity, and an increase in blood flow can be demonstrated radiographically. The increased blood flow may be associated with "localized osteoporosis" in the region surrounding the infarct.

Eventually there is a variable amount of bone resorption with a more consistent deposition of new bone on the surface of necrotic trabeculae ("creeping substitution") (Figure 3). As remodeling progresses, the infarct reaches its point of least mechanical stability and collapse may occur. During this phase the infarct becomes radiolucent, and articular collapse with associated degenerative and reparative changes may complicate the histologic

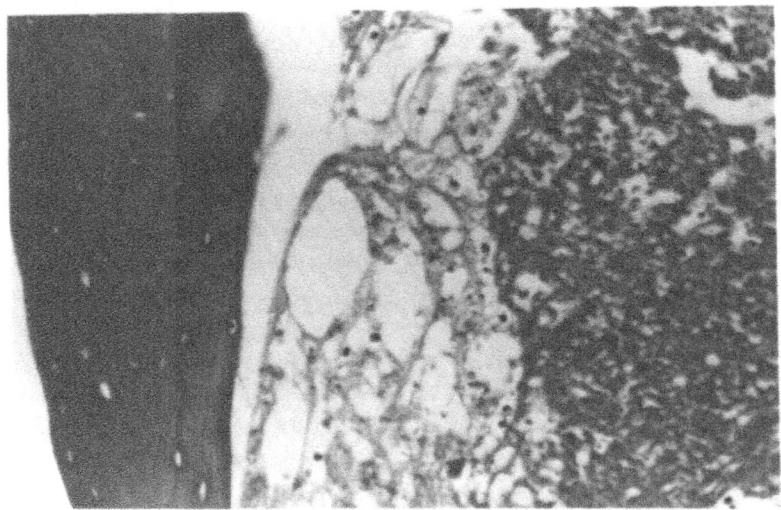

Figure 2. Easily recognizable "coagulative" osteonecrosis, with global loss of osteocytes and necrotic debris in the marrow space.

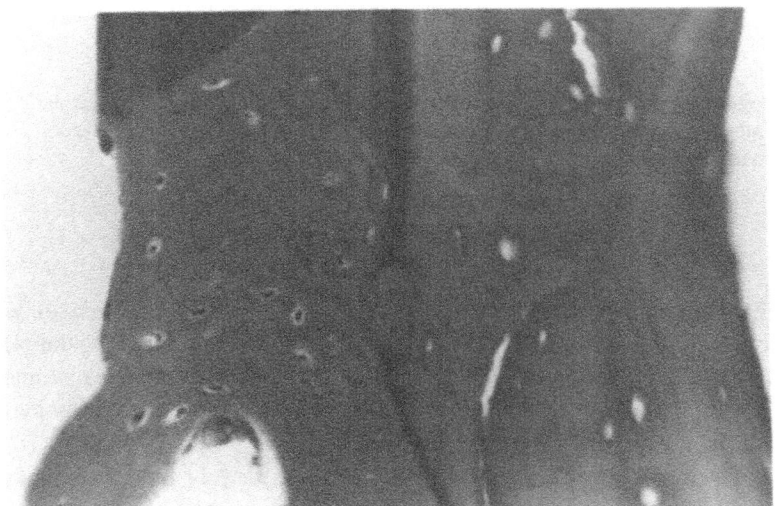

Figure 3. Photomicrograph of "creeping substitution." A geographic region of empty lacunae is surrounded by new, viable bone in which osteocytes can be identified.

picture. As remodeling continues, small bone infarcts may eventually be completely replaced by viable new bone. Larger infarcts, however, are unlikely to be completely replaced. Remodeling peripherally leads to local osteosclerosis that is evident radiographically as a sclerotic rim. Dystrophic calcification of the fibrotic marrow results in punctate calcification that radiographically mimics a cartilaginous tumor.

TISSUE PROCESSING AND HISTOLOGIC ARTIFACTS

In order to prepare microscope slides of bone biopsies, the tissue must be suitably fixed, dehydrated, embedded, sectioned and stained. Protocols for processing bone tissue are available in standard histotechnology texts (Anderson, 1982), but bone biopsies are viewed as difficult specimens by technologists in most busy hospital laboratories. While undecalcified processing with plastic embedding results in the best final result, this is impractical for routine pathology specimens. Satisfactory methods for testing decalcification, embedding and staining must be worked out for each individual laboratory. There are many opportunities to damage bone biopsies, and some of these artifacts make the diagnosis of necrosis difficult. Selected types of artifacts are summarized below:

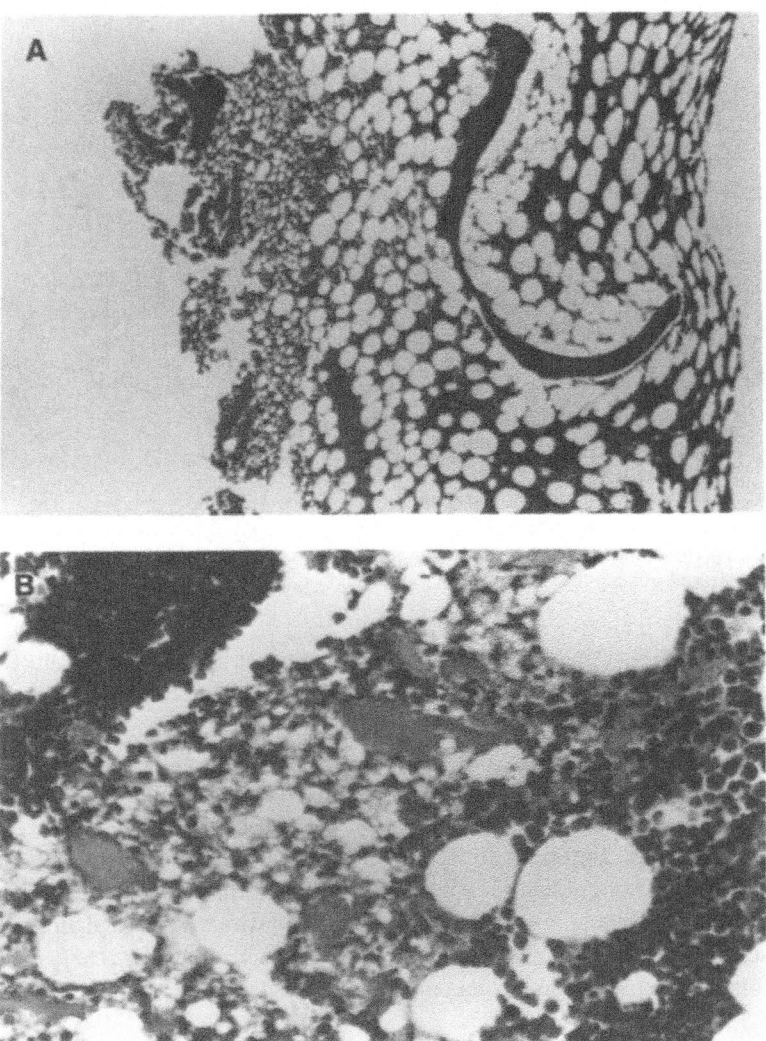

Figure 4. A) Low magnification photomicrograph of the edge of a core biopsy specimen. A zone of thermal artifact along the edge of the specimen has been induced by the biopsy procedure. B) At higher magnification, the change to marrow caused by heat could be mis-interpreted as ischemic necrosis.

287

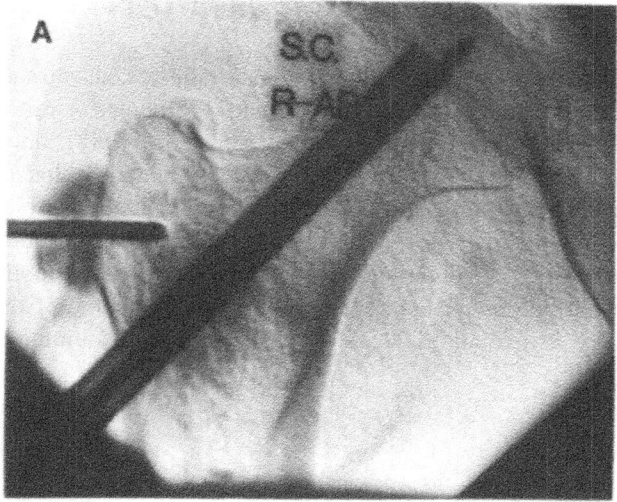

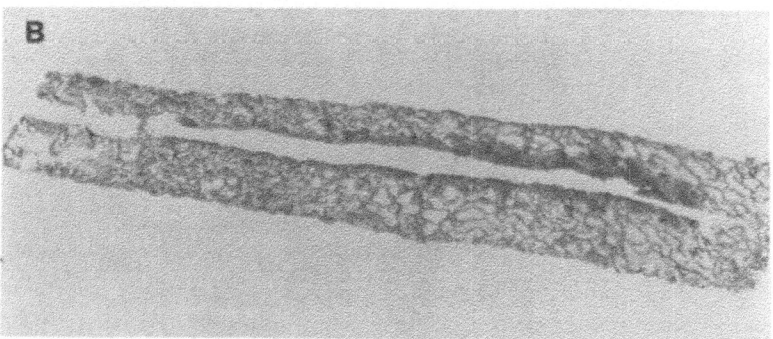

Figure 5. Low magnification photomicrograph of a core biopsy specimen. The longitudinal space in the center of the biopsy represents the placement of a guide wire prior to obtaining the core biopsy. A) Radiograph demonstrating biopsy needle and guide pin. B) Biopsy with visible space of guide wire.

Thermal Artifact. Power saws or drills may increase the temperature of adjacent tissue, especially when drilling sclerotic bone. Diffusion of heat into the adjacent marrow may cause coagulative changes that resemble the "microvescicular fatty change" of early ischemic necrosis (Figure 4), while higher levels of heat completely obscure cellular detail. Identifying the edge of the biopsy tract can be helpful in recognizing this as an artifact.

Guide Wire Tract. Before drilling the core biopsy, a thin wire may be inserted into the femoral head for radiographic localization. This wire tract is subsequently included in the biopsy specimen (Figure 5). Although the crush artifact produced does not usually resemble ischemic necrosis, awareness of the biopsy technique helps explain the mechanical disruption present in the center of the biopsy specimen.

Saw Artifact. Using a band saw to trim specimens may also generate thermal artifacts. In addition, the saw tends to force fragments of bone dust into the adjacent marrow, often resembling necrotic debris (Figure 6). Lightly brushing the cut surface of bone with a toothbrush helps minimize this artifact.

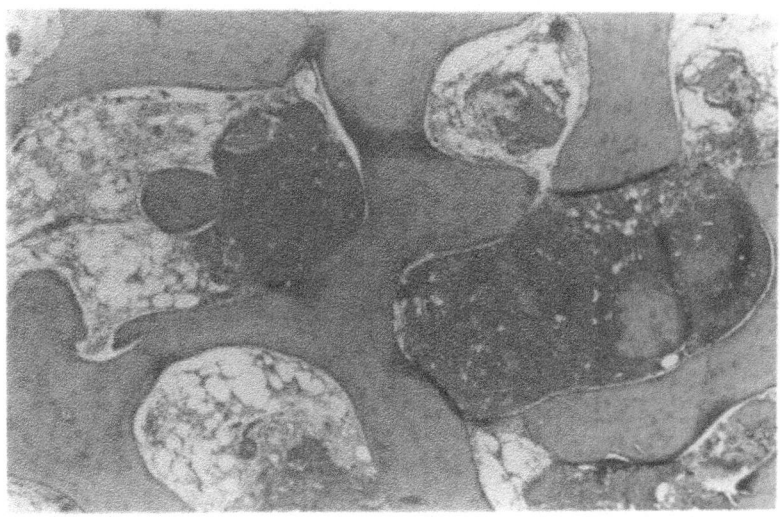

Figure 6. Bone dust and other debris can be forced into the bone by either the biopsy drill or a saw used in the laboratory. This artifact can be minimized by gently rinsing the cut surface of the bone under running water before further histologic processing.

Fixation. Underfixation will occur if the pieces of bone are too big, the volume of fixative too small, or if there is insufficient time allowed for fixation. Fixation artifacts are most prominent near the center of the specimen, and may be responsible for the apparent loss of osteocyte nuclei within the center of some trabeculae. The absence of a relation between the empty lacunae and cement lines may suggest fixation artifact rather than osteonecrosis.

Fixation is an integral part of the entire dehydration and embedding sequence, and inadequately fixed specimens tend to be tough, even if completely decalcified. This hardness may be misinterpreted as inadequate decalcification by inexperienced technicians, but re-processing with additional decalcification may make the block even harder. In our experience, inadequate fixation is the most common error in processing bone biopsies.

Decalcification. Underdecalcification prevents satisfactory sectioning. The resulting distortion prohibits recognizing cellular detail. Calcium is removed from bone specimens using strong acids, and any acid will have some effect on the stainability of the tissue. This influence increases with the degree of acidity and the length of the decalcification period. Overdecalcification inhibits the staining of nucleic acids by hematoxylin and other basic dyes. The staining of collagen with eosin is also influenced, resulting in a flat, uniform brick red or magenta color with few differential shades. Severe overdecalcification will completely remove nuclei from bone, with obvious consequences for the pathologist.

Dehydration and Embedding. Inadequate dehydration results in thick sections that may show severe cracking or spreading when placed on the flotation bath. Spreading causes separation of the marrow from trabeculae of bone. Over-processing may excessively harden the tissue, making sectioning and staining more difficult.

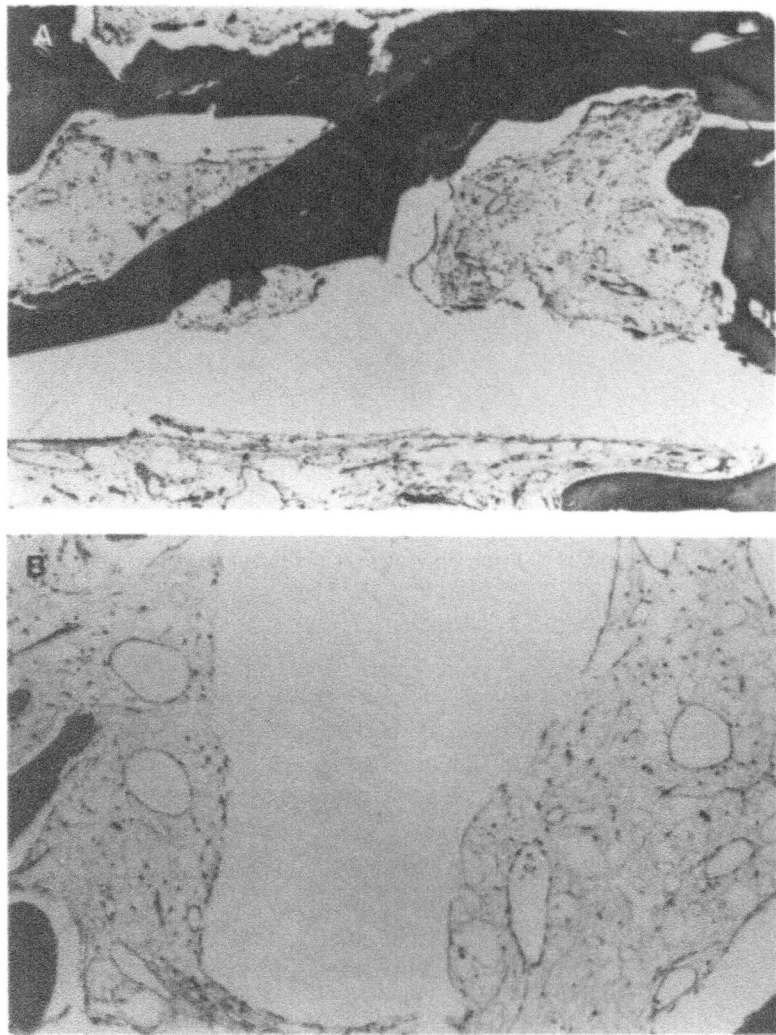

Figure 7. A) Low magnification photomicrograph of a poorly dehydrated biopsy. There is separation of the marrow from the edge of the bone, and a trabeculum has folded away from the surface of the microscope slide. A rim of osteoblasts remains adherent to the glass slide. B) Higher magnification of the apparent space created by this artifact could be mis-interpreted as a blood vessel or even a cavernous hemangioma.

Sectioning. Most sectioning artifacts are easily recognized. These include shattering due to a dull knife, length-wise scratches due to defects in the cutting edge or dragging particles of calcified bone across the section, and "venetian blind" effect when a thick fibrous specimen is cut with a dull, thin knife. When sections are placed onto the microscope slide, the bone may not adhere to the glass as well as adjacent marrow. When trabeculae of bone fold away from the slide, a rim of surrounding osteoblasts may remain, creating the false impression of a vascular space lined by endothelium (Figure 7). This common artifact needs to be recognized and distinguished from a vascular lesion.

290

Staining. Most staining artifacts are easily recognized, including stain bubbles beneath the section, hematein precipitation from incompletely dissolved hematoxylin, and bubbles due to water in the alcohol or xylene.

Exogenous Material. When biopsy specimens are placed on dry gauze, filaments may become embedded in the tissue and be difficult to remove. Gauze filaments can be recognized in histologic sections by their strong polarization and the lack of cellular reaction. Glove powder also polarizes strongly, and may adhere to fatty tissue more strongly than to other tissues. Although not usually a problem after tissue processing, glove powder can be misinterpreted as foreign material or crystal deposition in a frozen section. The source is often the laboratory technician who prepares the slide using new gloves that have not been rinsed with water.

SUMMARY

All laboratories experience technical difficulties with bone specimens, and there is no substitute for a dedicated technologist. Nevertheless, recognition of common artifacts may increase the sensitivity and specificity of diagnosing early osteonecrosis.

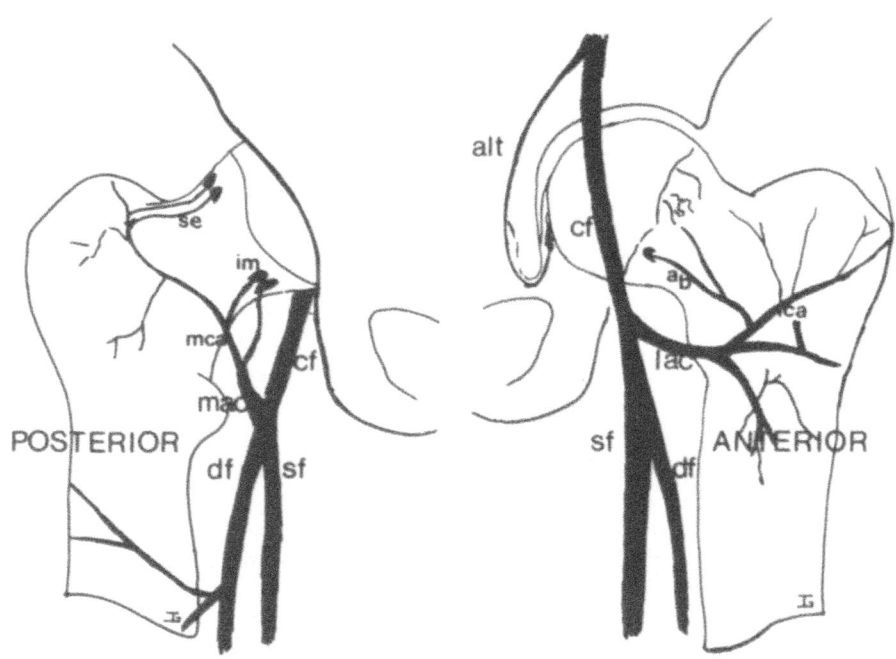

Normal vascular anatomy of the hip.
cf = Common Femoral Artery ; df = Deep Femoral Artery ; sf = Superficial Femoral Artery ; mac = Medial Circumflex Artery ; lac = Lateral Circumflex Artery ; mca = Medial Collum (Neck) Artery ; lca = Lateral Collum (Neck) Artery ; im = Inferior Metaphyseal Artery ; se = Superior Epiphyseal Artery ; alt = Teres Ligament Artery ; ab = Ascending branches of arterial ring of femoral neck.
(after Gardeniers).

REFERENCES

Anderson C.: Manual for the Examination of Bone. CRC Press, Inc. (1982).

Bauer T.W., H.C. Chiao, B.N. Stulberg: Steroid effects on adipocytes and intraosseous pressures in the canine femoral head. Lab Invest 56: 5A (1987).

Downey D.J., P.A. Simkin, W.L. Lanzer, F.A. Matsen: Hydraulic resistance: A measure of vascular outflow obstruction in osteonecrosis. J. Orthop. Res. 6: 272-278(1988).

Ficat R.P.: Idiopathic bone necrosis of the femoral head. Early diagnosis and treatment. J. Bone Joint Surg. 67B: 3-9 (1985).

Ficat R.P., J. Arlet: Ischemia and Necrosis of Bone. Williams and Wilkins Publ., Baltimore 1980.

James J., G.L. Steijn-Myagkaya: Death of osteocytes. Electron microscopy after in vitro ischemia. J. Bone Joint. Surg. 68(B): 620-624 (1986).

Kawai K., A. Tamaki, K. Hirohata: Steroid induced accumulation of lipid in the osteocytes of the rabbit femoral head. J. Bone Joint. Surg. 67A: 755-762 (1985).

Levine M., T.W.Bauer, B.N. Stulberg: Histologic evaluation of the femoral head following failed core decompression for osteonecrosis. Trans. Orthop. Res. Soc. 12: 499 (1986).

Saito S., A. Inoue, K. Ono: Intramedullary haemorrhage as a possible cause of avascular necrosis of the femoral head. J. Bone Joint. Surg. 69(B): 346-351 (1987).

Sissons H.A., M.A. Nuovo, G.C. Steiner: Pathology of osteonecrosis of the femoral head. Skeletal Radiol. 21: 229-238 (1992).

Spencer J.D., S. Humphreys, J.R. Tighe, R.R. Cumming: Early avascular necrosis of the femoral head. J. Bone Joint. Surg. 68 (B): 414-417 (1986).

Stulberg B.N., T.W. Bauer, G.H. Belhobek: Making core decompression work. Clin. Orthop. Rel. Res. 261: 186-195 (1990).

Stulberg B.N., A.W. Davis, T.W. Bauer, M. Levine, K. Easley: Osteonecrosis of the femoral head. A prospective randomized treatment protocol. Clin. Orthop. Rel. Res. 268: 140-151 (1991).

Stulberg B.N., M. Levine, T.W. Bauer, G.H. Belhobek, W. Pflanze, D.H. Feiglin, A.I. Roth: Multimodality approach to osteonecrosis of the femoral head. Clin. Orthop. Rel. Res. 240: 181-193 (1989).

Wang G.J., D.E. Sweet, S.I. Reger: Fat cell changes as a mechanism of avascular necrosis in the femoral head of cortisone-treated rabbits. J. Bone Joint. Surg. 59A: 729-735 (1977).

Zizic T.M., C. Marcoux, D.S. Hungerford, M.B. Stevens: The early diagnosis of ischemic necrosis of bone. Arthritis and Rheum. 29: 1177-1186 (1986).

BONE ARTERIOGRAPHY OF THE FEMORAL HEAD OF HUMANS IN NORMAL AND PATHOLOGICAL CONDITIONS

Takashi ATSUMI

Department of Orthopaedic Surgery
Fujigaoka Hospital
Showa University School of Medicine
1-30 Fujigaoka, Japan

Human femoral head has a specific vascular anatomy. Many studies were reported for investigations about this condition using angiography in vivo (Hipp, 1968; Mussbichler, 1956) and microangiography of autopsy (Chung, 1976; Sevitt, 1965; Trueta and Harrison, 1953; Trueta, 1957). However, it is difficult to demonstrate these arteries clearly in vivo, especially in the pathological conditions. Superselective angio-graphy of the hip was reported by Theron (1977) and this method was able to demonstrate the vascular changes of the femoral head in vivo (Atsumi et al., 1990, 1992). Microangiography of affected femoral head which was resected at operation (Atsu-mi et al., 1989) could be effective for observation of the details of vascular changes of the femoral head in pathological conditions. I investigated the vascular changes of the femoral head in normal and pathological conditions using superselective angiography of the femoral medial circumflex artery and microangiography of affected femoral head which were resected at prosthetic replacement.

SUPERSELECTIVE ANGIOGRAPHY IN ADULTS

Normal Hips

Perifemoral Head Vascular Pattern was obtained from 23 normal adult hips without risk factor for osteonecrosis. It is generally accepted that the femoral medial circumflex artery originates from the deep femoral artery or from the common femoral artery, and the posterior collum artery branches from the medial circumflex artery. Posterior collum artery was found to originate from the medial circumflex artery in 12 hips (52 %), from both medial circumflex artery and a branch of internal iliac artery in 10 hips (43 %). In the remaining hip (5 %), this artery originated directly from the internal iliac artery.

Bone Circulation and Vascularization in Normal and Pathological Conditions
Edited by A. Schoutens *et al.*, Plenum Press, New York, 1993

Vascular Distribution in the Femoral Head. Superior retinacular arteries are the most important nutrient arteries for supplying blood to a large area including the weightbearing portion of the femoral head. They branch off from the posterior collum artery. These arteries were demonstrated to stem from posterior collum artery at its distal portion and to enter the head, with main trunks delineating an arch, situated proximal to the epiphyseal scar, and extending to the center of the head. However, small arteries branching from the main trunks were not demonstrated clearly (Figure 1). The inferior retinacular arteries were seen in small regions of the medial subcapital area, but were very thin it was impossible to observe them in the intraosseous portion of the femoral head. The ligamentum teres arteries were also not demonstrated in the head.

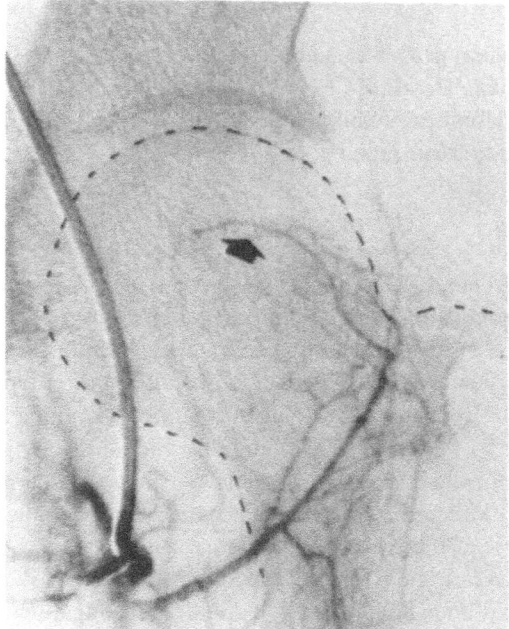

Figure 1. Superselective angiography of the femoral medial circumflex artery in the normal hip of a 38 year old male who has no risk factors for osteonecrosis. Superior retinacular arteries (Arrow) are well demonstrated up to the center of the femoral head.

Early Osteonecrosis

Angiographies were obtained from 16 hips with osteonecrosis diagnosed on bony scintigraphy, while X-rays were normal. Osteonecrosis was confirmed during the 2 to 16 months follow up by radiological examinations in 12 hips and by biopsy in 4. The main branches of the superior retinacular arteries in their extension from extraosseous area to the intraosseous portion of the femoral head were never seen. Small arteries penetrated into the femoral head.

In contrast, angiographies of 44 radiographically normal hips without abnormal findings suggesting osteonecrosis with bony scintigraphy showed a normal vascular

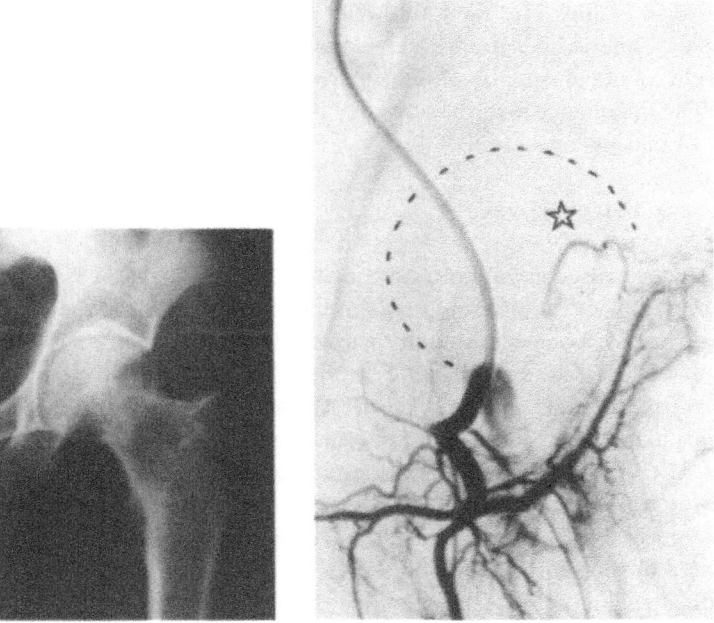

Figure 2. Superselective angiography of a hip with osteonecrosis in a 56 year old male suffering from alcoholic abuse. Moderate collapse is seen on radiograph and the superior retinacular arteries are not stained in the affected femoral head (Star).

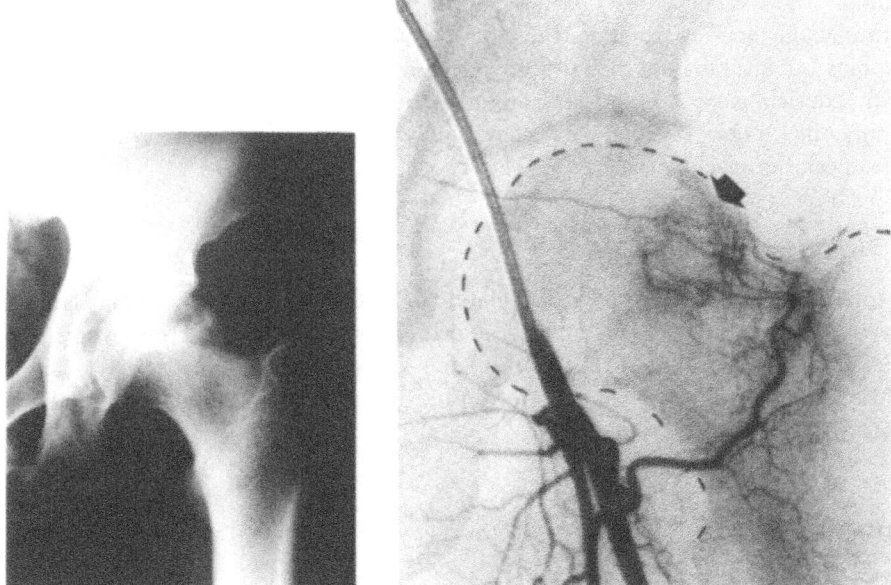

Figure 3. Superselective angiography of a hip with osteonecrosis in a 40 year old male. Slight collapse is seen on radiograph. Angiography demonstrates erratic formed small arteries penetrating into the affected head (Arrow).

pattern in 7 of these 44 hips (16 %). In the remaining 37 hips, the real superior retinacular arteries were not demonstrated. Penetration of small arteries was seen at the lateral margin of the femoral head in 21 hips. In the radiological follow up of mean 26 months, 7 out of 12 hips with small arteries penetration developed osteonecrosis. These result indicate that the impairment of blood supply of the superior retinacular arteries occurs in preradiological stage. The circulatory disturbances may develop in contralateral normal hips or normal hips of patients on steroids.

Vascular Changes in Osteonecrosis

In order to study revascularization in osteonecrosis, superselective angiography of 10 hips was performed twice at different times in the natural course of the disease. At the time of first angiography, bone scintigraphy showed increased uptake suggesting osteonecrosis in 9 out of 10 hips. The second angiography took place after the radiological evidence for osteonecrosis became evident. The first angiography showed the lack of the real superior retinacular arteries in the extraosseous area and small vessels penetrating into the head. The second angiography demonstrated arteries penetrating into the femoral head which were not seen in the first angiography and these arteries increased in size. These findings indicate that spontaneous revascularization after ischemia was initiated by the neoformation of small vessels of increasing diameter.

Superselective angiographies of 122 hips with advanced stage of ON were performed for studying the different conditions of revascularization. Revascularization originated from the newly formed superior and inferior retinacular arteries, while the ligamentum teres arteries were not seen in the intraosseous portion (Figures 2, 3). The main trunks of the real superior retinacular arteries were never demonstrated. The characteristic angiographic findings of advanced stage of ON were that the superior retinacular arteries appeared abnormal in both the extraosseous and intraosseous areas. Even though the overall status of revascularization varied from case to case and the inferior retinacular arteries penetrated to the demarcation area and increased in size. In angiographies of 88 hips out of 94 (93.6 %) with advanced osteonecrosis, the inferior retinacular arteries were found to extend to the demarcation area. This finding may indicate that the inferior retinacular arteries play an important role to compensate the impaired superior retinacular arteries.

Osteoarthritis

43 hips with osteoarthritis were obtained for superselective angiography and vascular changes were studied. Both superior and inferior retinacular arteries were hypertrophic and well demonstrated in depth of the femoral head. No evidence of impairment of the blood supply from the superior retinacular arteries was seen, as was the case in osteonecrosis.

PERTHES' DISEASE

In selective angiographies of 5 normal hips in children, superior retinacular arteries originated from the posterior collum artery and penetrated into the center of the bony epiphysis. Small branching of these arteries was not clearly seen. The superior metaphyseal arteries penetrated downward to the metaphysis. However the main trunks of these arteries were not well demonstrated. The inferior metaphyseal arteries were shown to penetrate from the inferior region of the metaphysis, but vessels were very thin and not clearly observable in the metaphysis. The medial epiphyseal arteries were retrogradely stained, but not observed in the bone.

Superselective angiography was performed on 23 hips with Perthes' disease. Lateral epiphyseal arteries were not seen in the early and advanced stages.

In the early stage with no evident repair on radiography, the lateral epiphyseal arteries were interrupted at their origin in the extraosseous area and revascularization consisting in small arteries was seen in a small lateral portion on the non-weightbearing portion of the femoral head. In the age period where Perthes' disease prevalence is high, the absence of blood supply by the medial epiphyseal arteries was reported by Trueta (1957). In Perthes' disease, these arteries were hypertrophied and divided into numerous small arteries. However, vascularization was not found in the bony epiphysis in the necrotic stage.

Characteristic findings of the late (repair) stage were the abnormally appearing arteries both in their extraosseous and intraosseous portions. Marked increase of vascularity arose from newly formed small lateral arteries and developed into the revascularized area of the bony epiphysis (Figure 4). No artery similar to the vascular trunks observed in normal hips was found. Hypervascularization consisting in numerous small arteries from medial epiphyseal arteries was disclosed on a medial small portion of the epiphysis. However, the load-bearing epiphyseal area remained avascular.

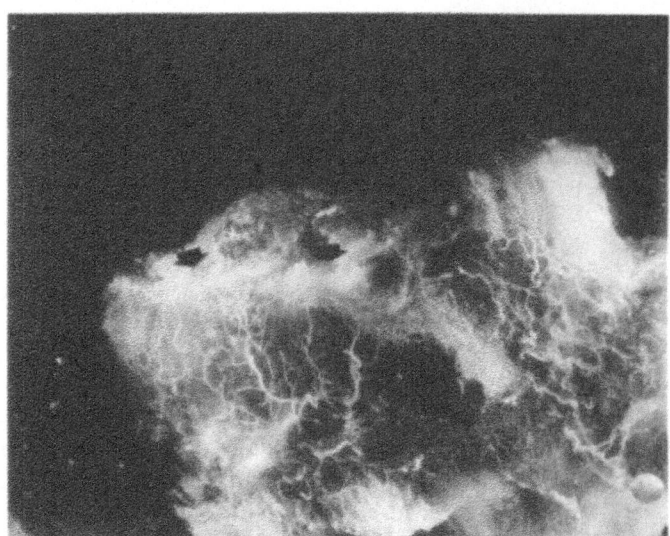

Figure 4. Superselective angiography of Perthes' disease in the repairing stage. Repair is obvious on the lateral aspects of the epiphysis. Angiography demonstrates numerous small arteries in the repairing region (Arrow).

MICROANGIOGRAPHY IN OSTEONECROSIS AND OSTEOARTHRITIS

Microangiography of 35 femoral heads with advanced osteonecrosis was performed to investigate the details of vascular distribution. In the posterior lateral area where the superior retinacular arteries normally exist, an obstruction of the arteries was seen in the extraosseous area. Numerous arteries penetrated into the affected femoral head and were also demonstrated at the area of demarcation surrounding the necrotic lesion. However, these arteries were interrupted by subchondral fracture and collapse in the load-bearing portion of the head (Figure 5). The vascularization in the non-weightbearing area was uninterrupted and penetrated in the subchondral area. From these findings it was assumed that the real superior retinacular arteries were obstructed in the extraosseous area and that subsequent revascularization occurred. However the revascularization was blocked during the repair process by collapse and subchondral fracture.

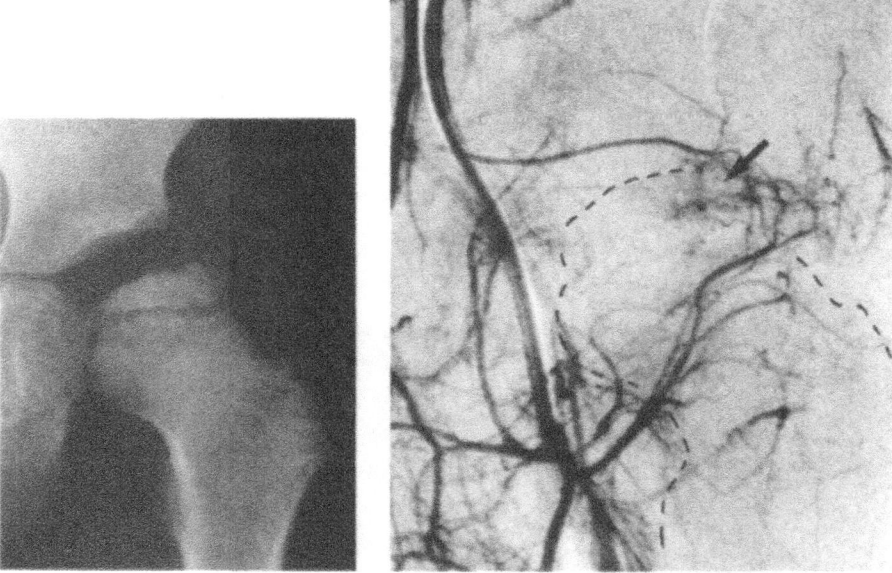

Figure 5. Microangiography of advanced stage of osteonecrosis with marked deformity. Numerous hypertrophic arteries of the proximal portion were interrupted at the site of collapse.

No evidence of impairment of blood supply of the superior retinacular arteries was seen in the extraosseous area of 53 hips with end stage osteoarthritis. In hips with large cystic lesions in the depth of the femoral head, the nutrient arteries were blocked on the surface of the cyst submitted to load-bearing. Nutrient arteries were distributed to subchondral area and there were no signs of impairment of blood supply in hips free from significant cyst.

DISCUSSION AND CONCLUSION

From these findings using superselective angiography in adults, although several patterns of blood supply surrounding the hip joint can coexist, vascular impairment of blood supply by the superior retinacular arteries was consistently found in the extraosseous area in preradiographical stage of osteonecrosis. Blockage of revascularization was found with microangiography. However, these pathological conditions were not seen in hips with osteoarthritis. The real superior retinacular arteries were abnormal in early necrosis. This circulatory disturbance may be induced by venous abnormality (Arlet, 1971), intravascular coagulation after fat embolism (Jones, 1985), or increase in size of fat cells (Wang, 1977). In the angiography of Perthes' disease, pathological findings were basically similar as those in adults osteonecrosis. It is assumed therefore that impairment of blood supply of the superior retinacular arteries is the most important single factor resulting in osteonecrosis. In some of affected hips, the penetration of newly formed vessels is prominent. However, in some cases, revascularization will aborte, one of the possible causes being the interference by subchondral fracture and collapse (Salter, 1984) caused by mechanical stress. Interference of revascularization is a real challenge in the clinical outcome of osteonecrosis.

REFERENCES

Arlet, J.: Pertrochanteric phlebography in primary necrosis of the femoral head in the initial stage (stage 1). In: W.N. Zinn (ed.): Idiopathic Ischemic Necrosis of the Femoral Head in Adults. pp. 152-157. Georg Theim. Publ., Sttutgart 1971.

Atsumi, T., Y. Kuroki, K. Yamano: A microangiographic study of idiopathic osteonecrosis of the femoral head. Clin. Orthop. 246:186-194(1989).

Atsumi, T., Y. Kuroki: Vascular changes of idiopathic necrosis of the femoral head. In : J. Arlet, B. Mazières (eds.): Bone Circulation and Bone Necrosis. pp. 311-315. Splinger-Verlag, Berlin Heidelberg 1990.

Atsumi, T., Y. Kuroki,: Impairment of the hemodynamics and revascularization in Perthes' disease. In: J. Arlet, B.Mazières (eds.): Bone Circulation and Bone Necrosis. pp. 295-299. Splinger-Verlag, Berlin Heidelberg 1990.

Atsumi, T. et al.: Role of impairment of blood supply of the femoral head in the pathogenesis of idiopathic osteonecrosis. Clin. Orthop. 277: 22-30 (1992).

Chung, S. M. K.: The arterial supply of the developing proximal end of the human femur. J. Bone and Joint Surg. 58-A: 961-970 (1976).

Hipp, H.E.: Das rontogenologische und angiographische bild der spontanen huftkopfnekrose des erwachsenen. Ver Dtsch. Orthop. Ges., 54: 236-244 (1968).

Jones, J.P., Jr.: Fat embolism and osteonecrosis. In: J .E. Kenzora (ed.): Idiopathic Osteonecrosis. pp. 595-633. Ortho. Clin.North, 1985.

Mussbichler, H.: Arterial supply to the head of the femur, an arteriographic study in vivo of lesions attending fracture of the femoral neck. Acta Radiol., 46:533-546 (1956).

Salter, R.B., G.E. Thompson: Legg-Calve-Perthes' disease. The prognostic significance of the subchondral fracture and a two group classification of the femoral head involvement. J. Bone and Joint Surg. 66-A: 479-489 (1984).

Sevitt, S., R.G. Thompson: The distribution and anastomosis of arteries supplying the head and neck of the femur. J. Bone and Joint Surg., 47-B: 560-573 (1965).

Theron, J.: Superselective angiography of the hip. Radiology, 124: 649-657 (1977).

Trueta, J., M.H. Harrison: The normal vascular anatomy of the femoral head in adult man. J. and Bone Joint Surg. 35-B:442-461 (1953).

Trueta, J.: The normal vascular anatomy of the human femoral head during growth. J. Bone and Joint Surg. 39-B: 358-394 (1957).

Wang, G.J., D.E. Sweet, S.I. Reger: Fat-cell changes as a mechanism of avascular necrosis of the femoral head in cortisone-treated rabbits. J. Bone and Joint Surg. 59-A: 729-735(1977).

COMPARED MICROANGIOGRAPHIC IMAGES OF OSTEONECROSIS OF THE FEMORAL HEAD AND OSTEOARTHRITIS OF THE HIP

KENJI OHZONO, NOBUHIKO SUGANO, NOBUO NAKAMURA, KEIRO ONO

Department of Orthopaedic Surgery, Osaka National Hospital, Osaka University Medical School, Osaka Teishin Hospital
Osaka, Japan

INTRODUCTION

It is important to elucidate the pathomechanism of hip diseases. Some investigators have studied vascular involvement in osteonecrosis by using selective angiography, however, very fine vascular imaging could not be obtained (Atsumi, 1983; Hipp, 1973). A few reports on microangiographic studies of osteonecrosis have been published, but the site of vascular obstruction was not clarified (Atsumi et al., 1989; Merle d'Aubigne et al., 1965). We have been investigating morphologic abnormalities of vascularity of the femoral head in osteonecrosis (Ohzono et al., 1992), osteoarthritis and rapidly destructive coxopathy (Postel et al, 1970). We believe that the accumulated microangiographic data will help to understand pathogenesis and pathophysiology of the various hip diseases.

MATERIALS AND METHODS

Sixty-three patients, 70 femoral heads with osteonecrosis, osteoarthritis or rapidly destructive coxopathy were studied by microangiographic and histologic technique. Eight patients, 9 heads with osteoarthritis, 12 patients, 13 heads with RDC and 43 patients, 48 femoral heads with osteonecrosis were examined. There were seven females and one male in osteoarthritis, ten females and two males in RDC, and twenty-three females and twenty males in osteonecrosis. The patient's age was 56 years in osteoarthritis, 61 years in RDC and 40 in osteonecrosis on average.

During prosthetic replacement surgery, the femoral heads were obtained along with the posterolateral joint capsule including the nutrient arteries. Fifty per cent of Barium sulphate suspension was infused through the posterior collum branch of the medial femoral circumflex artery, after irrigation with saline. Each specimen was fixed in formalin, cut into 10 mm thick slice and decalcified in a 25 % formic acid and 25 % citric

Bone Circulation and Vascularization in Normal and Pathological Conditions
Edited by A. Schoutens *et al.*, Plenum Press, New York, 1993

301

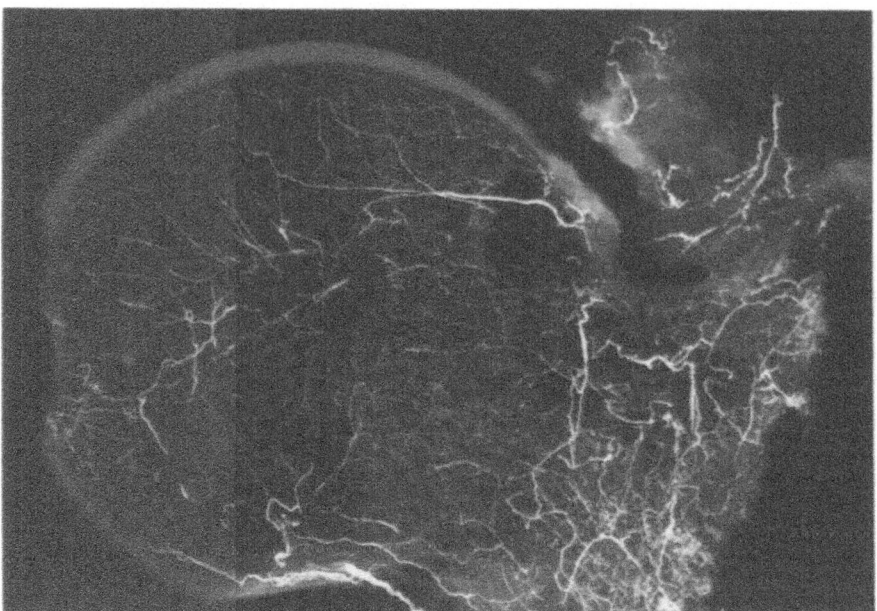

Figure 1. Normal microangiographic image of the femoral head in 70 year old female with femoral neck fracture. The arterial arch of the lateral epiphyseal artery and the inferior retinacular artery are visible.

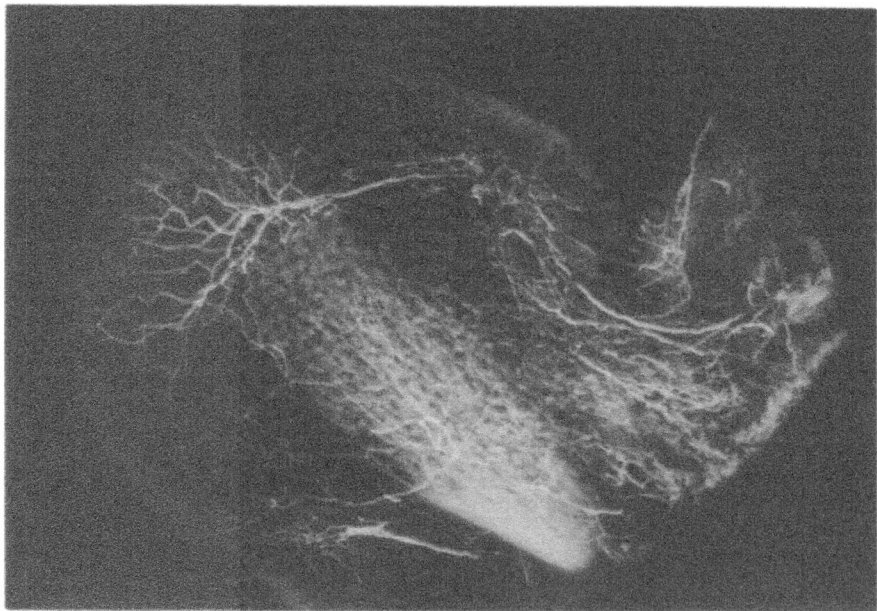

Figure 2. Microangiographic image of osteoarthritis in a 56 year old female. Typical arterial arch is preserved and hypervascularity in the subchondral region is observed.

acid solution for two weeks. Each slab was radiographed by low density X-ray, embedded in paraffin wax, cut into 5 μm slices and stained with hematoxylin and eosin.

Many reports (Müssbichler, 1965; Sevitt, 1965; Trueta et al., 1953; Tucker, 1949; Wertheimer et al., 1971) about the normal vascular anatomy of the femoral head served as reference for this study.

RESULTS

In the normal hip, the lateral epiphyseal artery, the inferior metaphyseal artery and the ligamentum teres artery are visible. Vascular anastomoses, such as typical superoinferoligamental arch (Wertheimer et al., 1971), are evident (Figure 1).

Osteoarthritis of the Hip

In the femoral head with osteoarthritis, basic vascular architecture is the same as in normal hip. Typical arterial arch and anastomoses are preserved (Figure 2). Hyperemic changes around cysts and meandering vessels could be seen (Figure 3). We could not observe any avascular area in these specimens with osteoarthritis.

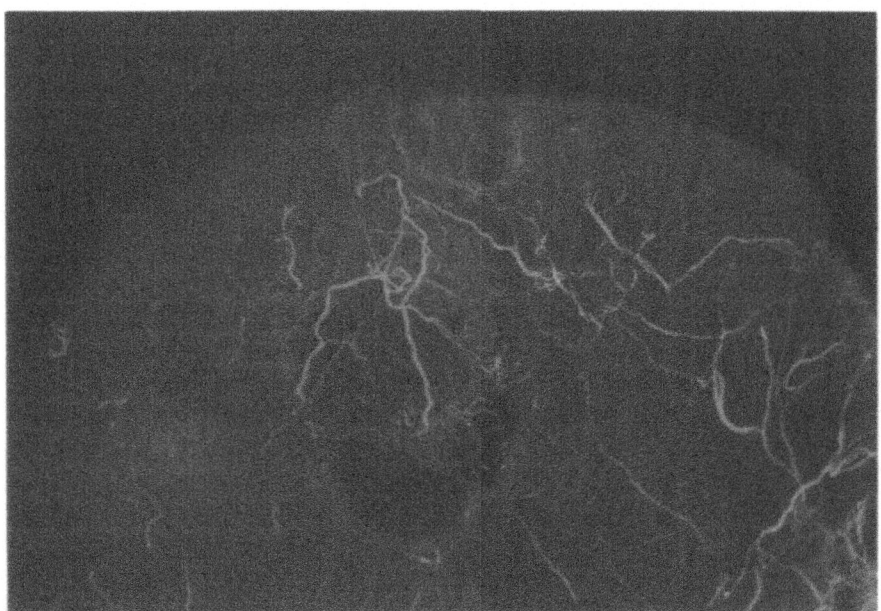

Figure 3. Subchondral cyst in a femoral head with osteoarthritis. Small vessels are surrounding the cyst and the meandering of the vessels is seen.

Rapidly destructive coxopathy

Clinically, rapid progession of destruction of the hip joint occurs within twelve months in relatively aged patients (Figure 4). No infection could be documented. Rheumatoid arthritis and neuropathic arthropathy were excluded.

In rapidly destructive form of osteoarthritis characteristic vascular patterns were observed by microangiography. Hyperemic changes in the femoral head and massive vascular budding onto the articular surface were evident (Figure 5). Very localized avascular areas were also observed. At the site of vascular budding, histology showed aggressive granulation with many blood vessels involving the articular surface (Figure 6). These findings must be closely related to the rapid progression of this disease.

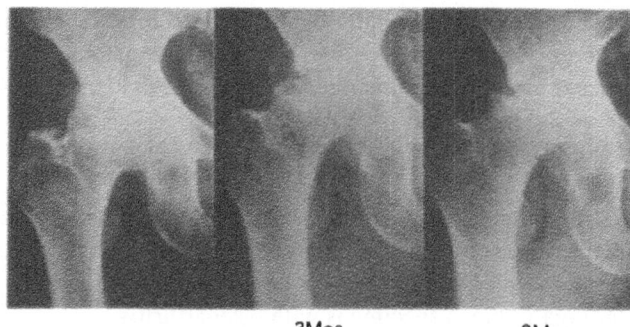

3Mos   6Mos

Figure 4. 70 year old female, rapidly destructive coxopathy of the right hip joint progressing in less than 1 year.

Osteonecrosis of the femoral head

In case of idiopathic osteonecrosis of the femoral head characteristic findings were observed. Figure 7 shows a microangiographic image of stage II osteonecrosis before decalcification. Just beneath the demarcation line, normal nutrient arteries can be seen. Just above the demarcation line no vessels were visualized.

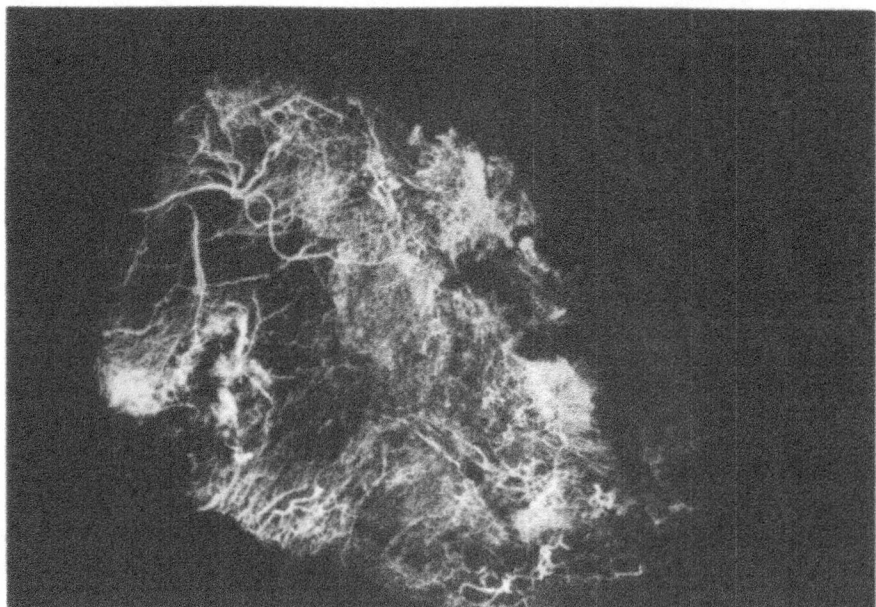

Figure 5. Microangiographic image shows severe hyperemic changes of the femoral head and evident vascular budding to the articular surface.

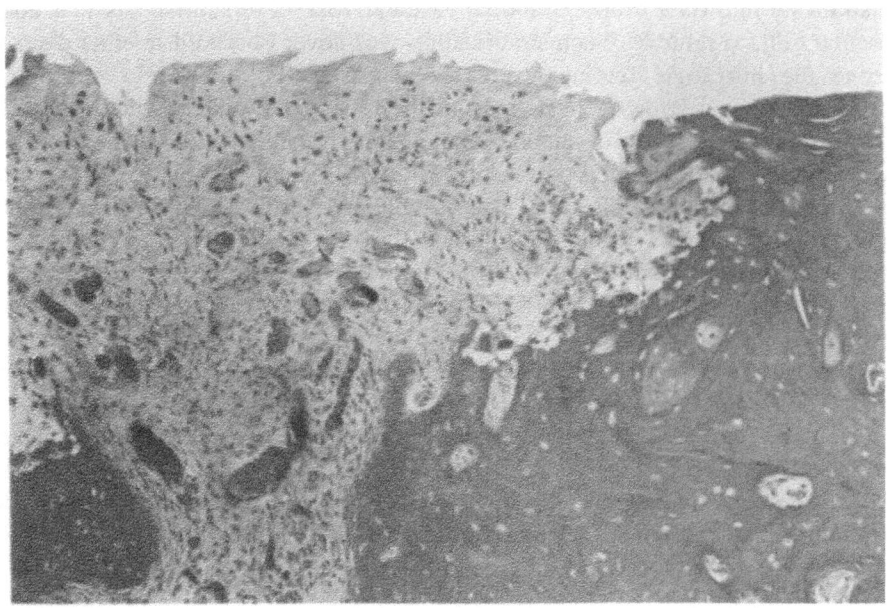

Figure 6. Histologic section shows aggressive granulation with many blood vessels involving the articular surface (Haematoxylin and Eosin).

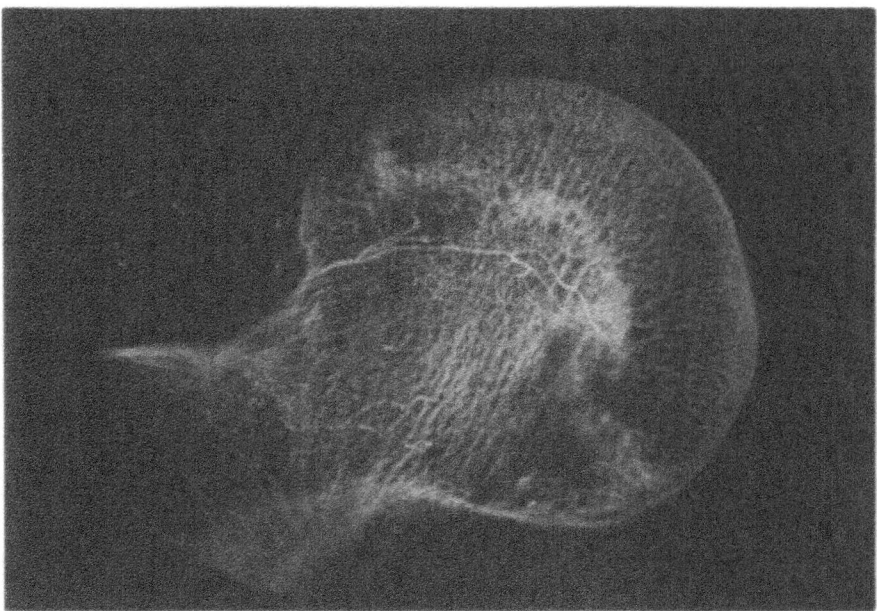

Figure 7. Microangiographic image of stage II (without collapse) osteonecrosis before decalcification in a 37 year old female. A sclerotic demarcation line is evident. Beneath the demarcation line, unaffected nutrient arteries can be seen. Above the demarcation line, no vessels are observed.

According to microangiographic images, the femoral head with osteonecrosis of stage III can be stratified into three zones; a normal vascular zone, a reparative vascular zone and an avascular zone (Figure 8). Such stratification was never observed in other diseases. In each zone the histologic findings were characteristic. In the normal vascular zone trabeculae and bone marrow were normal. In the reparative vascular zone dense fibrous tissue and reparative bone formation on the dead trabeculae were observed. In the avascular zone complete necrosis of trabeculae and bone marrow was evident (Figure 9).

In stage II of osteonecrosis, reparative vasculature was poor and in stage III, massive reparative vessels could be seen. Partly normal shape of the lateral epiphyseal artery was observed both in stage II and III (Figure 10). In case of extensively affected osteonecrosis, most part of intraosseous nutrient arteries were involved. On the other hand, in case of circumscribed osteonecrosis, vascular involvement was localized and vascular anastomosis was preserved (Figure 11). A typical superoinferoligamental arch could be observed. The ligamentum teres artery is also visualized through the anastomoses. The normal part of the lateral epiphyseal artery can be seen both in extensive and circumscribed osteonecrosis.

These findings suggest that the final site of vascular interruption in osteonecrosis may be the intraosseous stem of the nutrient artery of the femoral head.

DISCUSSION

Although normal vascular anatomy of the femoral head is well known, pathological vascular condition of the hip such as osteoarthritis, a rapidly destructive form of osteoarthritis, and even osteonecrosis of the femoral head have not been clarified.

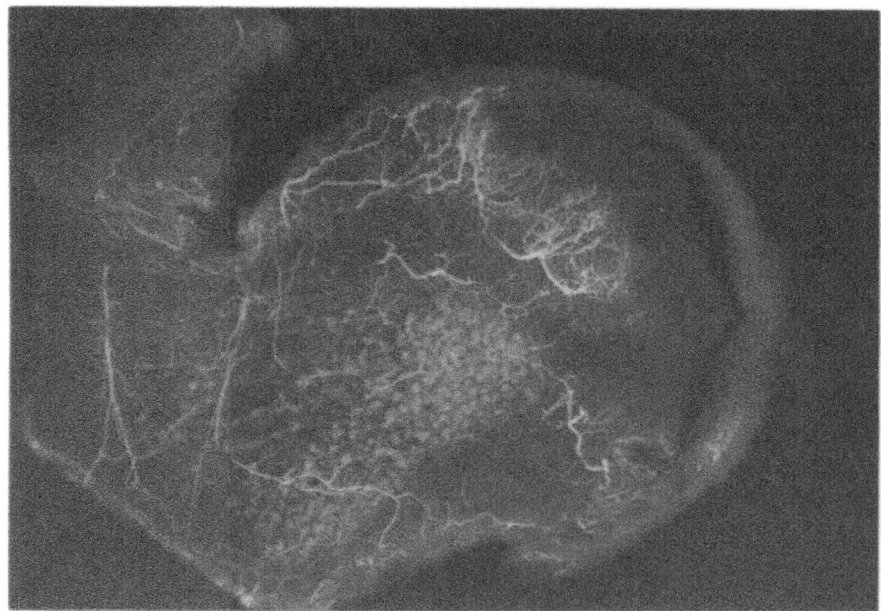

Figure 8. Microangiogram of the femoral head with osteonecrosis of a 39 year old male. In the center of the head and in the neck the vascular pattern is normal (Normal Vascular Zone). The subchondral area is avascular (Avascular Zone). Between the two zones there is interposition of repairing fine vascular network. (Reparative Vascular Zone).

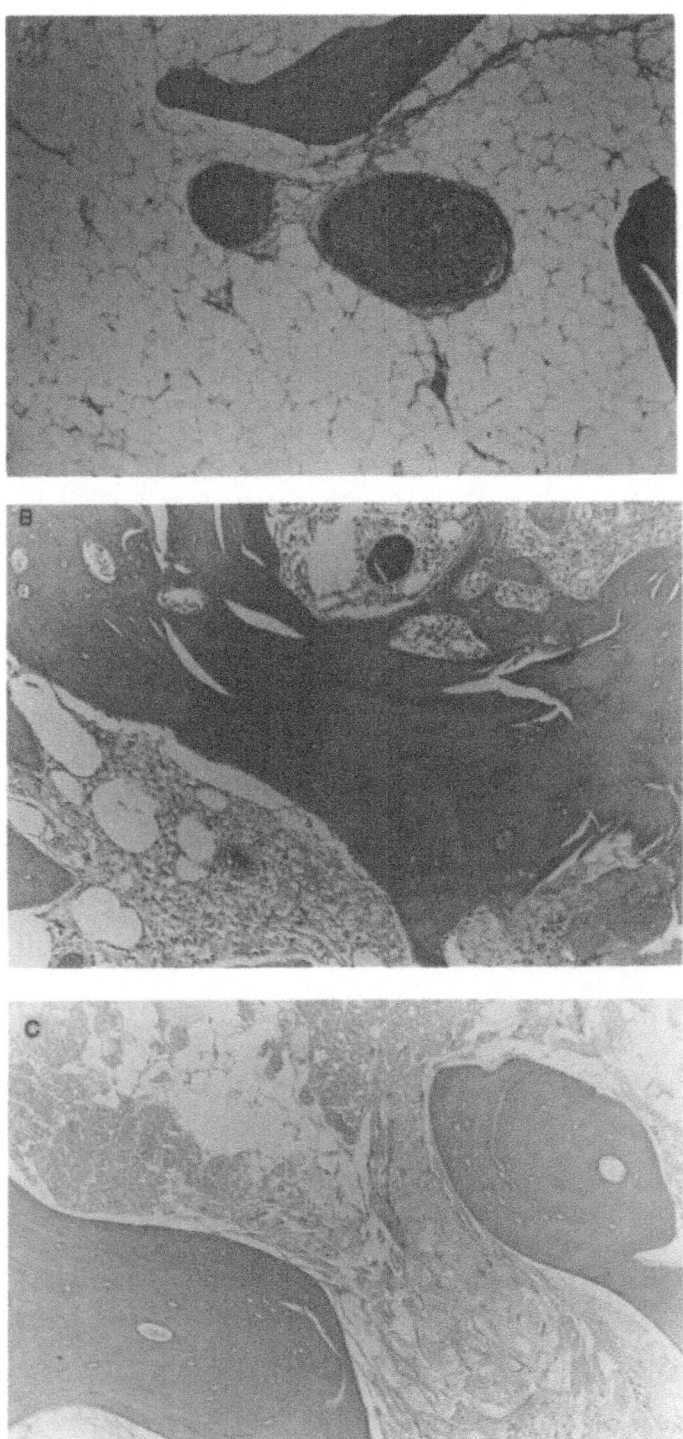

Figure 9. A) In the Normal Vascular Zone trabeculae and bone marrow are normal. B) In the Reparative Vascular Zone appositional bone formation on the dead trabeculae with empty lacunae and dense fibrous tissue are observed. C) In the Avascular Zone complete necrosis of trabeculae and bone marrow is observed (Haematoxylin and Eosin).

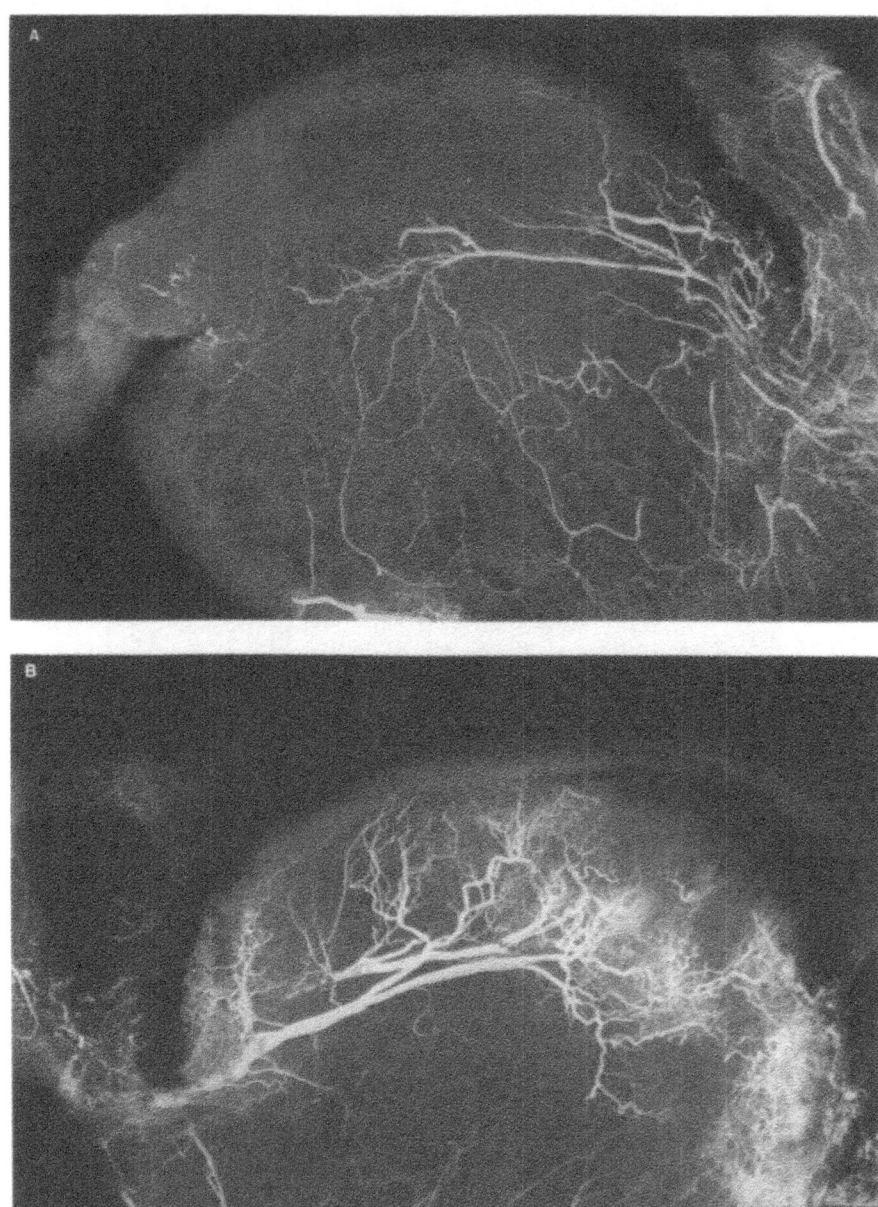

Figure 10. A) Microangiogram of stage II osteonecrosis. Unaffected trunk of the nutrient artery and poor repairing vessels can be seen. B) Microangiogram of stage III osteonecrosis. A massive proliferation of repairing vessels is evident. The unaffected trunk of the nutrient artery is also visible.

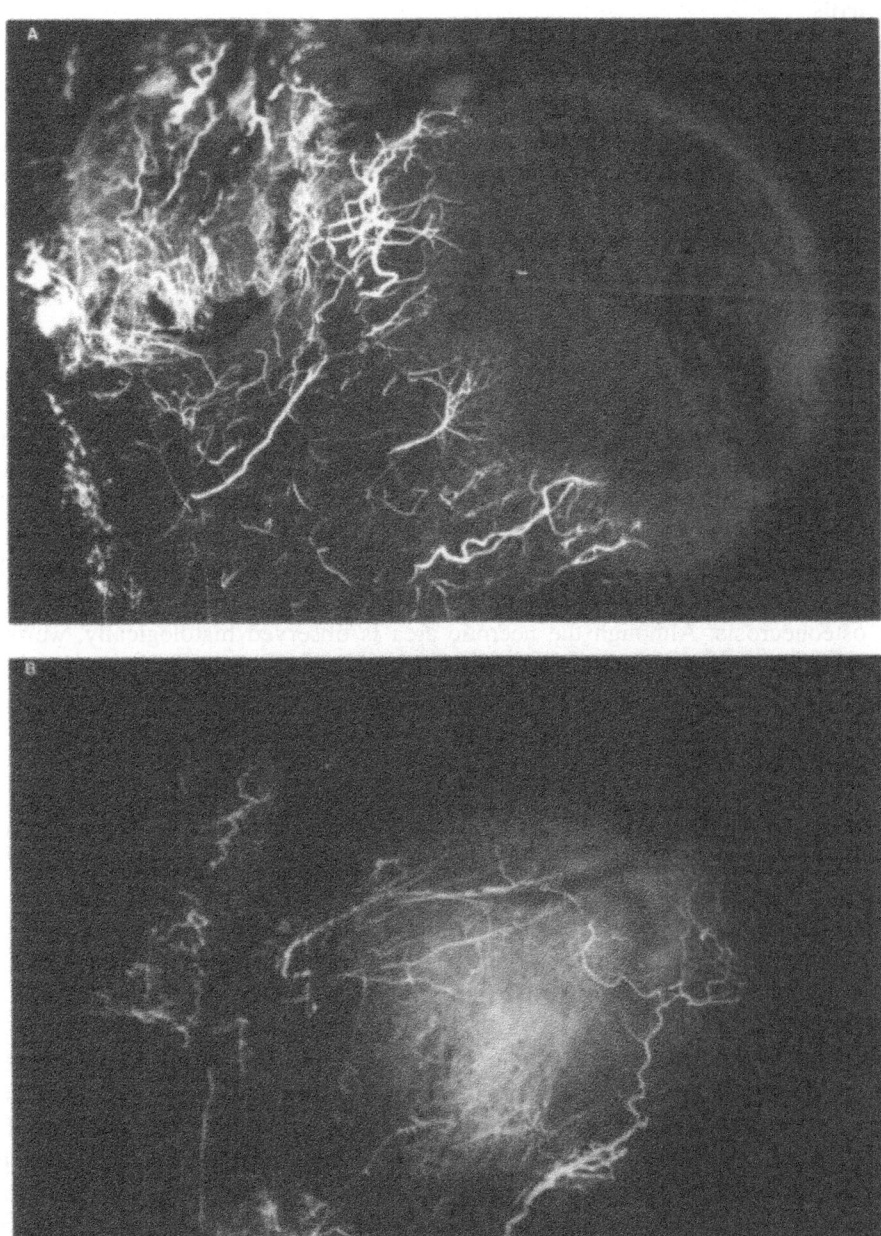

Figure 11. A) Microangiogram of an extensively affected hip with osteonecrosis in stage III. Most part of the intraosseous nutrient arteries are involved. B) The microangiogram of limited osteonecrosis in stage III. The vascular involvement is localized and vascular anastomoses are preserved. The typical superoinferoligamental arch can be observed.

Based on microangiographic study on these conditions, we report many important findings in each disease.

Osteoarthritis

Most important finding in this condition is that basic vascular architecture is the same as that of the normal hip. Osteoarthritic change is primarily degeneration of the articular cartilage. Therefore, vascular abnormality is secondary due to osteophyte formation and cystic degeneration. Hypervascularity from the stem of the nutrient aretry into the site where osteophyte is growing was characteristic. Hyperemic change surrounding the subchondral cyst was often noted, but is a secondary change.

Rapidly destructive coxopathy

This condition has been considered to be a special form of osteoarthritis, a rapidly progressive form of osteoarthritis. We found some characteristic vascular abnormalities in this condition. The first finding is tremendous vascular budding to the articular surface. According to histological findings of this vascular budding, aggressive granulation with many small vessels invading through the subchondral area was observed. The pathomechanism of this aggressive granulation is not known, but this proliferation of granulation is closely correlated to clinically very prompt destruction of the hip joint. The second finding is presence of a localized avascular area. This area is not as extensive as found in osteonecrosis. Although the necrotic area is observed histologically, we think that the cause of localized necrosis is not ischemia but may be aseptic inflammatory environment.

Osteonecrosis

Microangiographic findings of different necrotic size and stage of osteonecrosis disclosed two important points concerning the pathomechanism of osteonecrosis of the femoral head. The first is that reparative arteries developed from the trunk of the intraosseous nutrient arteries, whether the necrotic area is extensive or circumscribed. The second point is that the trunk of the nutrient artery remained partly unaffected within the femoral head. The initial site of vascular interruption is not known, but it might be the intraosseous part of the nutrient arteries. Initial vascular obstruction may occur in the microcirculation in the subchondral region and retrogradely cause vascular obstruction of the trunk of the nutrient arteries, such as the lateral epiphyseal artery and the inferior metaphyseal artery. According to size and number of the involved nutrient arteries and the amount of vascular anastomoses it might be possible to determine the size and location of osteonecrosis. However most of our cases are progressed too far to provide answers to pathomechanism of the osteonecrosis. But much evidence was obtained suggesting that the intraosseous arteries were indeed interrupted. Further investigation and experimental studies in the early stages of osteonecrosis are urgently needed.

REFERENCES

Atsumi, T.: Hemodynamic study of the idiopathic necrosis of femoral head using superselective angiography. Nippon Seikeigeka Gakkai Zassi. 57: 353 (1983).

Atsumi, T., Y. Kuroki, K. Yamano: A microangiographic study of idiopathic osteonecrosis of the femoral head. Clin. Orthop. 246: 186-194 (1989).

Hipp, E.: Angiographic studies in idiopathic osteonecrosis of the femoral head. In: Ficat, P., Arlet, J. (eds.): Proceedings of the First International Symposium on Circulation of Bone, pp. 189. Bris, Editions Inserm Publ., 1973.

Merle d'Aubigne, R., M. Postel, A. Mazabraud, P. Massias, J. Gueguen: Idiopathic necrosis of the femoral head in adults. J. Bone Joint Surg. 47B: 612-633 (1965).

Müssbichler, H.: Angiography of the hip region. Acta. Radiol. 11: 593-603 (1971).

Ohzono, K., K. Takaoka, M. Saito S. Saito, M. Matsui, K. Ono: Intraosseous arterial architecture in non-traumatic avascular necrosis of the femoral head. Microangiographic and histologic study. Clin. Orthop. 277: 79-88 (1992).

Postel, M., M. Kerboull: Total prosthetic replacement in rapidly destructive arthrosis of the hip joint. Clin. Orthop. 72 : 138-144 (1970).

Sevitt, S.: The distribution and anastomoses of arteries supplying the head and neck of the femur. J.Bone Joint Surg. 47-B: 560-573 (1965).

Trueta, J., M.H.M. Harrison: The normal vascular anatomy of the femoral head in adult human. J. Bone Joint Surg. 35-B: 442-461 (1953).

Tucker, F.R.: Arterial supply to the femoral head and its clinical importance. J. Bone Joint Surg. 31-B: 82-93 (1949).

Wertheimer, L.G., S.D.L. Fernandes Lopez: Arterial supply of the femoral head: A combined angiographic and histological study. J. Bone Joint Surg. 53-A: 545-556 (1971).

VALUE OF QUANTIFIED MRI TO PREDICT LONG-TERM PROGNOSIS OF EARLY STAGE AVASCULAR NECROSIS OF THE FEMORAL HEAD

P.C. ACQUAVIVA, P. LAFFORGUE

Service de Rhumatologie, Hôpital Timone
Marseille, France

INTRODUCTION

There is a general consensus about the need for diagnosis of femoral head osteonecrosis (ON) as early as possible, but the indications and efficacity of treatment modalities at early stages remain controversial. This results for part from the poor knowledge of factors influencing the natural course of the disease. It is now more currently thought that the development of collapse is closely related to the extent and location of ON, rather than to its radiographic stage (Beltran, 1990; Chan, 1991; Ohzono, 1991).

Many investigators have already demonstrated the usefulness of Magnetic Resonance Imaging (MRI) for early diagnosis of osteonecrosis (Beltran, 1988; Hauzeur, 1989; Markisz, 1987; Mitchell 1987; Sarrat, 1988; Seiler, 1989). Adequate correlation has also been shown between MRI appearance and histopathology of ON (Bassett, 1987; Hernigou, 1989; Lang, 1988; Mitchell, 1987; Takatori, 1987). In particular, MRI distinctly shows the limits of the necrotic area within the femoral head, even when the lesion is poorly or not seen in conventional roentgenography.

The aim of this study was to define with MRI quantitative parameters reflecting the size and location of early stage osteonecrosis of the hip, in an attempt to correlate them with long-term behavior of the disease.

METHODS

Patients

31 hips with radiographic Stage I or II avascular osteonecrosis in 24 patients were included. All had typical signs of ON at MRI. Diagnosis was confirmed in all cases either by histopathology in the patients who underwent further surgery (core decompression or total hip replacement, i.e. 17 cases), or by later occurrence of typical roentgenographic changes (14 cases).

Bone Circulation and Vascularization in Normal and Pathological Conditions
Edited by A. Schoutens *et al.*, Plenum Press, New York, 1993

313

Initial radiographic staging according to Ficat's classification showed 14 stage I ON (i.e. normal plain X-ray films) and 17 stage II ON (i.e. mottled sclerosis and/or radiolucencies with no crescent sign, collapse, or joint narrowing). Computed tomography performed on 16 hips also ascertained the absence of subchondral fracture.

Outcome Assesment

All the patients had at least a two-year follow-up (mean = 46 months). At the last survey, clinical evaluation was done using the first two parts of the Merle d'Aubigné scale as modified by Charnley. Good and very good results using this scale were considered as a satisfactory clinical outcome. Radiographs were available for 27 of the 31 hips and were staged according to Ficat's classification. Radiological outcome was considered as good if the last radiographs showed Stage I or II and poor when the ON had progressed to Stage III or more.

12 hips underwent core decompression soon after the initial MRI examination, while the 19 others had only symptomatic medications and partial weightbearing suppression.

MRI Procedure

MRI was performed at the time of radiographic staging with supraconducting units (Magniscan CGR in 21 hips and Magnetom Siemens in 10 hips) operating respectively at 0.5 and 1.5 Tesla.

Coronal sections with T1-weighted spin-echo sequences (short repetition time TR ranging from 400 to 650 ms/ echo time TE ranging from 15 to 40 ms) were obtained in all patients. Sections were contiguous, section thickness varied from 4 to 7 mm.

The parameters studied were the proportion of the weightbearing cortical bone involved by the necrosis (WB), in percentage, and the proportion of the femoral head surface affected (FH) in percentage (Figure 1). WB was calculated by dividing the entire weight-bearing area by the portion of femoral weightbearing cortex involved by ON. The affected femoral cortex was considered as the area comprised between the medial and lateral raccordments of ON with the cortex. When the necrotic portion was entirely circumscribed and spared the cortical bone, WB was estimated as 0 %. FH was calculated considering the femoral head as a circle and the necrotic process as an ellipse.

Each parameter for each patient was the average of measurements made by 2 readers, blinded to the clinical and radiological history, on the contiguous slides throughout the median portion of the femoral head totaling a width of 20 mm (i.e. 3 sections of 7 mm, 4 sections of 5 mm, or 5 sections of 4 mm thickness). The results of individual values were displayed according to the outcome, i.e. satisfactory or poor, and also using subgroups of ARCO's classification: types 1, 2, 3 indicate the weightbearing area involved by necrosis (1<33 %, 2=33-66 %, 3>66 %) and types A, B and C correspond to the femoral head area involved (A<15 %, B=15-30 %, C>30 %).

Statistical Analysis

The non parametric Mann-Whitney U-test was used for comparisons of WB and FH according to the outcome. Receiving Operator Curve (ROC) analysis was used to define the more accurate parameter for poor outcome prediction (the more accurate diagnostic tool with this method is the one with the highest area under the curve (AUC). An AUC of 1 designates a perfect criterion while 0.5 indicates a chance occurrence).

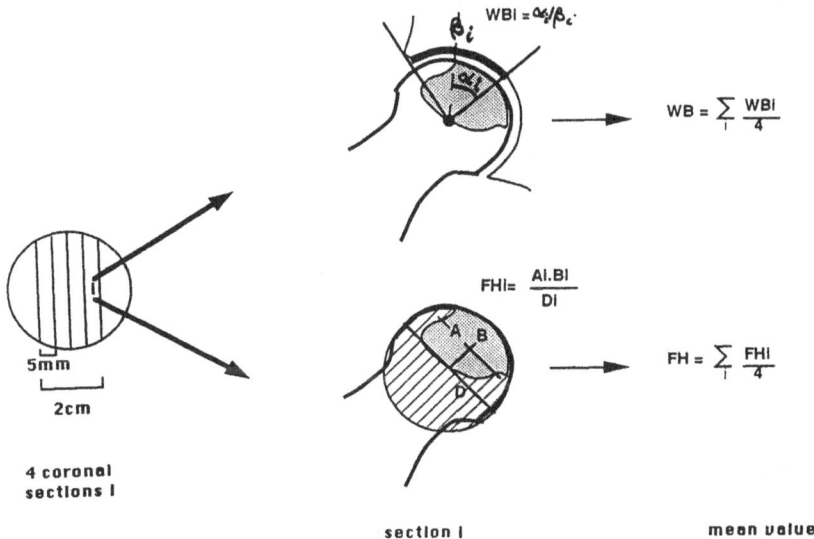

Figure 1. Measurement of quantitative parameters on MRI: the adequate contiguous MRI slices are determined, i.e. slices corresponding to the central 2 cm portion of the femoral head (4 slices of 5 mm thickness in this example). For each section i, the portions of the weightbearing cortex (WB) and of the femoral head surface (FH) involved are measured. The final value for one hip is the average of the values obtained in each section.

RESULTS

There was no difference in age, sex, initial radiographic staging and symptoms, or etiology versus the final outcome (Table 1). Of the 31 hips, 35.5 % had a good clinical outcome and 33.3 % had a final radiographic staging at I or II. Clinical and radiological evolutions were highly correlated ($p \leq 0.0001$).

The values of the parameters greatly differed in groups with good clinical/radiological outcome versus poor outcome (Table 2). The involved weightbearing cortex WB and the necrotic surface FH were very significantly higher in patients with poor prognosis. Moreover, the diagrams of repartition of the values showed very few overlap between the values of patients with good and poor outcome. ROC showed the proportion of involved weight bearing cortex involved (WB) to be the more reliable parameter for outcome prediction (Figure 2). This was especially true for the necroses which entirely spared a cortical subchondral margin. The ARCO's classification subgroups were only partially adequate. Considering the weightbearing area involvement, types 1 and 3 were clearly associated with either satisfactory or poor outcome, but there was considerable overlap in the intermediate type 2, which had poor predictive value. Considering the percentage of femoral head involvement, types B and C were indicative of poor outcome but about 30 % of type A also had a bad course. In fact, the critical value of weightbearing involvement appears to be 45 %, with satisfactory positive and negative predictive values of an unsatisfactory outcome (respectively 0.95 and 0.91).

Core decompression gave poor results. This might be reasonably explained by the extent of the osteonecrose, which was generally important in the cored group. In fact, it appeared from the repartition of values in cored and non operated hips, that the evolution

315

Table 1. Main features of the osteonecrosis.

clinical behavior	good outcome (n=11)	poor outcome (n=20)
sex ratio	4F/7M	5F/15M
age	46.8±17.5	43±13
initially asymptomatic	3 (27.3 %)	5 (25 %)
Initial radiographic	5 Stage I (45.5 %)	9 Stage I (45 %)
staging	6 Stage II (54.5 %)	11 Stage II (55 %)
Etiology:		
corticosteroids	5	9
alcoholism triglycerids	2	7
pregnancy	1	0
diving	1	0
none	1	3
cored hips	1	11
previous contralateral ON with Stage≥ III	4 (36.4 %)	6 (30 %)

Table 2. Means and standard deviations of WB and FH according to further outcome.

	clinical course		radiographic course	
	Good (n=11)	Bad (n=20)	Stage I or II (n=9)	stage ≥ III (n=18)
WB (%)	25.5±29.6	86.5±22.3	19.9±20.6	91.9±13.
	P ≤ 0.0001		P ≤ 0.0001	
FH (%)	8.7±8.4	30.9±15.6	7.4±5.2	34.2±14.6
	P ≤ 0.0004		P ≤ 0.0003	

was closely related to the size of the lesion and the weightbearing area involvement and not to treatment schedule.

COMMENTS

Determining the prognosis of avascular necrosis of the femoral head is of importance in order to decide if surgery is needed and which procedure is the most adequate. The factors influencing the course of the disease are probably multiple ones, however clinical experience indicates that the size and site of osteonecrosis are of special importance. The attempt of grading the extent of ON is not new. Steinberg already recommended a subdivision of each radiographic stage into three classes according to the extent of the lesion (Steinberg, 1984). More recently, Ohzono et al. 1991 showed that the natural course of osteonecrosis in non operated hips was closely related to its location within the femoral head. Those osteonecroses affecting less than one-third of the weightbearing area and those sparing a subchondral margin had no further radiographic progression, while the others generally experienced further collapse (Ohzono, 1991). However, such radiographic systems do not apply to cases with normal radiographs and are somewhat approximative in many cases. Since MRI is more sensitive than plain radiography, it is presumably more accurate for such a measurement. Recently, ARCO committee has proposed to use this subclassification for MRI quantitation too. A semi-quantitative approach with MRI was also used by Beltran et al. 1990 to assess the results of core decompression in a prospective series of 34 Stage I and II hips. They used the combination of the coronal and the axial sections showing the greatest involvement. In this study, results of core decompression were invariably good for osteonecroses involving less than 25 % of the femoral head and worsened as osteonecrosis had a larger extent.

The method of measurement used for this study is original in the way that values were the average of extent and location of ON assessed on multiple sections in a critical site. The zone of interest was determined as the coronal sections throughout the 2 centimeters central portion of the femoral head (such measurements would not be valid in the most anterior and posterior processes since the lesion in these areas would be tangent to the plane of investigation). Indeed, we think that the use of several sections appraises the reality more closely than the use of the only section showing the widest involvement.

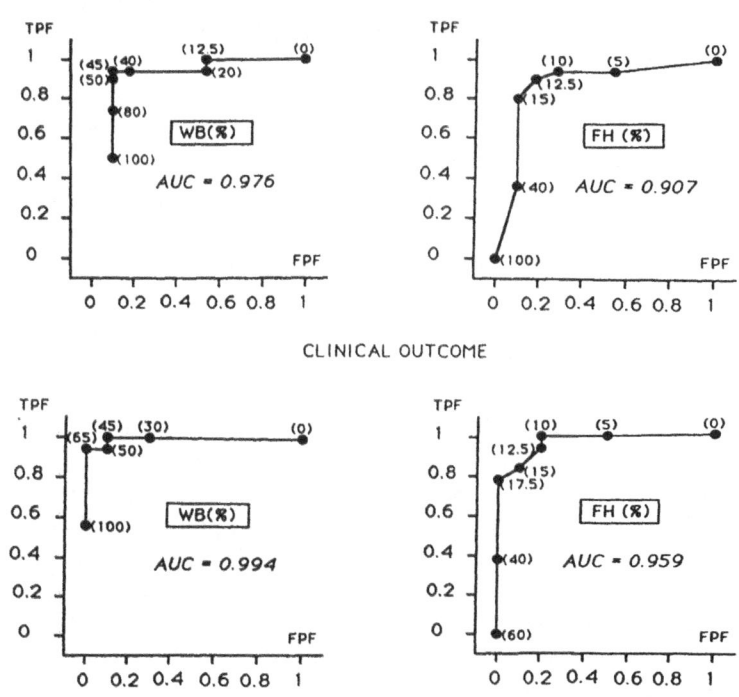

Figure 2. Receiving Operator Curves (ROC) for each parameter, according to later occurrence of unsatisfactory outcome. AUC: area under the curve. Theoretical maximal value for AUC is 1. TPF: true positive fraction. FPF: false positive fraction.

The age, sex ratio, number of initial "silent hips", initial radiographic staging and prevalence of contralateral advanced osteonecroses seemed not to play a role. On the contrary, the quantitative parameters assessed were very significantly higher in the patients who experienced chronic discomfort or radiological collapse. Such a result is not surprising but the differences observed were even more pronounced than expected. Though the differences were very significant with the two parameters, the more accurate was in our experience the proportion of the weightbearing cortical bone involved by the necrosis (WB). Furthermore, very little overlap occurred between the values of WB in each group, resulting in good predictive and negative values. This is of practical importance since it may enable prediction of the likelihood of outcome in individuals. The prognosis was consistently poor for ON involving more than 45 % of weightbearing cortex and good under this threshold. Thus, the classification into three categories (0-33 %, 33-66 %, 66-100 %) apppeared less adequate than the use of only two groups approximately corresponding to more or less than 50 % of weightbearing involvement. The percentage of femoral head involved was in our experience more difficult to measure and there was no critical threshold: the ON affecting less than 10 % or more than 20 % of the femoral head had invariably either good or poor radiological outcome, but there was considerable overlap in the intermediate group.

This method of evaluation appeared satisfactory but requires making some remarks. Firstly, the differences observed between good versus poor behavior might be less marked with a greater sample of patients. Secondly, we used the coronal plane which was most commonly performed at our institution and which allows the study of both hips. The exclusive use of this plane, however, does not permit a pertinent evaluation of the most anterior part of the femoral head, which is commonly affected by osteonecrosis. Thus, a combination of several incidences, especially coronal and sagittal planes, should be recommended.

Of the non-operated hips, about half had no or very mild symptoms and 40 % did not experience collapse nor joint space narrowing. This results are better than the 68-95 % of poor results of natural course previously reported (Hungerford, 1980; Ohzono, 1991; Patterson, 1964; Saito, 1988). This might be imputed to the use of MRI, whose sensitivity allows detection of small osteonecroses. These small-sized lesions are precisely those with the better behavior and might have been overlooked with conventional investigations used in the past. On the opposite side, the poor result of core decompression seems due to the larger extent of the necrotic area in the cored patients. As core decompression and non-surgical managing were not randomized, this does not allow a pertinent comparison between the two methods, but our data raise question about the effective role of core decompression. The controversy about the ability of core decompression to influence long-term outcome of avascular necrosis of the hip is not new. Some authors published highly satisfactory results (Ficat, 1985; Hungerford, 1980; Tooke, 1988), while more recent studies failed to show significant benefit from this procedure (Camp, 1986; Hopson, 1988; Learmonth, 1990; Seiler, 1989) and others obtained mitigated results (Chan, 1991; Saito, 1988; Warner, 1987). Recent studies have shown that positive results of core decompression occurred especially in necrosis of small size(Beltran, 1990; Chan, 1991), and that spontaneous good outcome was also observed in these kinds of patients as well (Ohzono, 1991). Any evaluation of any treatment of osteonecrosis of the hip should now include MRI as a major criterion. Firstly, addition of MRI would certainly improve the ascertainment of diagnosis, since none of the other diagnosing modalities has perfect sensitivity and specificity at early stage of the disease, including histopathology obtained by core biopsy. Secondly, MRI proves to be very accurate in assessment and quantification of the extent and location of osteonecrosis, especially regarding the involvement of the weightbearing area. Thus, it could help in determining the relationships between the treatment, the size and site of the necrosis, and the progression of the disease.

In summary, MRI proved to be an accurate way for quantifying the extent and location of osteonecrosis of the hip, even in cases with normal radiographs. The initial MR examination enabled a meaningful approach to the likelihood of further course of the disease, whether or not a core decompression was carried out.

REFERENCES

ARCO-Committee on Terminology and Classification. ARCO News. 4: 41-46 (1992).

Bassett, L.W., J.M. Mirra, A. Cracchiolo, R.H. Gold: Ischemic necrosis of the femoral head: correlation of Magnetic Resonance Imaging and histologic sections. Clin. Orthop. 223: 181-187 (1987).

Beltran, J., L.J. Herman, J.M. Burk, W.A. Zuelzer, R.N. Clark, J.G. Lucas, L.W. Weiss, A. Yang: Femoral head avascular necrosis: MR Imaging with clinical-pathologic and radionuclide correlation. Radiology. 166: 215-220 (1988).

Beltran, J., C.T. Knight, W.A. Zuelzer, J.P. Morgan, L.J. Schwendeman, V.P. Chandnani, J.C. Mosure, P.B. Shaffer: Core decompression for avascular necrosis of the femoral head: correlation between long-term results and preoperative MR staging. Radiology. 175: 533-536 (1990).

Camp, J.F., C.W. Colwell: Core decompression of the femoral head for osteonecrosis. J. Bone Joint Surg. 68-A: 1313-1319 (1986).

Chan, T.W., M.K. Dalinka, M.E. Steinberg, H.Y. Kressel: MRI appearance of femoral head osteonecrosis following core decompression and bone grafting. Skel. Radiol. 20: 103-107 (1991).

Ficat, R.P.: Idiopathic bone necrosis of the femoral head: early diagnosis and treatment. J Bone Joint Surg. 67-B: 3-9 (1985).

Hauzeur, J.P., J.L. Pasteels, A. Schoutens, M. Hinsenkamp, T. Appelboom, I. Chochrad, N. Perlemutter: The diagnostic value of magnetic resonance imaging in non-traumatic osteonecrosis of the femoral head. J. Bone Joint Surg. 71-A: 641-649 (1989).

Hernigou, P., M.C. Voisin, M. Marichez, E. Despres, D. Goutallier: Confrontation de l'imagerie par résonance magnétique nucléaire et de l'histologie dans les nécroses des têtes fémorales. Rev. Rhum. 56: 74 1-744 (1989).

Hopson, C.N., S.W. Siverhus: Ischemic necrosis of the femoral head: treatment by core decompression. J. Bone Joint Surg. 70-A: 1048-1051 (1988).

Hungerford, D.S., T.M. Zizic: The treatment of ischemic necrosis of bone in systemic lupus erythematosus. Medicine. 59: 143-148 (1980).

Lang, P., H.E. Jergesen, M.E. Moseley, J.E. Block, N.I. Chafetz, H.K. Genant: Avascular necrosis of the femoral head: high-field strength MR imaging with histologic correlation. Radiology. 169 : 517-524 (1988).

Learmonth, I.D., S. Maloon, G. Dall: Core decompression for early atraumatic osteonecrosis of the femoral head. J. Bone Joint Surg. 72-B: 387-390 (1990).

Markisz, J.A., R.J.R. Knowles, D.W. Altchek, R. Schneider, J.P. Whalen, P.T. Cahill: Segmental patterns of avascular necrosis of the femoral head: early detection with MR Imaging. Radiology. 162: 717-720 (1987).

Mitchell, D.G., V.M. Rao, M.K. Dalinka, C.E. Spritzer, A. Alavi, M.E. Steinberg, M; Fallon, H.Y. Kressel: Femoral head avascular necrosis: correlation of MR imaging, radiographic staging, radionuclide imaging, and clinical findings. Radiology. 162: 709-715 (1987).

Mitchell, D.G., P.M. Joseph, M. Fallon, W. Hickey, H.Y. Kressel, V.M. Rao, M.E. Steiberg, M.K. Dalinka: Chemical-shift MR imaging of the femoral head: an in vitro study of normal hips and hips with avascular necrosis. AJR. 148: 1159-1164 (1987).

Ohzono, K., M. Saito, K. Takaoka, K. Ono, S. Saito, T. Nishina, T. Kadowaki: Natural history of nontraumatic avascular necrosis of the femoral head. J. Bone Joint Surg. 73-B: 68-72 (1991).

Patterson, R.J., W.H. Bickel, D. C. Dahlin: Idiopathic avascular necrosis of the head of the femur. J Bone Joint Surg. 46-A: 267-282 (1964).

Saito, S., K. Ohzono, K. Ono: Joint-preserving operations for idiopathic avascular necrosis of the femoral head. Results of core decompression, grafting and osteotomy. J Bone Joint Surg. 70-B: 78-84 (1988).

Sarrat, P., P.C. Acquaviva, P. Lafforgue, M. Lopez-Vasquez, P. Bernard, B. Bouscarle: Ostéonécrose aseptique de la tête fémorale: apports de l'imagerie par résonance magnétique. Ann. Radiol. 31: 133-139 (1988).

Seiler, J.G., M.J. Christie, L. Homra: Correlation of the findings of magnetic resonance imaging with those of bone biopsy in patients who have Stage I or II ischemic necrosis of the femoral head. J. Bone Joint Surg. 71-A: 28-32 (1989).

Steinberg, M.E., C.T. Brighton, D.R. Steinberg, S.E. Tooze, G.D. Hayken: Treatment of avascular necrosis of the femoral head by a combination of bone grafting, decompression, and electrical stimulation. Clin Orthop. 186: 137-153 (1984).

Takatori, Y., M. Kamogawa, T. Kokubo, T. Nakamura, S. Ninomiya, K. Yoshikawa, H. Kawahara: Magnetic resonance imaging and histopathology in femoral head necrosis. Acta Orthop. Scand. 58: 499-503 (1987).

Tooke, S.M.T., P.J. Nugent, L.W. Bassett, P. Nottingham, J. Mirra, R. Jinnah: Results of core decompression for femoral head osteonecrosis. Clin. Orthop. 228: 99-104 (1988).

Warner, J.J.P., J.H. Philip, G.L. Brodsky, T.S. Thornhill: Studies of nontraumatic osteonecrosis: the role of core decompression in the treatment of nontraumatic osteonecrosis of the femoral head. Clin. Orthop. 225: 104-127 (1987).

OSTEONECROSIS
3. Treatment

CONSERVATIVE, NON INVASIVE TREATMENT OF OSTEONECROSIS OF THE FEMORAL HEAD

B. MAZIERES

Service de Rhumatologie
CHU de Rangueil
Toulouse, France

Is osteonecrosis of the femoral head (ON) spontaneously curable? One believes that it is, but one cannot yet confirm it. Of course, at collapse stage, healing is impossible. However, as long as the subchondral shell is intact, there is a possibility of healing. On the other hand, the evolution of controlateral hip, which goes to collapse in more than 50 % of the cases, invited us to strongly manage these patients without any hope in a spontaneous healing of the disease.

Is there a medical treatment of ON? The goal in the treatment of ON is to preserve, not replace, the femoral head. Although many methods have been proposed, none has proved completely satisfactory .

PREVENTION

When the risk of ON is high, and when an etiological factor is open to treatment, prevention is the best way:

- Most decompression accidents can be avoided if all decompressions follow the rules which have been established (slow resurfacing with staged decompression);
- Hyperlipemia, diabetes, alcohol intake should be treated or minimised;
- Each time corticosteroid treatment is undertaken in moderate or high doses, particularly greater than 20 mg of prednisone-equivalent daily for more than one month, the risk of bone necrosis must be recognised and weighted against the benefits to be derived from this therapy. Steroid use should be kept as low as possible.

Bone Circulation and Vascularization in Normal and Pathological Conditions
Edited by A. Schoutens *et al.*, Plenum Press, New York, 1993

323

THE CONSERVATIVE TREATMENT

It includes several methods, from the classical non weightbearing procedure to the modern electromagnetic fields which are still in progress.

Discontinuing weightbearing

Discontinuing weightbearing on the affected hip from one to several months, using crutches or cane, gives bad results in all series. Although there are few published reports describing the results of this approach, it appears that without some type of intervention, inevitable progression will occur once the condition is established.

Evaluating the effectiveness of core decompression in patients with lupus associated osteonecrosis, Hungerford and Zizic (1980) paired their surgical patients with a similar group of conservatively managed patients. In this later group, 21 of 22 hips (95 %) with Ficat stage II disease deteriorated symptomatically, 19 (86 %) progressed radiographically and 15 (68 %) required total hip replacement in a mean of 31 months. For stage III hips, 11 of 11 were worse symptomatically and 7 (68 %) came to total hip arthroplasty in an average of 28 months.

A retrospective study of 341 patients with ON of the femoral head allowed to select 36 patients with 50 hips available for a clinical and radiographic conservative managed follow-up (Musso et al., 1986). Using the Merle d'Aubigné-Postel 18-point grading system for clinical assessment, the clinical grade at first presentation was 13.9 and at the last follow-up examination it was 10.3. The mean time of clinical follow-up was 16 months. Radiographically (35 hips in 25 patients for whom serial radiographs were available), most of the hips were worse, increasing of one class in Ficat grading (figure 1).

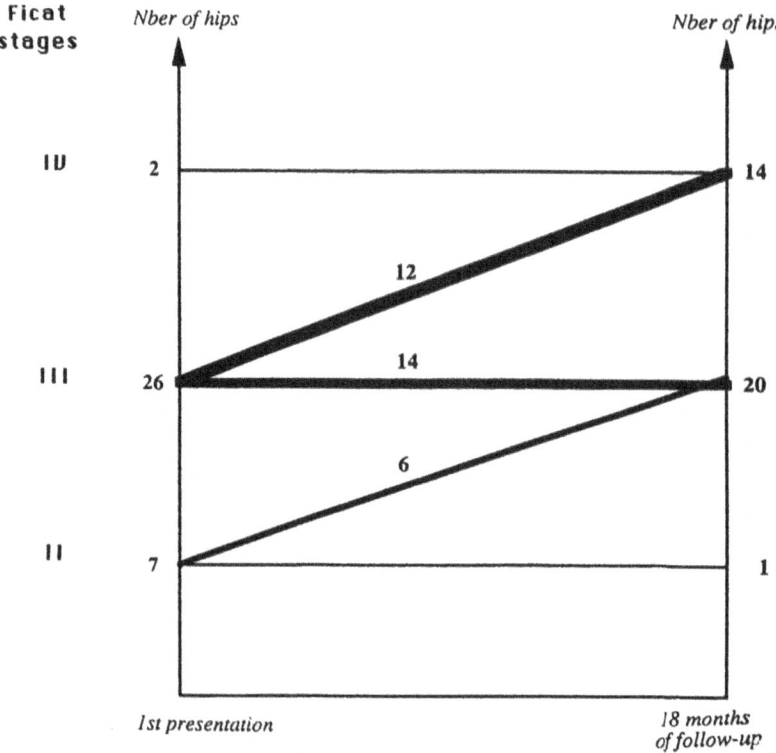

Figure 1. Radiographical follow-up of conservatively managed ON of the femoral head (Musso et al., 1986).

324

In a review of more than 250 cases of ON of the femoral head seen between 1970 and 1980 at the University Hospital of Pennsylvania, Steinberg et al. (1984) obtained 55 cases where there was a sufficient duration of follow-up prior to replacement surgery to allow a retrospective analysis with regard to the progression of the disease according to the stage of involvement at the time that the hip was first seen. Whatever the initial stage, 87 % of the hips were worse at the end-point (table I). The progression in 48 of these 55 hips, treated conservatively, was noted. Forty-four (92 %) showed definite progression and required femoral head replacement or total hip arthroplasty. No differences were noted between those allowed full weightbearing and those kept on partial or non-weight-bearing. In this study, the follow-up time is not known and clinical assessment is not available.

In conclusion, once a definitive diagnosis is made, progressive changes will occur in virtually all cases if untreated. Even if early stages are diagnosed at the first presentation, (pre-collapsed stages), evolution is similar. Partial weightbearing or non-weightbearing using bed rest, crutches or canes, analgesics and antiinflammatory agents are ineffective.

Table 1. Spontaneous radiographic follow-up of osteonecrotic hips according to the stage at the time the hip was first seen (Steinberg, 1984)

Stage at the beginning	N. of hips	better	same	worse
Before collapse	16	1	2	13
Cresent sign without flattening	11	0	2	9
Flattening ± OA changes	28	0	2	26
Total	55	1	6	48

Vasoactive Drugs

Vasoactive drugs may play a role in treating early cases of ON. Although it is difficult to prove the efficacy of so-called peripheral vasodilatators, most of them appear to be useful for relief of pain. These drugs are extensively used in black Africa in the treatment of sickle-cell crises. Furthermore we also have considerable experience with these drugs, which have the effect of rapidly reducing symptoms in patients who have not responded to either analgesics or NSAIDs. ON is considered to be of ischemic origin. This decrease of blood supply within bone tissue cannot be directly measured in humans, but indirect evidence of circulatory changes can be seen even in the early stages of ON: venous stasis (observed by intramedullary venography) and intramedullary hyperpressure, reflect the decrease in bone blood flow. Such an increase in intramedullary pressure could provide an explanation for patients with painful ON (Arnoldi's engorgement pain syndrome), especially in cases where pain is increased by coughing which also temporarily increases the intramedullary pressure (IMP).

Some presumed vasoactive drugs such as Hydergine, Vincamine of Naftidrofuryl have an antalgic effect on the pain of ON. This is a pragmatic finding without any assessment, but it was interesting to test the short-term effects of such drugs on the increased medullary pressure. We injected 40 mg of naftidrofuryl by i.v. route during a 30-60 s period in nine patients with a definite ON with a sustained increase in IMP (Arlet et al., 1990). IMP was recorded during the next 15 min. In six out of the nine patients, there was a stable and statistically significant decrease of IMP, while systemic pressure was not modified (Table 2).

Table 2. Intramedullary pressure before and after i.v. injection of 40 mg naftidrofuryl (Praxilène®) in patients with definite osteonecrosis of the femoral head

Patient no.	Intramedullary pressure (mm Hg)	
	Before injection	15 min after injection
1	60	40
2	40	25
3	55	40
4	55	45
5	65	45
6	60	40
mean ± SD	55.8 ± 8.6	39.2 ± 7.4

Furthermore, Nifedipine - a calcium inhibitor which has a myorelaxant activity through its pharmacological properties of relaxation of the arterial tonus - was tested in the treatment of hip pain of ON (Laroche et al., 1990). Ten mg of nifedipine are orally given to 18 patients with ON of the femoral head and to 7 patients with osteoarthritis or other non-inflammatory diseases of the hip. Pain was assessed by the visual analog scale (VAS) 30 and 60 min after drug intake. In the ON group, decrease of pain is significant until 30 min and goes on decreasing significantly at 60 min, while there is no change in the VAS score throughout the follow-up in the control group.

Of course, all these studies on the vascular aspect of the disease, need further well controlled investigations, but they may be an interesting way for future.

Lipid-Clearing Agents

Are the lipid-clearing agents a putative treatment? Wang and his team (1990) have demonstrated for a long time ago that steroid-treated rabbits are developing accumulation of fatty emboli within the intraosseous arterioles, especially in the subchondral region of the femoral head, hypertrophy of fat cells, and bone marrow hyperpressure with decreased bone blood flow. All these lesions are reduced or suppressed when rabbits are treated with clofibrate. No clinical trial is available with this medication.

Pulsing Electromagnetic Fields

Promising results were recently reported using pulsing electromagnetic fields (PEMF) applied externally. Rationale for using PEMF is: 1) in ON a coupled process of resorption of dead bone and replacement with living bone has been shown to occur, 2) exposure of bone to electrical currents of various configurations and time-varying magnetic fields has been shown to result in enhancement of bone repair and alteration in bone turnover clinically and experimentally. In ON, this bone turnover raises the possibility for a conservative treatment to influence the rates of resorption and repair and to modify the loss of structural integrity and conserve the femoral head.

In the experience of Aaron et al. (1989) PEMF were applied 8 hours per day for 12 to 18 months and the results compared with those of core decompression. The mean follow-

Table 3. Comparison of the results of core decompression and PEMF in osteonecrosis of the femoral head (Aaron et al., 1989)

Ficat grading	PEMF				Core decompression			
	N. of hips	clinical success	clinical failure	X-rays progression	N. of hips	clinical success	clinical failure	X-rays progression
II	23	20	3	6	26	16	10	16
III	33	18	15	16	24	6	18	16
Total	56	38	18	22	50	22	28	32

Table 4. Steinberg stage at start of treatment and at the most recent follow-up evaluation (from Bassett et al., 1989)

Steinberg stages at 1st presentation		N. of hips	Progress better	same	worse
Precollapse	A*	1	0	1	0
stage	B	2	1	1	0
	C	9	6	3	0
Subchondral	A	0	0	0	0
collapse	B	3	2	1	0
without	C	0	0	0	0
flattening					
Flattening	A	7	0	6	1
	B	31	0	24	7
	C	41	0	31	10
Flattening	A	13	0	12	1
with joint	B	3	0	3	0
narrowing	C	5	0	5	0
Total		115	9	87	19

* A, B and C subdivide the lesion in mild, moderale and severe (Steinberg et al., 1991)

up time was three years (minimum: 24 months). The PEMF used was a single-pulse configuration at a frequency of 72 Hz. The coil was held in place over the greater trochanter in specially fabricated shorts. Clinical success (assessed by pain relief using a modified Merle d'Aubigné scale) was observed in 68 % of the 56 cases treated with PEMF versus 44 % of the 50 hips operated. Rœntgenographic progression (defined as an increase in Ficat grade or collapse more than 2 mm compared to pretreatment rœntgenograms) was 26 % in Ficat grade II and 48 % in grade III with PEMF, compare to respectively 62 and 67 % in the operated group. The clinical and rœntgenographic results of PEMF treatment were superior to those of core decompression, particularly in hips with Ficat II lesions (Table 3).

Bassett et al. (1989), use the same PEMF procedure, with 3 types of coils which produced identical field values at the level of the femoral head. They quantified their results with the same clinical scale and the more sensitive Steinberg rating for rœntgenograms and the follow-up since the treatment began was 49 months. They observed 7 better rœntgenograms out of 12 pre-collapse cases and 81 out of 100 stable grade in hips with collapse (table 4). Clinically, using Merle d'Aubigné-18 points scores, the hips were 12.8 at start and 15.8 at the most recent evaluation.

These two studies demonstrate that treatment with PEMF can retard the usual roentgenographic progression of precollapse and postcollapse ON when compared with the natural evolution of the disease or evolution after core decompression. These results seem to be independent from the etiology of ON. Furthermore, the frequency of the need for surgical salvage has been reduced in patients whose average age is not optimal for joint arthroplasty.

On an other hand, using decompression and bone grafting with (20 patients) or without (20 patients) electrical stimulation, Steinberg et al. (1990) were unable to observe that the addition of capacitive coupling gave better results. This technique is still on evaluation and development.

REFERENCES

Aaron R.K., D. Lennox, G.E. Bunce, T. Ebert: The conservative treatment of osteonecrosis of the femoral head. A comparison of core decompression and pulsing electromagnetic fields. Clin. Orthop. 249: 209-218 (1989).

Arlet J., B. Mazières, M. Thiéchart, G. Vallières: The effect of i.v. injection of naftidrofuryl (Praxilène®) in intramedullary pressure in patients with osteonecrosis of the femoral head. In: J. Arlet and B. Mazières (eds): Bone Circulation and Bone Necrosis, pp. 405-406. Springer-Verlag, Berlin 1990.

Bassett C.A.L., M. Schink-Ascani, S.M. Lewis: Effects of pulsed electromagnetic fields on Steinberg ratings of femoral head osteonecrosis. Clin. Orthop.. 246: 172-185 (1989).

Hungerford D.S., T.M. Zizic: II. The treatment of ischemic necrosis of bone in systemic lupus erythematosus. Medicine 59: 143-148 (1980).

Laroche M., J.M. Jacquemier, Ph. Montané de la Roque, J. Arlet, B. Mazières: La nifédipine per os améliore les douleurs de l'ostéonécrose de la tête fémoral. Rev. Rhum. 57: 669-670 (1990).

Musso E.S., S.N. Mitchell, M. Schink-Ascani, C.A. Bassett: Results of conservative management of osteonecrosis of the femoral head. A retrospective review. Clin. Orthop. 207: 209-215 (1986).

Steinberg M.E., G.D. Hayken, D.R. Steinberg: The "conservative" management of avascular necrosis of the femoral head. In: Arlet J., Ficat R.P., Hungerford D.S. (eds): Bone circulation, pp. 334-337. Williams & Wilkins Publ., Baltimore 1984.

Steinberg M.E., C.T. Brighton, R.E. Bands, K.M. Hartman: Capacitive coupling as an adjunctive treatment for avacular necrosis. Clin. Orthop. 261: 11-18 (1990).

Steinberg M.E., D.R. Steinberg: Evaluation and staging of avascular necrosis. Seminars in Arthroplasty 2: 175-181 (1991).

Wang G.J., R.E. Fechner, J.P. O'Hara, N. Stoffel, W.G. Stamp: Improvement of femoral head blood flow in steroid-treated rabbits using lipid-clearing agents. In: J. Arlet and B. Mazières (eds): Bone Circulation and Bone Necrosis, pp. 395-398. Springer-Verlag Publ. Berlin 1990.

LONG TERM RESULTS IN ELECTROMAGNETIC FIELDS (EMF) TREATMENT OF OSTEONECROSIS

M. HINSENKAMP[1], J.P. HAUZEUR[2], and S. SINTZOFF Jr.[3]

Services d'Orthopédie Traumatologie[1], de Rhumatologie[2] et de Radiologie[3]
Hôpital Erasme, Université Libre de Bruxelles
Bruxelles, Belgique

INTRODUCTION

Recent studies based on histological and MRI examination have improved the understanding of the pathogenesis of osteonecrosis of the femoral head (ONFH) (Hauzeur, 1989). The early stages have been more systematically explored. Until the presence of typical radiological lesions with the crescent sign and collapse, radiology could be doubtful or normal. In such cases bone scintigraphy is positive in 51%, MRI and biopsy in 100 %. Excessive intraosseous pressure is found in 80 % of the early stages. This data led to the division of the evolution of ONFH into three phases. First, the necrotizing process produces areas of bone weakening or bone thickening. Such remodelling induces modification in the mechanical properties of the bone. The second phase results from the bad outcome of the first one when a fracture of the subchondral bone arises, usually in the weightbearing portion of the head. The third and last phase is osteoarthritis produced by acquired dysplasia due to the fracture.

The therapeutic strategy has to take into account this data: it must try to reverse the necrosis and avoid bone mechanical inhomogeneity during the first phase, it must restrict the evolution of the dysplasia during the second phase and treat the osteo-arthritis during the last phase. The effect of time varying EM fields on the acceleration of the ossification (Hinsenkamp, 1985, 1987, 1990) makes this treatment suitable for the two first phases corresponding to Ficat I, II and III.

MATERIAL AND METHODS

The EMF stimulation is applied by EBI (ElectroBiology Inc., P.O. Box 682, West Caldwell, New-Jersey 07006, USA) devices composed of a pair of Helmholtz

Bone Circulation and Vascularization in Normal and Pathological Conditions
Edited by A. Schoutens *et al.*, Plenum Press, New York, 1993

331

coils or a single coil and the generator unit. The induced single pulse, measured in a probe centered in the target area as described by Bassett (1984), is a quasi rectangular pulse of 380 microsecondes wide repeated with a frequency of 72 Hz. The coils were centered by the patient over the greater trochanter and activated for at least 12 hours per day.

The population consisted of a series of 19 patients with 29 ONFH and 2 osteonecroses of the femoral condyles. Among these 31 sites, 28 were treated. The mean age was 46 years with equal distribution between the thirties, forties and fifties. There were 12 men and 7 women. The etiological distribution was idiopathic (4), steroids (11), alcoholism (4), trauma (2) and hyperlipemia (2). In our series, steroids had been used for long term treatment in 58 % of the cases. Unilateral lesions were relatively rare (3 on the left and 6 on the right side) in comparison with the bilateral involvement which repre-sented 58 % of the cases (n=11).

At the beginning of the treatment, the first screening included: standard X-rays (4 incidences: anteroposterior, lateral, 2 obliques 45° and - 45°); bone biopsy ; CT scan and MRI. During the treatment standard X-rays were repeated every two months in the first year and then every three months. CT scans and MRI were repeated every six months.
At first screening, staging was Ficat I (1), II (8), III (19) and IV (3). In each case, a bone biopsy performed with a thin 3 mm trephine confirmed the presence of a true avascular necrosis. During this procedure, pressure measurements in the distal epiphysis before and after such biopsy proved the absence of a significant variation in the medullary pressure inside the femoral head induced by the biopsy (Hauzeur, 1992).
In the various examinations the comparison was made between the status at the beginning of the treatment, after 3 years and at the term of a follow-up ranging from 40 to 68 months. Scores were simplified to two categories corresponding to improvement or no alteration, and failure if any degeneration appeared.

RESULTS

Results are presented for the clinical evaluation, X-rays, CT scan and MRI. Non-stimulated heterolateral hips or follow-up shorter than 36 months still under treatment were excluded from the study unless they resulted in surgical procedure for a total hip prosthesis.

Clinical Evaluation

The clinical results are based on 28 osteonecroses evaluated after 36 months of treatment and 13 after a mean period of 54 months. One case of Ficat I remained asymptomatic after 54 months. The Ficat II and Ficat III evolution are presented in Table 1. At three years all Ficat II had improved as they had after prolonged follow-up, mean 58 months (range 47 to 68). No-one required surgery. With Ficat III, 50 % improved or showed no degradation after three years while 50 % required surgery. Failure increased to 69 % after a mean period of 54 months (range: 47-68 months). Table 2 gives the incidence of surgical procedure (total hip prothesis, THP) during the follow-up.
For the Ficat IV, two required surgery after 7 and 8 months treatment and one was still improving after 54 months.

Table 1. Clinical evolution

	Ficat II		Ficat III	
Follow up in months	36	58 (mean)	36	54 (mean)
Improved or status quo	6	4	9	5
Failure	0	0	9	2
Total	6	4	18	7

Table 2. Incidence of surgery in Ficat III

Year of Follow up	Number of THP	% of the remaining population
1	4	22
2	3	21
3	2	18
4	1	14
5	1	16
Total	11	69

X-rays Evaluation

The X-rays results are based on 22 osteonecroses after 36 months and 13 after a mean period of 54 months (range: 47-68 months).
One case of Ficat I remained invisible on X-rays after 54 months. The Ficat II and Ficat III evolution is presented in Table 3. At 36 months in 6 cases and after a mean period of 58 months (range: 47-68 months) in 4 cases all the Ficat II had improved or showed no modification. In Ficat III after 3 years 50 % of the X-rays lesions were altered and this increased to 64 % after a mean period of 53 months (range: 40-68 months). After 54 months the patient with Ficat IV showed an unexpected absence of degradation in the X-rays parameters.

Table 3. X-rays evolution

	Ficat II		Ficat III	
Follow up in months	36	58 (mean)	36	53 (mean)
Improved or status quo	6	4	7	5
Failure	0	0	7	2
Total	6	4	14	7

CT Scan and MRI Evolution

CT Scan and MRI examinations are indispensable for the classification at the beginning of the treatment. They identified or revealed the extent of the structural bone lesion in the CT-scan and the vascular disturbance under MRI examination. During the evolution in this series they reflected roughly the X-rays evolution.

Histological Examination

Histological examination was possible on 8 out of the 11 femoral heads removed for total hip arthroplasty. All the 8 heads exhibited an unusual, intense osteoblastic response on the margin of the necrotic area.

DISCUSSION

For ethical reasons we did not carry out a double blind study because of the long and cyclic evolution of the necrosis and the slow effect of the treatment. We chose to analyse well-documented cases and to compare them to the natural evolution presented in the literature.

The high percentage of bilateral lesions found in this series does not seem abnormal considering that most heterolateral lesions are asymptomatic and are discovered during the thorough screening prescribed for the symptomatic side. This led us to the assumption that the usual 30 % of bilateral lesions often found in the literature is probably underevaluated due to incomplete screening (Hauzeur, 1987).

The bibliography gives only a few series where the natural evolution of the osteonecrosis is selectively analysed for each Ficat stage and no-one to our knowledge has presen-ted a detailed follow-up up to a mean period of at least three years. In our results the difference observed between the evolutions of Ficat II and III after three years and four and a half years of treatment makes this selective analysis mandatory.

Table 4. X-rays evolution of Ficat II after conservative treatment

Authors	Year	N. of hips	% Alteration	Follow up months
Hungerford	1980	22	86	21
Steinberg	1982	11	82	21?
Musso	1986	7	86	16

For Ficat II, table 4 gives the radiologic evolution after conservative treatment in three series found in the literature. Compared to the results of the present study where not one case in six degenerated after 36 months, the effect of EMF appears obvious. Considering recourse to surgery for Ficat II, Hungerford's series (1980) of 24 hips revealed 68 % requiring surgical procedure after 31 months while after 36 months of EMF treatment, in our series no surgical procedure was required. In his series of 7 Ficat II treated conservatively, Stulberg (1991) found a median survival period before surgery of 6 months.

Table 5 gives the X-rays evolution for conservative treatment of Ficat III. The present results of which only 50 % are aggravated after 36 months and 64 % after 53 months, compared to the results and the follow-up of the conservative treatment, are in favour of EMF stimulation. With a view to a statistical evaluation of the effect of the treatment, we regrouped the stages Ficat II and III and considered separately the cases improved or not altered versus the cases with a roentgenologic failure, in the literature after a mean period of 20 months (16-28 months) and after 3 years in our series. Table 6 presents this comparison and the result of the Chi^2 test which reveals a significant difference (a < 0.001).

Table 5. X-rays evolution of Ficat III after conservative treatment

Authors	Year	N. of hips	% Alteration	Follow up months
Hungerford	1980	11	91	28
Steinberg	1982	10	90	21
Musso	1986	26	46	16

Table 6. Comparison of the X-rays results after F II and F III regroupment in the literature and in the present series ($Chi^2 = 11.65$, ddl = 1, a < 0.001)

Treatment	Conservative	EMF	Total
Improvement or Status quo	22	13	35
Degradation	65	7	72
Total	87	20	107

Regarding the surgical outcome of Ficat III, on 11 hips, Hungerford (1980), observed that 64 % required surgery after 28 months compared to 50 % after 36 months in the present study and 59 % after 54 months. In his series of 10 Ficat III treated conservatively, Stulberg (1991) found a median survival period before surgery of 5 months. It is interesting to note that throughout the five years of follow-up under treatment the incidence of recourse to surgery did not increase and even seemed to decrease. With reservations regarding the small number of cases in this series and the presumed similarity of the populations found in the literature the present results of EM stimulation in Ficat II and III are encouraging. The results show less positive influence on Ficat III which can be expected as the activation of the ossification observed as the main effect of the EMF stimulation (Hinsenkamp, 1985, 1987, 1990) cannot restore the anatomical geometry of the joint surfaces when they are deformed. However, in those cases the surgical procedure may be delayed which can be valuable for young patients providing they are able to sustain normal activity.

Larger series describing the natural evolution of osteonecrosis and the evolution under EM treatment during periods of three and five years should be performed to identify the indication and the true efficiency of EMF treatment for each stage. Long periods of follow-up are required to establish the dynamics of the evolution which in this pathology is not well known even in its natural progression. These studies should include a thorough screening and differenciate the various stages. The surgical outcome in the more advanced cases allows histological studies which should be performed to identify the mechanisms of action of the electromagnetic fields on the repair processes of osteonecrosis.

REFERENCES

Bassett C., M. Scink, S. Mitchell: Treatment of osteonecrosis of the hip with specific, pulsed electromagnetic fields (PEMFs): a preliminary clinical report. In: J. Arlet, R. Ficat and D. Hungerford (eds): Bone circulation, pp. 343-354. Williams and Wilkins Publ., 1984.

Hauzeur J.P., J.L. Pasteels, S. Orloff: Bilateral non-traumatic aseptic osteonocrosis in the femoral head. J. Bone Joint Surg. 69-A (8): 1221-1225 (1987).

Hauzeur J.P., J.L. Pasteels, A. Schoutens, M. Hinsenkamp, T. Appelboom, I. Chochrad, N. Perlmutter: The diagnostic value of magnetic resonance imaging in non-traumatic osteonecrosis of the femoral head. J. Bone Joint Surg. 71-A (5): 641-649 (1989).

Hauzeur J.P.: Contribution à l'étude de l'ostéonécrose aseptique non traumatique de la tête fémorale: apport de nouvelles méthodes d'investigation au diagnostic et à l'étude de l'évolution de cette affection. Thèse d'Agrégation, U.L.B.: pp. 71 (1991).

Hinsenkamp M., M. Rooze, M. Noorbergen, B. Tuerlinckx: Topography of EM exposure and its relationship to biological effects on tissues. In: A. Chiabrera, C. Nicolini and H. Schwan (eds): Interactions between electromagnetic fields and cells, pp. 557-567. Plenum Press Publ., 1985.

Hinsenkamp M., M. Rooze, B. Tuerlinckx: Biophysical and biological intervention: future directions. In: J. Lane (ed): Fracture Healing, pp. 267-275. Churchill Livingstone Publ., 1987.

Hinsenkamp M: Stimulation électromagnétique de la croissance et de la consolidation osseuse. Thèse d'Agrégation, U.L.B.: pp. 381 (1990).

Hungerford D.: Treatment of ischemic necrosis of bone in systemic lupus erythematosis. Medicine 59: 143-148 (1980).

Musso E., S. Mitchell, M. Schink-Ascani, C. Bassett: Results of conservative mangement of osteonecrosis of the femoral head. A retrospective review. Clin. Orthop. 207: 209-215 (1986).

Steinberg M., G. Hayken, D. Steinberg: The "conservative" management of avascular necrosis of the femoral head. In: J. Arlet, R. Ficat and D. Hungerford (eds): Bone Circulation, pp. 334-337. Williams and Wilkins Publ., 1984.

Stulberg B., A. Davis, T. Bauer, M. Levine, K. Easley: Ostenecrosis of the femoral head. Clin. Orthop. 268: 140-151 (1991).

EFFECTS OF ELECTRIC AND ELECTROMAGNETIC FIELDS ON BONE FORMATION, BONE CIRCULATION AND AVASCULAR NECROSIS

Marvin E. STEINBERG

Department of Orthopaedic Surgery
University of Pennsylvania School of Medicine
Philadelphia, Pennsylvania, U.S.A.

INTRODUCTION

For over 150 years it has been recognized that electrical signals are involved in a number of physiological processes. In 1812 a London surgeon by the name of Birch reportedly healed a tibial nonunion with galvanic current (Brighton, 1991; Canady et al., 1991). In the 1950s Yasuda, Fukada and associates, and other Japanese investigators rekindled interest in the relationship between electricity and bone. They demonstrated that when bone was mechanically deformed it generated an electrical potential, and that electricity could induce bone formation. During the next three decades several investigators demonstrated that the appropriate electrical signal could influence a number of tissues by a variety of mechanisms. These include not only bone, but also cartilage, tendon, nerve, blood vessels, skin and soft tissue. This signal can enhance healing of delayed union and non-union of fractures, osteotomies, spine fusions, and chronic skin ulcers. It may also impede tumor growth, retard the development of osteoporosis, stimulate angiogenesis, prevent vasoconstriction in skin flaps, enhance new bone formation, and retard progression in avascular necrosis of bone.

THE ELECTRICAL SIGNAL

At the present time three types of electrical and electromagnetic signals are of interest: direct current (D.C.), inductive coupling or pulsing electromagnetic fields (PEMFs) and capacitive coupling (C.C.). This paper will examine the effects of these signals on bone formation, bone circulation, and avascular necrosis.

Not only are there different methods for producing an electrical signal in bone, but within each method there are several factors which can be varied, thus altering the characteristics of the signal. These factors are quite important as they can dramatically alter the response in living tissue. In addition, a signal which is too "weak" may have no

Bone Circulation and Vascularization in Normal and Pathological Conditions
Edited by A. Schoutens *et al.*, Plenum Press, New York, 1993

337

demonstrable effect, whereas one which is too "strong" may cause cell death and tissue necrosis. Thus considerable time and effort have been spent to determine the optimum signal to be used under different circumstances.

Initially interest focused on direct current (D.C.) signals and attempts were made to duplicate the electrical wave forms recorded from mechanically stressed bone. It was felt by many that electricity might well be the means by which Wolff's Law might operate: mechanical stress causes bone deformation that generates an electrical signal which, in turn, stimulates bone formation. Areas under compresssion are electronegative in relationship to areas under tension. When bone is exposed to the appropriate external signal, new bone formation takes place about the negative electrode or cathode (Yasuda et al., 1955; Yasuda, 1977).

C.A.L. Bassett (1982) reported that bone formation and repair could also be stimulated by exposure to a pulsing electromagnetic field. Similar effects were observed with capacitive coupling, another method for producing electrical fields. (Brighton, 1991; Farmer et al., 1991.)

Direct Current (D.C.). A direct electrical current can be delivered by invasive or semi-invasive techniques. In the invasive technique a flexible cathode wire is placed at the site to be stimulated by an open operative procedure, either with or without a supplemental bone graft. This electrode is connected to a small self contained battery pack and generating unit which is placed in a subcutaneous position close to the site of entry into the bone. This power pack is later removed (Figure 1). In the semi-invasive technique, rigid wires are drilled percutaneously into the bone under direct image intensifier guidance. The metallic electrode is exposed at the tip where the stimulation is to take place, whereas the rest of the wire is insulated. These wires are then connected externally to the negative pole of a generating source. Each of these methods has advantages and disadvantages. These devices are designed to deliver a constant direct current of approximately 20 microamperes in the face of changing resistance at the electrode - tissue junction (Friedenberg et al., 1974).

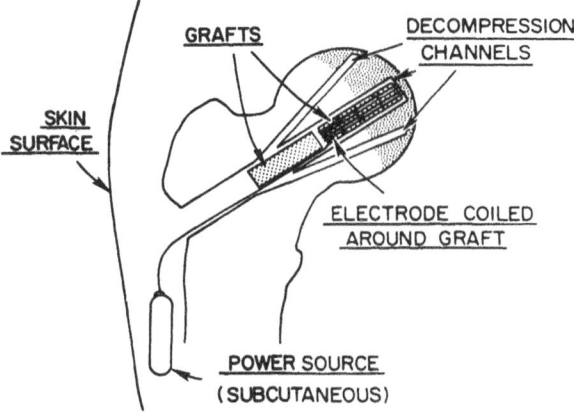

Figure 1. Direct Current (D.C.): decompression channels, bone graft, electrode and power source.

Capacitive Coupling (C.C.). In capacitive coupling capacitor plates or electrodes are placed opposite each other on the skin surface over the site to be stimulated (Figure 2) Either a continuous or a variable signal can be used. In practice the signal now generally used is a 20 to 200 KHz sine wave with a peak to peak potential of one to ten volts. At our institution a signal of 60 KHz and 5 volts is normally employed. This has been calculated to produce an electrical field in the tissues between the electrodes of 3.3 V/cm. with a current density of 300 MA / cm^2. The unit generates a large field which is 2.5 times the diameter of the electrode plates in all directions. The electrodes are connected to a small portable power source worn around the waist and kept in continuous operation (Steinberg et al., 1990).

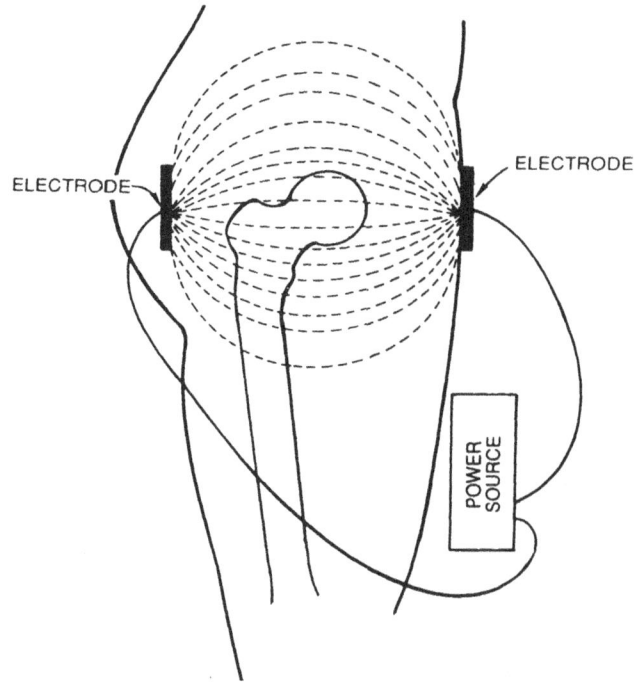

Figure 2. Capacitive Coupling (C.C.): placement of electrodes, electrical field, and power source.

Inductive Coupling or Pulsing Electromagnetic Fields (PEMF's). Electrical fields can also be produced in bone and other tissues by inductive coupling, a technique which uses an external coil or coils through which is passed a time-varying or pulsed electrical signal. This generates an electromagnetic field which in turn induces a secondary electrical field in the underlying bone (Figure 3). In this technique the tissue is exposed to both an electrical and a magnetic field, but it is generally felt that the electrical field is the most important in regard to biological stimulation. The signal can be varied in many regards and different signal characteristics are employed, for example, when treating fracture non-unions than when treating avascular necrosis. In this latter condition, the signal presently used has a single-pulse configuration at a frequency of 72 Hz. Each pulse is quasirectangular and has a duration of 380 microseconds. The pulses have a peak magnetic field of 35 Gauss, a dB/dT of 6.5 T/s and an electrical field of 0.1 MV/cm (Aaron et al., 1991).

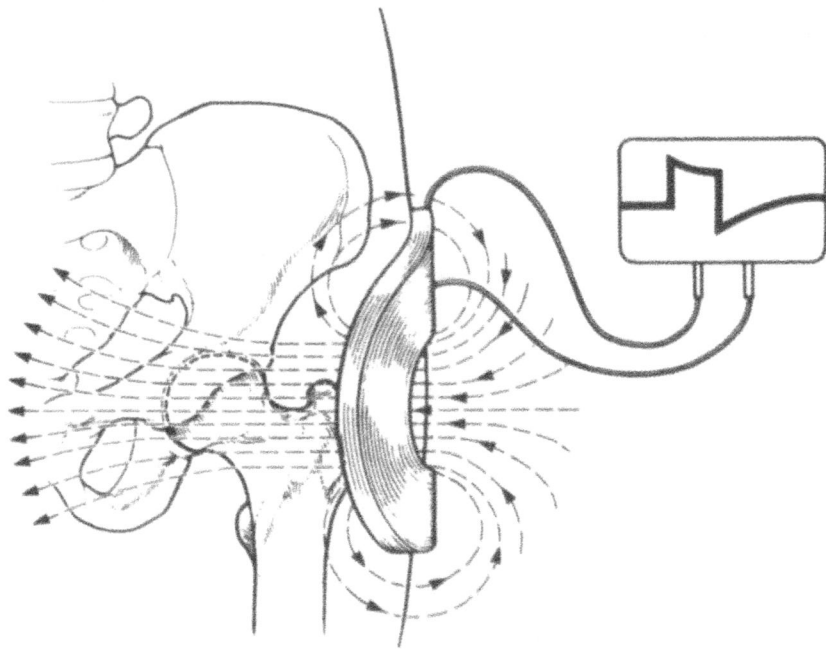

Figure 3. Inductive Coupling (PEMFs): single curved coil over lateral aspect of hip, electromagnetic fields, and external current generator.

MECHANISM OF ACTION

There are many effects common to the three different types of electrical stimulation which are presently being employed. For example, they seem to be equally effective in stimulating the healing of non-unions and delayed unions. On the other hand, these signals must not be considered identical and each type of signal may have variable biological effects as certain of its characteristics are changed. Much work has been done to characterize these effects selectively, yet much more still needs to be done. In this section we will discuss the common physiologic effects of these electrical and electromagnetic fields, regardless of the specific techniques used to produce them.

All the mechanisms of action of electrical and electromagnetic fields have not yet been fully delineated, but it appears that they function at several levels from the molecular to the macroscopic. One of the most important sites of action is at the cell membrane. Here a number of effects have been observed. These include an alteration in ion channels, membrane permeability, and ion binding, and an activation of cell membrane receptors

340

and enzymes, including Na, K-ATPase. The net effect would seem to be an alteration in the rate at which ions enter or leave the cell, notably Ca^{++}, Na^+ and K^+. This in turn can influence a number of intracellular processes. These include the overall metabolism of the cell, the activity of cAMP, the type of macromolecules synthesized, (including DNA, RNA, collagen and proteoglycans), the activity of various enzyme systems, hormones and growth factors, and cell proliferation and differentiation. There also appears to be an inhibition of parathyroid homone (PTH) and a reduction in osteoclastic activity leading to a decrease in bone resorption (Tenforde, 1991; Canady, 1991).

EFFECT ON BONE CIRCULATION

Very little has been reported on the specific effects of electric or magnetic fields on bone circulation per se. A number of studies have demonstrated that electrical stimulation can speed the healing of chronic skin ulcers and soft tissue defects, many of which are associated with poor circulation. They also increase the survival of skin flaps, in part due to improving local circulation by inhibiting vasoconstriction. In experimental animals PEMF's have been able to limit the size of myocardial infarction produced by coronary artery ligation. Although the mechanisms are uncertain, it has been suggested that they act by decreasing inflammation and the release of vasoconstricting agents.

In 1988 Ameia Yen-Patton and associates reported that PEMF's had a significant effect on venous and arterial endothelial cells grown in cell culture. In addition to stimulating cell proliferation, they caused morphological changes such as cell elongation and formation of a "sprouting" pattern. In certain areas there was reorganization of cells into vessel-like structures. Discrete stages of neovascularization were observed, quantitatively similar to stages of angiogenesis observed in vivo. Braun and Lemons noted that PEMF's seemed to increase the penetration of small vessels into devascularized bone (Bassett et al., 1989).

Nannmark et al. (1988) directly investigated the influence of electrical stimulation on bone vascular supply. They used an optical chamber inserted into the metaphysis of rat tibias to observe the effects of a constant direct current on bone blood flow, vascular exchange, and microvascular behavior. They found that after one hour of stimulation there was macromolecular leakage around capillaries and venules. After two to eleven weeks there was increase in both the size and number of capillaries, simultaneous with an increase in bone formation. No changes in net blood flow were found. From these studies it was not possible to determine whether any of the vascular changes caused the increased bone formation, or whether these were two separate and independent effects of electrical stimulation.

NEW BONE FORMATION

In the early 1950s Yasuda and other investigators in Japan reported on the production of new bone in response to a D.C. electrical stimulus (Yasuda, 1977); This was confirmed by a number of other investigators using direct curent, inductive coupling (PEMF's) and capacitive coupling. (Friedenberg et al., 1974; Brighton et al., 1991; Matsunga et al., 1990; Bassett et al., 1982). A variety of in vivo and in vitro experimental models were used. These observations laid the foundation for further experimentation and clinical application of electrical signals to several conditions. These included fracture healing, stimulation of osteotomies and surgical fusions, prevention of osteoporosis, and treatment of avascular necrosis.

FRACTURE HEALING

The area where electrical fields have received the most attention is that of fracture healing. All three types of signal have shown positive effects in a number of studies performed by several different investigators. In animal studies they demonstrated their effectiveness in healing osteotomies of the long bones and both delayed union and non-union. They have also been used in a number of clinical studies involving human patients. After some initial skepticism due to a lack of prospective, blinded controls, sufficient evidence regarding their effectiveness has been accumulated to make these devices now readily available for clinical use.

Each of the methods for electrical stimulation has advantages and disadvantages, as discussed earlier. Whichever technique is selected, it must be used in addition to and not in place of optimum fracture care. Although there have been some data suggesting that electrical fields can accelerate normal fracture healing initially, there is no definitive evidence that the final result differs whether electricity is or is not applied. The value of electrical stimulation lies in treatment of the small percent of fractures which fail to heal normally and go on to either delayed union or non-union. Once this has been established and the situation found amenable to electrical stimulation (e.g. the absence of a synovial pseudarthrosis, a wide fracture gap, a large segment of dead bone, etc.) this technique may be of significant benefit. At times it may be used in place of open surgical procedures, and at time as an adjunct to surgery and bone grafting.

During the past twenty years a number of studies have been published and several thousand patients with delayed union or non-union have been treated. The results were quite consistent whether D.C., C.C., or PEMFs were used. They indicated an overall success rate in the range of 75 to 80 %, although differences were noted, in part depending on location of the fracture. (Brighton, 1991; Bassett et al., 1982; Cundy, 1990; Hinsenkamp, 1991; Scott and King, 1991).

MISCELLANEOUS CONDITIONS

Electrical fields have been used in a number of other conditions, a few of which will be mentioned here.

Osteoporosis. Both PEMFs and C.C. have shown their ability to inhibit the development of osteoporosis in experimental models (Bassett L.S., 1979; Brighton et al., 1989). The details for its effective application in human patients have not yet been worked out, although this is an area of major interest and major importance.

Surgical Fusions. A small number of studies have evaluated the effect of D.C. stimulation as an adjunct to open surgery with bone grafting in producing a stabilization or fusion of the spine and other areas. In general these studies showed a small but definite increase in the incidence of fusion (approximately 10 %) when electricity was used. This technique is gaining in popularity, particularly in regard to fusions of the spine. (Brighton, 1991).

Growth Plate Stimulation. Initial experiments showed that both cartilage cells in culture and animal growth plates in vivo could be stimulated by electrical fields. The early reports were met with considerable enthusiasm because of the clinical importance of this observation. To date a long term in vivo increase in growth has not been established, and more work is needed in this area. (Armstrong and Brighton, 1986; Kuhlman and Brighton, 1991).

TREATMENT OF ATRAUMATIC OSTEONECROSIS

Osteonecrosis (ON) of the femoral head (ONFH) is a frustrating condition which, if untreated, results in femoral head collapse and the need for hip replacement in the majority of cases. After the initial vascular insult and death of bone and marrow a process of repair is initiated. Unless the lesion is quite small this is ineffective.
Osteoblastic new bone formation cannot keep pace with osteoclastic resorption and the other processes which remove necrotic tissue. The net result is a marked weakening of the subchondral bone with subsequent stress fractures and collapse of the articular surface. If the processes of resorption could be retarded and those of bone repair accelerated, much of the strength of the subchondral bone could be preserved and the incidence of segmental collapse decreased.
A number of surgical procedures have been used to retard or reverse the progression of ONFH. Although the results have generally been better than with symptomatic treatment alone, they leave much to be desired. Electrical fields have been shown to stimulate bone formation, retard bone resorption, and promote healing of fractures. A small number of investigators therefore applied various types of electrical stimulation to femoral heads with ONFH, either alone or as a supplement to a surgical procedure, in an attempt to improve the results of treatment.

Inductive Coupling (PEMFs). In 1982 Bassett et al. presented a preliminary report on the positive results of PEMFs in treating ONFH of the femoral head. (Bassett et al., 1989) (Figure 3). This work was continued by other investigators and a multi center study was initiated. Recently Aaron reported on a series of patients with a mean follow up of six years and compared these to treatment with core decompression. The results with PEMFs were superior to those with core decompression and were significantly more durable. (Aaron and Steinberg, 1991; Aaron et al., 1991).
This topic will be covered in greater detail by Professor Hinsenkamp later in this book.

Direct Current (D.C.). Stimulation with direct current can be accomplished by semi-invasive and invasive tehcniques as described earlier, and can be used by itself or combined with surgical treatment. We elected to use it as an adjunct to core decompression with bone grafting in a series of patients with avascular necrosis of the femoral head. In a preliminary study the signal was delivered by multiple rigid wire electrodes placed circumferentially about the central core and graft and traversing the soft tissues to emerge from the skin surface. Because of the motion of the soft tissues about the hip, these wires were not well tolerated and this technique was abandoned. A second technique was then employed.
One hundred sixteen hips with ONFH underwent core decompression and bone grafting. Seventy-four of these were also treated with D.C. electrical stimulation by means of a negatively charged electrode, the cathode, coiled about the cancellous graft which

343

was inserted into the center of the necrotic lesion. This was connected to a constant direct current power pack and electrical stimulator which was placed subcutaneously in the thigh ("Orthofuse", DePuy, Warsaw, Indiana, U.S.A.) (Figure 1). This had an estimated life of six months and was removed at the end of that time. Patients were kept on partial weight bearing for six months. They were followed every three months for the first year, every six months for the next year, and annually thereafter. Hips were evaluated clinically by means of the Harris hip rating system and radiographically by a quantitative method of evaluation and staging (Steinberg et al, 1989). Results were determined by comparing the Harris score and radiographic stage preoperatively to those at the latest follow up, and by determining the number of hips which required subsequent total hip replacement (THR). Patients were included in this protocol if they had non-traumatic ONFH of one or both femoral heads, regardless of etiology. Hips were in Stage I (pre-radiographic, with a positive bone scan and/or MRI) through Stage IV-A (a small amount of femoral head flattening) at the start of treatment. They were randomly assigned to electrical or non-electrical groups. These groups were well matched by radiographic stage, Harris score, etiology, age and gender. Follow up time was two to eight years in both groups. Mean follow up was 33 months in the non-stimulated group and 44 months in the stimulated group. The results in these two groups were also compared to the results in a group of 55 hips treated non-operatively and evaluated retrospectively prior to the start of this present study.

The 55 hips managed symptomatically served as controls and indicated the natural progression of ONFH without surgical intervention. These hips were followed from six months to ten years with a mean follow up of twenty-one months. This mean follow up time was short since a large number of hips required arthroplasty during the first two years. Radiographic progression and the need for arthroplasty are broken down by stage and are shown in Table 1.

Table 1. Progression of ON with non-operative management (6 months to 10 y follow- up).

STAGE	RADIOGRAPHIC PROGRESSION	ARTHROPLASTY REQUIRED
O-II	81%	69%
III	82%	82%
IV	100%	90%
V & VI	94%	94%
ALL HIPS	92%	84%

Hips treated with decompression and grafting alone showed mean radiographic progression of 1 1/3 stages with 21 % improved or unchanged and 79 % more advanced. There was a mean loss of 3 points in the Harris score. 43 % required total hip replacement (THR) regardless of the preoperative stage.

Hips treated with decompression, grafting and D.C. stimulation showed mean radiographic progression of 2/3 stage with 30 % improved or unchanged and 70 % more

Table 2. Results of decompression and grafting with and without D.C. stimulation (stage I-IV ONFH).

RADIOGRAPHIC PROGRESSION:	WITHOUT D.C.	WITH D.C.
BY STAGE	1 1/3	2/3
% OF HIPS	79%	70%
CHANGE IN HARRIS SCORE	↓ 3	↑ 5
ARTHROPLASTY REQUIRED	43%	35%

advanced. A small clinical improvement was also noted with a mean increase of 5 points in the Harris score. Overall 35 % of hips required THR. However, only 25 % required THR if treatment began prior to surface flattening (Stages I-III), whereas 47 % required THR when flattening was already present at the start of treatment (Stage IV) (Table 2).

In conclusion, hips treated by decompression and grafting, with or without supplemental electrical stimulation, did significantly better than hips treated symptomatically (Figure 4). The addition of D.C. electrical stimulation by the technique described gave better results by all parameters of evaluation than decompression and grafting alone. The differences between these two groups were, not enough to be statistically significant, however there was a definite suggestion that D.C. ws a useful adjunct to decompression and grafting, especially when instituted prior to the start of femoral head collapse.

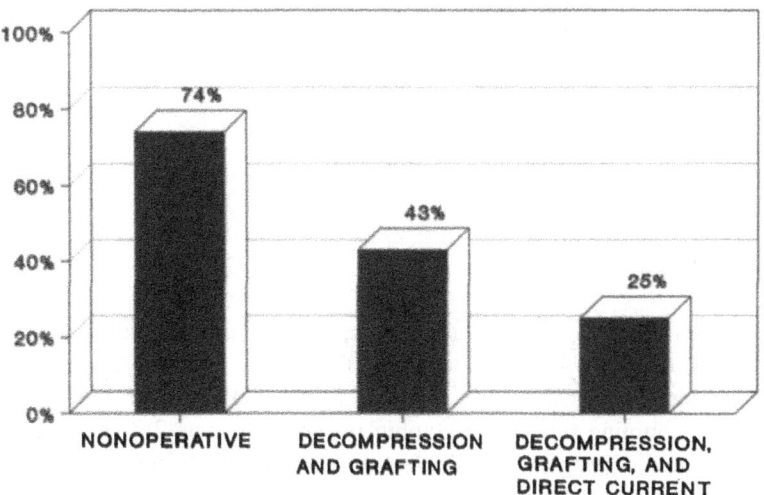

Figure 4. Percent of hips requiring arthroplasty after non-operative treatment, decompression and grafting alone, and decompression and grafting with supplemental D.C. stimulation. (Stages I-III at start of treatment).

Capacitive Coupling (C.C.). We were encouraged by the results obtained with direct current and sought a more effective method for electrical stimulation of the femoral head in ONFH. Inductive coupling (PEMFs) had already been evaluated. Because of the positive results with capacitive coupling (C.C.) both in laboratory experiments and in treating patients with delayed union and non-union, we sought to evaluate this technique in treating avascular necrosis. C.C. has certain advantages over D.C. as used here. It generates a much larger electrical field and can thus stimulate viable cells at the margins of the necrotic region which might be beyond the reach of the fields generated by the internal D.C. coil. It is a non-invasive technique and could lend itself to future use without surgery if it proved effective as an adjunct to surgery.

Forty patients were entered into a prospective, randomized, double blind study monitored by the United States Food and Drug Administration (FDA). The study parameters were similar to those described for D.C. stimulation, however no patients were included in Stage IV (femoral head flattening). After decompression and grafting all patients wore a battery pack and signal generator connected to skin electrodes placed anteriorly and posteriorly about the hip (Figure 2). One half the units were active and one half were inactive. Neither the patient nor the examiners knew which were the active units. These were worn for six months. Patients were followed from two to four years with a mean follow up time of 31 months. Results were again evaluated by determining change in clinical status (Harris score), radiographic progression or improvement, and the need for total hip replacement. All evaluations were done before the code was broken to separate active from inactive units. At this time the two groups of patients were compared and were found to be closely matched by all parameters.

Table 3. Results of decompression and grafting with and without capacitive coupling (Stage I-II ONFH).

	WITHOUT C.C.	WITH C.C.
RADIOGRAPHIC PROGRESSION:		
BY STAGE	1	2/3
% OF HIPS	50%	58%
CHANGE IN HARRIS SCORE	↑ 1	↓ 12
ARTHROPLASTY REQUIRED	20%	25%

The non-stimulated or control hips showed mean radiographic progression of one full stage with 50 % improved or unchanged. There was no change in the mean clinical status as determined by Harris score. 20 % required THR. In the hips treated with the active capacitive coupling units radiographic progression was 2/3 of a stage with 42 % showing no evidence of progression. There was a loss of 12 points in the Harris score. 25 % of hips required THR (Table 3).

In conclusion, there was no difference in outcome between hips treated with or without active capacitive coupling. As in previous studies, results in both groups were significantly better than in hips treated symptomatically (Figure 5). As mentioned earlier there are many ways in which an electrical signal or field can be varied and these variations may produce quite different effects when applied to living tissues. It is therefore quite possible that an optimum capacitive coupling signal would be effective in treating ONFH, although the signal used in these experiments was not (Steinberg et al., 1990). These results cannot be directly compared to those with direct current described earlier, as the studies were conducted at different times and there were differences in the extent of involvement and follow up time.

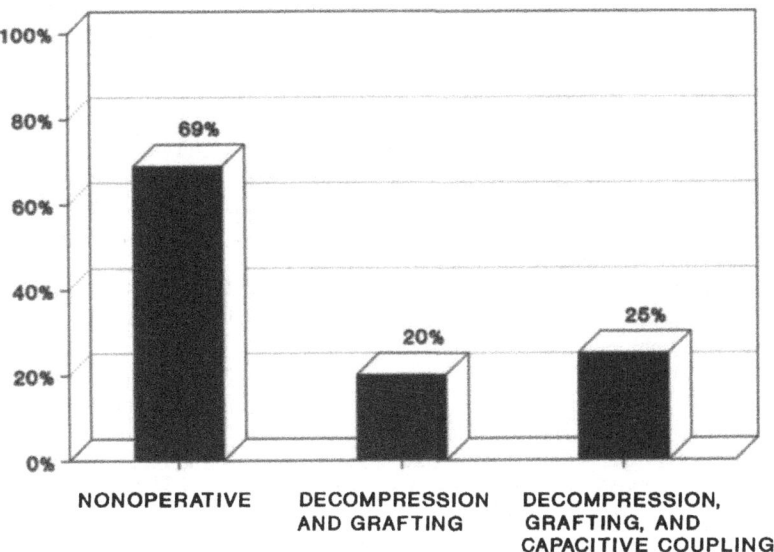

Figure 5. Percent of hips requiring arthroplasty after non-operative treatment, decompression and grafting alone, and decompression and grafting with supplemental capacitive coupling. (Stages I-II at start of treatment) .

SUMMARY AND CONCLUSIONS

During the past three decades there has been a great deal of interest in the effects of electrical and electromagnetic fields on living tissue. Several types of signal have been evaluated and a number of different effects have been observed at both the molecular and macroscopic level. The proper signal can stimulate cellular activity and promote bone formation and repair. Clinically promising results have been obtained in the treatment of delayed union and non-union of fractures and in healing skin and soft tissue lesions. Results in treating osteonecrosis have been mixed with some techniques apparently able to retard progression and others showing no definitive effect.

A great deal of work has yet to be done to determine which specific signal or electrical field will be most effective under specific conditions. We remain optimistic that in the future we will be able to add electrical stimulation to our armamentarium in combating a variety of disease processes, including some of those which involve the musculoskeletal system.

REFERENCES

Aaron, R.K., D. Lennox, D. Ciombor, G. Bunce, T. Ebert: Durability of outcome in osteonecrosis of the femoral head treated with pulsing electromagnetic fields or core decompression. In: C.T. Brighton, S.R. Pollack (eds.): Electromagnetics in Medicine and Biology, pp. 309-312. San Francisco Press, Publ., San Francisco 1991.

Aaron, R.K., M.E. Steinberg: Electrical Stimulation of Osteonecrosis of the femoral head. Seminars in Arthroplasty: Osteonecrosis of the Hip. 2: 214-221 (1991).

Ameia Yen-Patton, G.P., W.F. Patton, D.M. Beer, B.S. Jacobson: Endothelial cell response to pulsed electromagnetic fields: Stimulation of growth rate and angiogenesis in vitro. J.Cell. Physiol. 134: 37-46 (1988).

Armstrong, P.F., C.T. Brighton: Failure of the rabbit tibial growth plate to respond to the long term application of a capacitively coupled electric field. J. Orthop. Res. 4: 446-451 (1986).

Bassett, C.A.L., S.N. Mittchell, S.R. Gaston: Pulsing Electromagnetic field treatment in ununited fractures and failed arthrodeses. JAMA. 247: 623-628 (1982).

Bassett, C.A.L., M. Schink-Ascani, S. Lewis: Effects of Pulsed electromagnetic fields on Steinberg ratings of femoral head osteonecrosis. Clin. Orthop. Rel. Res. 246: 172-185 (1989).

Bassett, L.S., G. Tzitzikalakis, R.J. Pawluk, C.A.L. Bassett: Prevention of Disuse Osteoporosis in the rat by means of pulsing electromagnetic fields. In: C.T. Brighton, J. Black, S.R. Pollack (eds.): Electrical Properties of Bone and Cartilage: Experimental Effects and Clinical Applications, pp. 311-331. Grime & Stratton, Publ., New York 1979.

Brighton, C.T.: Advanced Clinical Applications of Electromagnetic field effects: Bone and Cartilage. In: C.T. Brighton, S.R. Pollack (eds.): Electromaqnetics in Medicine and Bioloqy, pp. 293-308. San Francicso Press, Publ., San Francisco, 1991.

Brighton, C.T., J.C. Farmer, B.J. Sennett, C.A. Hansen, J.P. Iannotti, J.R. Williamson, S.R. Pollack: Activation of the phosphoinositide cascade in the proliferation response of rat calvarial bone cells exposed to a capacitively coupled electrical field. Ibid., pp. 105-110.

Brighton, C.T., M.D. Luessenhop, S.R. Pollack, D.R. Steinberg, M.E. Petrik, F.S. Kaplan: Treatment of Castration-Induced Osteoporosis by a capacitively coupled electrical signal in rat vertebrae. J. Bone Joint Surg.: 71A: 228-236 (1989).

Canaday, D.S., R.C. Lee: Scientific basis for clinical applications of electric fields in soft tissue repair. In: C.T. Brighton, S.R. Pollack (eds.): Electromagnetics in Medicine and Biology. pp. 275-292. San Francisco Press Publ., San Francisco 1991.

Cundy, P.J., D.C. Patterson: A ten year review of treatment of delayed union and non-union with an implanted bone growth stimulator. Clin. Orthop. Rel. Res. 259: 216-221 (1990).

Friedenberg, Z.B., L.M. Zemsky, R.P. Pollis, C.T. Brighton: The response of non-traumatized bone to direct current. J Bone Joint Surg.: 56A: 1023-1030 (1974).

Hinsenkamp, M.: Realistic approach of electromagnetic stimulation in the treatment of non-union. Transactions of the Bioelectric Repair and Growth Soc. XI: S5 (1991).

Kuhlman, J.R., C.T. Brighton: Stimulation of bovine growth plate chondrocytes with pulsed capacitively coupled electric fields. In: C.T. Brighton, S.R. Pollack (eds.): Electromagnetics in Medicine and Biology, pp. 153-158. San Francisco Press Publ., San Francisco 1991.

Matsunaga, S., K. Sakore, K. Ijiri, M.A. Ali, E. Fukada, M. Date: Histochemical investigation of intramedullary callus by pulsed electromagnetic fields (PEMFs): Ibid. pp. 177-190.

Nannmark, U., F. Buch, T. Albrektson: Influence of direct currents on bone vascular supply. Scand. J. Plast. Reconstr. Surg. 22: 113-115 (1988).

Scott, G., J.B. King: A prospective double blind trial of capacitive coupling in the treatment of non-unions. Transactions of the Bioelectric Repair and Growth Society. XI: 42 (1991).

Steinberg, M.E., C.T. Brighton, R.E. Bands, K.M. Hartman: Capacitive coupling as an adjunctive treatment for avascular necrosis. Clin. Orthop. Rel. Res. 261: 11-18 (1990).

Steinberg, M.E., C.T. Brighton, A. Corces, G.D. Hayken, D.R. Steinberg, B. Strafford, S.E. Tooze, M. Fallon: Osteonecrosis of the femoral head: results of core decompression and grafting with and without electrical stimulation. Clin. Orthop. Rel. Res. 249: 199-208 (1989).

Tenforde, T.S.: ELF field interactions at the animal, tissue, and cellular levels. In: C.T. Brighton, S.R. Pollack, (eds.): Electromagnetics in Medicine and Biology, pp. 225-236. San Francisco Press Publ., San Francisco 1991.

Yasuda, I.: Electrical callus and callus formation by electret. Clin. Orthop. Rel. Res. 124: 53-56 (1977).

Yasuda, I., K. Noguchi, T. Sata: Dynamic Callus and Electric Callus. J Bone Joint Surg.: 37-A: 1292-1293 (1955).

THE PLACE OF CORE DECOMPRESSION IN THE TREATMENT OF OSTEONECROSIS OF THE FEMORAL HEAD

C. FICAT

Hôpital Beaujon, Université Paris VII
Paris, France

Core decompression of the femoral head for osteonecrosis (ON) is exactly the same procedure as core biopsy. The operative technique and the instrumentation are the same. Decompression of the bone marrow spaces with its analgesic effect is the fortuitous side effect of the bone biopsy.

In the early sixties, before bone scans were available, P. Ficat and J. Arlet realized that ON of the femoral head was preceded by a rather long period of absence of radiographic signs. They decided to remove a piece of bone from the suspected painful hips with a limitation of movements and normal X-rays in order to try to have a definite diagnosis and to detect the first tissue changes. They wished to understand the physiopathology of the disease. Early diagnosis, permitting early physiologic treatment, has always been the main concern of my father. This was the backbone of his philosophy and in this precise case, by chance, the diagnostic procedure proved also to be the treatment.

INDICATIONS FOR CORE DECOMPRESSION

Originally, core biopsy (or core decompression) was the third stage of a diagnostic procedure called "Functional Investigation of Bone" including, firstly, the measurement of the bone marrow pressure (BMP) and, secondly, intraosseous venography (IOV).

This functional investigation of bone was carried out in the case of a painful hip with limited movements, mostly internal rotation and normal X-rays. Some characteristics of the pain could suggest an ON: sudden onset, localization of pain in the groin, persistence at rest and during the night, impulsiveness on coughing, as well as the detection of a risk factor like alcoholism, cortisone treatment, hyperlipidemia, diabetes, sickle cell anemia, or a recent trauma.

Normally the core biopsy was done only in the cases of an excessive BMP and an abnormal IOV indicating a disturbance of the vascularization of the upper end of the femur.

Bone Circulation and Vascularization in Normal and Pathological Conditions
Edited by A. Schoutens *et al.*, Plenum Press, New York, 1993

351

Since those years where the Functional Investigation of Bone was the only way of making the diagnosis of ON at its pre-radiological stage, several reliable new investigations, less invasive and easier to prescribe, have progressively appeared. These are mainly: Isotopic Bone Scanning, Computerized Tomography Scanning, Magnetic Resonance Imaging. And now the question is: Is it still necessary to do this functional investigation of bone with a core biopsy/decompression?

Isotopic Bone Scanning, with Strontium 87m and now essentially with Technetium 99m usually gives an image of increased uptake in the femoral head which is not pathognomonic. An area of decreased uptake surrounded by a rim of increased uptake is more specific but is also less frequently observed. Isotopic Bone Scanning which, according to several authors gives positive information in 73 to 89 % of the cases at stage I, can be absolutely normal even at repeated intervals in early cases of ON.

Computerized Tomography Scanning (CT Scan) cannot demonstrate the early marrow lesions of the pre-radiographic stage, but it is very interesting for determining if the sphericity of the femoral head is still intact. This is a crucial landmark in the management of ON of the femoral head. CT Scans can demonstate a slight collapse of the subchondral bone or a rupture of the contour of the head when it is not yet clearly visible on conventional X-rays. The informations obtained by the CT Scan can change the radiological staging based on conventional X-rays. The crescent sign belongs definitely to stage III. It indicates a massive collapse of the cancellous bone of the femoral head. Its presence is due to the resilience of the strong intact articular cartilage attached to the thin shell of subchondral bone detached from the slightly collapsed sequestrum.

MRI (Magnetic Resonance Imaging) is the most recent, less invasive, and the most sensitive of the new diagnostic procedures. It can be pathognomonic. It can show important modifications of the marrow signal when the radiographs are normal. Generally an area of decreased signal within the femoral head on a T1 weighted sequences is seen. Its sensitivity could be 100 % pathognomonic according to some authors like Camp and Colwell (1986). In fact MRI can be non demonstrative at stage I and is not always totally reliable. According to Stulberg and al., it has a sensitivity of 86 %. Genez and al. from 11 cases histologically verified have observed 5 positive results and 6 negative.

So it is still necessary in some cases to do a Functional Bone Investigation with a core biopsy if any other investigation is not conclusive. The rationale for this attitude is simple. The diagnosis of ON is basically a histological one. ON means death of bone cells, i. e. necrosis of bone marrow cells with or without necrosis of osteocytes. The definite diagnosis has to be based on histological data, and the purpose of the core-biopsy is to provide the histological specimen.

On the other hand when the diagnosis is clear from all the information gathered by clinical examination, X-rays bone scan and MRI, and if the sphericity ot the femoral head appears to be preserved on X-rays and CT Scan, then the question is: Is it worthwhile to do a core decompression ? The answer is yes considering the natural evolution of the disease.

ON of the femoral head treated conservatively at stages 0, I, and II is progressive in 81 % of cases, with a gross collapse in 69 % according to Steinberg et al. (1984). The value of core decompression as a therapeutic measure in ON of the femoral head has been recently stressed by Stulberg et al. in a prospective randomized study of non-operative management versus core-decompression for 55 hips at radiological stages I, II and III in 36 patients. Non-operative treatment consisted in non-weightbearing during a minimum of six weeks.

For the 10 operated hips at stage I they have obtained a clinical success in 7 cases and on the 5 hips treated conservatively only 1 clinical success. At stage II a clinical success has been achieved in 5 cases of the 7 operated hips and 0 in the 7 non-operated hips. For stage III on the 11 operated hips a clinical success has been achieved in 8 cases and again on the 10 non-operated hips there is only 1 success, and the authors conclude that the use of core decompression for the surgical management of ON in stage I, II and early stage III hips appears justified.

However in cases that the hip is either asymptomatic or has just become asymptomatic abstention can be discussed and can be reasonable if the clinical context allows it and the classical measures of pressure relief (rest, use of crutches. complete or partial non-weight-bearing) and appropriate medical treatment are applied.

TECHNIQUE OF CORE DECOMPRESSION

Normally core decompression of the femoral head, which is also a core biopsy, is preceded by the measurement of the BMP and intraosseous venography. However, if the diagnosis is sufficiently assured, these two procedures are now not absolutely necessary.

The operation is done under general or epidural anesthesia. The patient lies supine on a radiotransparent operating table with an image intensifier placed just above the affected hip. A cushion is placed under the ipsilateral buttock to facilitate the approach by making the greater trochanter more prominent. The caudal part of the limb is dressed sterile in order to retain free movements in all directions.

The surgical approach is a small lateral incision of 5 to 8 cm starting from the apex of the great trochanter downward. The fascia lata is cut in the direction of its fibres and its borders are retracted.

The measurement of bone marrow pressure is simple (it can be recorded percutaneously under local anesthesia) but requires a precise technique. Bone marrow pressure must be recorded in the greater trochanter, near the base of the neck and also in the centre of the femoral head, because sometimes its level can be normal in the metaphysis but high in the head. Once the special trocar has been inserted into the bone, it cannot be removed anymore and placed elsewhere. This would create a leak at the first orifice leading to an incorrect recording of the intra medullary pressure. If the trocar is not in a good direction and has to be redirected one must let it at the same orifice and use another one driven carefully under the control of the image intensifier. The pressure is recorded by an electronic manometer giving an instant value of the BMP in millimeters of Hg. It is advisable to obtain a paper print of the pressure curve. If an electronic manometer is not available the pressure can be recorded with a simple Claude manometer giving a value in cm of H_2O.

The test is positive when the BMP is equal or superior to 30 mm Hg. One can also appreciate the amplitude of the pulsations. If the value recorded is normal or bordeline one can do a "stress test" consisting in an injection of 5 cc of normal saline solution. If the pressure increases more than 10 mm Hg 5 minutes after the injection the test is positive. The BMP, recorded in the trochanteric area, is abnormally elevated in 72 % of our cases, but after the stress test that percentage increases to 85 %. In the remaining 15 % of the cases, 10 % are abnormal if the recording is done in the center of the femoral head. Stulberg stated that the measurement of the BMP is an invasive procedure of less sensitivity than the other non invasive modalities. Nonetheless, in his experience, it was positive in 84 % of the cases.

Intraosseous venography (IOV) is done by the same trocar used for the recording of the BMP either in the metaphysis or in the femoral head. An Injection of 10 cc of a contrast medium (Hexabrix) normally fills the efferent veins of the superior part of the femur. Five minutes later no trace of contrast medium should remain in the bone. ON produces in 95 % of the cases a modification of the venogram, like a stasis of the contrast medium in the medulla five minutes after the injection or a filling of the central venous sinus or an amputation of the efferent veins (Arlet and Ficat, 1980). These images, like the recording of the BMP, give good indications about the circulatory disturbance but seem to have no prognostic value.

Normally the core biopsy is done only if an abnormality of the bone blood flow is established by these two investigations. It also can be done on the basis of information provided by non invasive methods such as bone scanning or MRI
The vastus lateralis is cut in the direction of its fibres which are retracted. The lateral cortex is exposed by two bone elevators placed on the anterior and posterior aspect of the femoral metaphysis. A small hole of the size of the trephine is made with a narrow gouge just below the line of insertion of the vastus lateralis on the great trochanter, at an equal distance of the anterior and posterior aspects, exactly in the middle of the metaphysis. As soon as the bone is open one can observe a dramatic fall of the BMP. Then the chosen trephine with its short mandrel is driven throuh the hole in the direction of the centre of the femoral head.
The trephine must have an appropriate size. There are three sizes available: 0.6 cm, 0.8 cm and 1 cm according to the hip morphology and to the type of drilling required. The mid-size of 0.8 cm is the one most currently used. The small one is used to make a second core if a larger decompression is needed or if the femoral head is small and the bigger ones cannot be used.
When the tip of the trephine reaches the intertrochanteric line at the base of the neck, its orientation in three planes must be precisely controlled with the image intensifier. Once the direction is established to be correct, the lower limb is placed in the frog leg (or number four) position and the trephine is pushed manually. Its progression from the middle of the neck towards the centre of the head is controlled continuously on the screen. In the head, the bone is often very hard and the progression by hand can be impossible. We hammer on the mandrel and give the trephine torsional movements between the blows. The progression is stopped approximately 5 mm from the subchondral plate. The subchondral plate must not be perforated. At this point the trephine is gently pulled out with some rotation movements, the mandrel left in place. The cylindrical piece of bone is extracted from the trephine by the introduction of a pusher. The specimen 5 to 7 cm long is put in a tube of a slightly larger diameter and sent to the pathology department. Its orientation in the tube is noted.
The aspect of the bone, its colour, its density, must be appreciated. Normally it is red in the neck but, in the head it can be yellowish with red spots and it is more dense.
Sometimes a second drilling with a smaller trephine can be indicated if the necrotic area is wide, or if an abnormal zone is not revealed by the first biopsy.

If a biopsy of cartilage is needed we use a special long "hip chondrotome" of 18 cm. A "chondrotome" is a very small trephine of a diameter of 3 mm specially made to remove a cylinder of cartilage with its subchondral bone. It is introduced inside the drilled canal and pushed by hand through the subchondal bone and the articular cartilage of the femoral head. By the same way it is also possible to take a biopsy of acetabular cartilage.
Just after the procedure one can determine that the BMP has returned to a normal level. The wound is closed plane by plane with a suction drainage for two days.
After a core decompression, the patient is normally allowed to walk the day after the operation, but according to the stage of the necrosis we recommend a variable period of total

or partial non-weightbearing, one month on the average. The risk of collapse of an eventual sequestrum is more than the risk of a fracture of the femur.

RESULTS OF CORE DECOMPRESSION

Histological Results (Core biopsy)

The value of core biopsy for the histological diagnosis of the ON in the femoral head according to several authors:

Table I

Authors	Date	Number of cases	Positive results
Camp et al.	1986	40	62 %
Zizic et al.	1986	259	94 %
Stulberg et al.	1987	54	75 %
Genez et al.	1988	11	100 %
Tooke et al.	1988	42	50 %
Hopson et al.	1988	21	95 %
Coleman et al.	1988	18	89%
Miller et al.	1989	16	94 %
Seiler et al.	1989	16	100 %
Robinson et al.	1989	23	78 %
Learmoth et al.	1990	41	68%

Zizic et al., with a technique similar to ours, have obtained 94 % positive results in 259 hips, Genez et al. 11 positive in 11 cases. One can expect nearly always valuable information from a core biopsy provided it is done correctly. In our experience histological information obtained by a core biopsy has always been of value and precise. However some papers in the literature do not corroborate this experience. Stulberg et al. describe 7 uninterpretable cases on 33 biopsies due to inadequate sampling or preparation. We think that besides the technique, the size of the core specimens is of utmost importance.

The advantage of to the development of this Functional Bone Investigation has been to reveal that 1) ON of the femoral head could exist with perfectly normal X-rays and 2) ON was not a limited disease as suggested by X-rays but a disease of the total proximal part of the femur. The characteristic triangular sequestrum being only the result of the mechanical stress due to the application of the resultant of the muscular forces and the body weight on that particular area of the femoral head, in most cases the anterosuperior quadrant. 3) In early ON there is a circulatory disturbance in the bone marrow spaces with histological stigmata such as plasmostasis, oedema and hemorrhages followed by the disappearance of the hematopoietic marrow, the destruction of the adipocytes (eosinophilic reticular necrosis) and finally the death of osteocytes. From the different biopsy specimens it was possible to hypothesize on the evolution of the disease and to predict the histological stage according to the classification of Arlet and Ficat. 4) There was no parallelism between the existing histological and radiological classifications. One can observe an ON of any histological grade with normal X-rays. This fact can be obvious peroperatively. After its extraction from the trephine the bone can look like normal cancellous bone to the naked eye or it can show two well defined parts, one distal, corresponding to the metaphysis, looking reddish and normal, and one proximal, corresponding to the head, entirely white and looking completely ischemic. Frequently, the cylindrical specimen breaks in two at the junction of these two parts. The core specimen can also look entirely white, thus ischemic.

Therapeutic Results (Core decompression)

The first effect of the drilling of the femur is to create a dramatic fall of the BMP. The BMP, which can be as high as 100 mmHg, returns instantaneously to a normal level after coring. This effect is essentially due to the trepanation of the lateral cortex of the femur for 90 % of the drop. The removal of the cervicocephalic core biopsy represents only 10 % of the drop (Ficat and Arlet, 1980). We agree with Stulberg who recommends a monitoring of the BMP during the procedure in order to be sure that a proper decompression of the bone has been realized. Normally the pain disappears immediately after the operation. The pain is related to the raised BMP and a persisting pain after a core decompression could therefore be attributed to an insufficient decompression.

The efficiency of core decompression is also demonstrated objectively by the normalization of Tc 99m uptake on bone scans done 6 months later or even earlier, and nowadays by a postoperative MRI. In some cases where a selective hip arteriography has been done before and after core decompression it has been observed that the intracapital superior arteries, which were not visible on the preoperative arteriogram, became visible again after the procedure (J. Théron, 1980). We hypothesize that an elevated BMP contributes to the mechanical obstruction of the blood flow coming from the small superior cephalic arteries and it prevents its penetration in the superior part of the femoral head. The mechanism is comparable to what is happening in a post-traumatic compartmental syndrome of the leg or of the forearm resulting in the Volkmann contracture. The raised pressure due to oedema inside the non-elastic osteomuscular compartment compresses the muscular capillaries and provokes ischemia and muscular necrosis. The aponevrotomy-like core decompression reduces the excessive pressure and the blood flow is restored. In most cases, probably, the lumen of the small intracephalic arteries remain permeable, in spite of some alterations of their wall (Arlet et al., 1991). If the blockade at the capillary level is suppressed the blood can pass again through these small intracapital arteries.

If the importance of the core biopsy to establish a precise diagnosis of early stages of ON is now generally recognized (despite some criticism), its therapeutic efficiency is still a matter of considerable controversy.

The therapeutic results of the core decompression of the femoral head have already been published by P. Ficat (1985) with a mean follow-up of 9 years 6 months (minimum: 5 years, maximum: 18 years).

In stage I very good and good clinical results were obtained in 93.9 % of the cases. The radiological results were in 86.6 % very good and good, 13.4 % were failures.

In stage II there is a drop of very good and good radiographic results to 66.7 %, but very good and good clinical results are still 82.4 %.

Table II. Clinical results after Core Decompression

Stage	Number of hips	Very good and good		Failure	
		Number	Per cent	Number	Per cent
I	82	77	93.9	5	6.1
II	51	42	82.4	9	17.6
Both	133	119	89.5	14	10.5

P. Ficat. J. Bone Joint Surg., 67-B: 3-9 (1985)

Table III. Radiographic results after Core Decompression

Stage	Number of hips	Very good and good		Failure	
		Number	Per cent	Number	Per cent
I	82	71	86.6	11	13.4
II	51	34	66.7	17	33.3
Both	133	105	78.9	28	21.1

P. Ficat. J. Bone Joint Surg., 67-B: 3-9 (1985)

From a pathophysiological point of view it is interesting to note that a narrowing of the joint space was observed in 9.8 % of the cases at stage I and in 9.6 % at stage II.

The disappointing results and complications of core decompression reported by some authors (Camp and Colwell, 1986; Hopson and Siverus, 1988) do not coincide with our experience.

Camp and Colwell have stated that core decompression was difficult and dangerous mainly on the basis of the occurence of secondary fractures of the femur. This complication is practically unknown in our experience which contains several hundred cases. The only fracture we have observed after a core decompression has occured after a car accident. The fracture was located at the point of penetration of the trepan on the lateral cortex of the femur. Camp and Colwell have observed 4 fractures in 40 core decompressions for stage I ON done by 13 different orthopaedic surgeons. Other authors have also mentioned the possibility of such a complication.

Table IV. Clinical results of core decompression in the literature

Authors	Date	Cases	Stage	Success
Zizic et al.	1985	49	I	69 %
		90	II	60 %
Camp et al.	1986	18	I + II	40 %
Tooke et al.	1986	10	I	100 %
		26	II	60%
Hopson et al.	1988	20	I + II	40 %
Colemanetal.	1988	18	I + II	94 %
Seiler et al.	1989	10	I	67 %
			II	34 %
Benoit et al.	1989	9	0	89 %
		3	I	67 %
		7	II	29 %
Lausten et al.	1990	11	I	63 %
		11	II	36%
Learmoth et al.	1990	12	I	25 %
		29	II	17 %

(The results have been expressed in percentage in order to allow an easier comparison. Stage 0: ON with normal X-rays and asymptomatic.)

Table V. Complications of Core Decompression

Authors	Fractures of the femur	Core decompression (Number of cases)
Zizic et al. (1985)	2	211
Camp et al. (1986)	4	40
Tooke et al. (1988)	1	45
Hopson e al. (1988)	1	17

We think that this type of complication can be related to an inappropriate technique. Core decompression is neither difficult nor dangerous if correct and minimal precautions are taken. The choice of the point of penetration of the trephine in the middle of the lateral side of the femoral metaphysis is important. Penetration should be made just under the upper limit of the vastus lateralis fibres where the cortex is thin and not through the thick cortex of the upper diaphysis. This causes a dangerous weakening of the femur.

The discordance in the therapeutic results of core decompression according to the different authors can have several reasons. It could be a question of 1) diagnosis, or 2) indication or 3) technique or 4) a question of population of patients.

An accurate radiological staging of the cases is not always easy, but it is certainly essential. If the quality of the images is poor, a stage II can be classified as a stage I, and a stage III as a stage II. The CT Scan is very useful for a precise staging. We would like to emphasize that a crescent sign as discrete as it can be is definitely a stage III. A slight flattening of the head which can be frequently overlooked is also a stage III.

For Beltram et al. (1990) MRI can have a prognostic value according to the percentage of bone involved in the upper half of the femoral head.

The second point is that in clinical practice core decompressions are frequently performed too late, at early stage III, because the diagnosis has not been made at the very beginning of symptoms. This could contribute to the fact that the procedure is ineffective. Core decompression is indicated if the femoral head has remained perfectly spherical, at the radiological stage I and II. If the decompression has an analgesic effect and presumably improves the bone circulation and therefore the bone metabolism, it cannot prevent the collapse of a sequestrum which already exists. If the pathogenesis of ON has some relationship with a compartmental syndrome, one must keep in mind that core decompression is the equivalent of the aponevrotomy which has to be performed as an emergency procedure in order to be effective. An aponevrotomy done at the stage of ischemia would be, of course, useless.

Technical problems may also play a role. A point of penetration too low on the lateral cortex, breaking the femoral neck cortex by a misguided trephine can produce a secondary fracture. The motorization of the drilling can be dangerous and needs great attention. We do not recommend it. An inadequate trephine of a too small diameter can be ineffective and a too large one can be dangerous. A routine penetration of the femoral head by an arthrotomy as advocated by Saito et al. is more complicated and hazardous. Learmoth et al. (1990) break the subchondral plate with trephine of a diameter of 1 cm and remove the articular cartilage with the piece of bone. We recommend not to go further than 5 mm from the subchondral plate. Core decompression must remain a simple and precise procedure.

The last point which could explain the discrepancies in the results is the fact that the different etiologies have not the same prognosis and their proportions can be different according to the different medical centres. Steroid associated ON is very common in the series published in the USA while it was relatively rare in Toulouse in the series of Arlet and Ficat. In our experience ON after renal transplant has the worst prognosis, followed by cortisone induced ON. Osteonecrosis induced by alcohol seems better tolerated, perhaps because of the particular psychology of the patients. Post-contusive and idiopathic cases seem to give the best results. One can observe that for these different etiologies the trabecular bone densities are most of the time abnormal and that the worst prognosis are associated with the lowest densities.

We propose the hypothesis that the prognosis of ON of the femoral head depends for a significant part on the anabolic capacities of the trabecular bone. "Bone formers" should have a better prognosis than "bone loosers". We now measure the bone density of the neck of the femur and spine, in order to have an idea of the strength of the cancellous bone of the femoral head and to undertake a sequential treatment by calcitonin or diphosphonate in order to prevent the microfractures of the trabeculae, which otherwise would lead to the collapse of the sequestrum and the fatal loss of sphericity.

REFERENCES

Arlet, J., M. Laroche., R. Soler, M. Tiéchart, M.T. Pieraggi, B. Mazières: Histopathology of the vessels of the femoral heads in specimens of osteonecrosis, osteoarthritis and algodystrophy. ARCO News. 3. 2: 108-112 (1991).

Beltram., J., C.T. Knight, W.A. Zueler, J.P. Morgan, L.J. Schwendeman, V.P. Chandnani, J.C. Mosure, P.B. Schaffer: Core decompression for avascular necrosis of the femoral head: Correlation between long-term results and preoperative MR staging. Radiology 175: 533-536 (1990).

Benoit, J., C. Got, P. Hardy, P. Videcoq: L'ostéonécrose de la tête fémorale chez le transplanté rénal. A propos de 82 hanches opérées. Rev. Chir. Orthop. 4: 216-227 (1989).

Camp, J.F., C.W. Colwell: Core decompression of the femoral head for osteonecrosis. J. Bone Joint Surg. 68 A: 1313-1319 (1986).

Coleman, B.G., H.Y. Kressel., M.K. Dalinka, M.L. Scheibler, L.B. Burk, E.K. Cohen: Radiographically negative avascular necrosis. Detection with MR imaging. Radiology 168: 525-528 (1988).

Conklin, J.J., P.O. Alderson, T.M. Zizic, D.S. Hungerford, J-Y. Dansereau, A. Gober, H.W. Wagner: Comparison of bone scan and radiograph sensitivity in detection of steroid-induced ischemic necrosis of bone. Radiology 147: 221-226 (1983).

Ficat, R.P.: Idiopathic bone necrosis of the femoral head. J. Bone Joint Surg. 67 B: 3-9 (1985).

Ficat, R.P., J. Arlet: Ischemia and Necrosis of Bone. D.S. Hungerford (eds). Williams & Wilkins, Baltimore 1980.

Ficat, R.P., J. Arlet: Diagnostic de l'ostéonécrose fémoro-capitale primitive au stade I (stade pré-radiologique). Rev. Chir. Orthop. 54: 637-648 (1968).

Genez, B.G., M. Wilson, R.W. Hauk, R.W. Weiland, H.R. Unger, N.N. Shields, K.S. Rugh: Early osteonecrosis of the femoral head: Detection in high risk patients with MR imaging. Radiology 168: 521-524 (1988).

Hopson, C.N., S.W. Siverus: Ischemic necrosis of the femoral head. J. Bone Joint Surg. 70 A: 1048-1051 (1988).

Lausten, G.S., B. Matheisen: Core decompression for femoral head necrosis. Prospective study of 28 patients. Acta Orthop. Scand. 51: 507-511 (1990).

Learmoth, I.D., S. Maloon, H. Dall: Core decompression for early atraumatic osteonecrosis of the femoral head. J. Bone Joint Surg. 72 B: 387-390 (1990).

Miller, I.L., C.G. Savory, D.W. Polly, G.D. Graham, J.M. Mc Cabe, J.J. Callaghan: Femoral head osteonecrosis. Detection by magnetic resonance imaging versus single-photon emission computed tomography. Clin. Orthop. 247: 152-162 (1989).

Robinson, H.J., P.D. Hartleben, G. Lund, J. Schrieman: Nuclear magnetic resonance imaging in femoral head osteonecrosis: In: J. Arlet, B. Mazières (eds): Bone circulation and bone necrosis, pp. 227-280. Springer Verlag, Heidelberg 1989.

Saito, S., K. Ohzono, K. Ono: Joint-preserving operations for idiopathic avascular necrosis of the femoral head. J. Bone Joint Surg. 70B: 78-84 (1988).

Seiler, J.G., M.J. Christie, L. Homra: Correlation of the finding of magnetic resonance imaging with those of bone biopsy in patients which have stage I or II ischemic necrosis of the femoral head. J. Bone Joint Surg. 71A: 28-32 (1989).

Steinberg, M.E., G.D. Hayken, D.R. Steinberg: The conservative management of avascular necrosis of the femoral head. Proceedings of the third International Symposium on Circulation in Bone, 1982, Toulouse, France. Williams & Wilkins, Baltimore 1984.

Stulberg, B.N., M. Levine, T.W. Bauer, G.H. Belhobek, W. Pflanze, D.H.I. Feiglin, A.I. Roth: Multimodality approach to osteonecrosis of the femoral head. Clin. Orthop. 240: 181-193 (1989).

Stulberg, B.N., A.W. Davis, T.W. Bauer, M. Levine, K. Easley: Osteonecrosis of the femoral head: A prospective randomized treatment protocol. Clin. Orthop. 268: 140-151 (1991).

Théron, J.: L'artériographie supersélective dans les nécroses primitives de la tête femorale. In: M. Lequesne, C. Massare (eds): Maladies de la hanche. VIème réunion annuelle du GETROA. Geigy, Paris 1980.

Tooke, S.M.T., P.J. Nugent, L.W. Bassett, P. Nottingham, J. Mirra, R. Jinnah: Results of core decompression for femoral head osteonecrosis. Clin. Orthop. 228: 91-104 (1988).

Zizic, T.M., C. Marcoux, D.S. Hungerford, M.B. Stevens: Radiography, scintigraphy and hemodynamics: Diagnosis of ischemic necrosis of bone (INB) (abstr.) Arthritis Rheum. 26: S37 (1983).

Zizic, T.M., D.S. Hungerford: Avascular necrosis of bone. In: W.N. Kelly, E.D. Harris, S. Ruddy, C.B. Sledge (eds): Textbook of rheumatology. W. B. Saunders Co. 1985.

Zizic, T.M., D.S. Marcoux, D.S. Hungerford, M.B. Stevens: The early diagnosis of ischemic necrosis of bone. Arthritis Rheum. 29: 1177-1186 (1986).

300 CASES OF CORE DECOMPRESSION WITH BONE GRAFTING FOR AVASCULAR NECROSIS OF THE FEMORAL HEAD
(Preliminary presentation. Full paper will be published elsewhere)

M.E. STEINBERG, W.B. HOSICK and K. HARTMAN

Department of Orthopaedic Surgery
University of Pennsylvania School of Medicine
Philadelphia, Pennsylvania

INTRODUCTION

Today we do not have a completely satisfactory method for treating avascular necrosis of the femoral head. One of the more popular approaches for treatment in the early stages is core decompression. Reports in the literature have varied considerably concerning the effectiveness of this treatment. Unfortunately, many of the reported studies have significant shortcomings, and it is difficult to compare the results reported by different authors.
We have modified the technique of core decompression as initially reported and have added to it a loosely fitting cancellous graft. This paper reviews the results in over 300 cases of avascular necrosis treated by this technique by the senior author and followed for up to eleven years. This is a prospective study in which the results were evaluated using objective clinical parameters and a quantitative method of radiographic measurement. It should therefore help dispel much of the contradiction and ambiguity of previous studies.

METHODS

Between 1981 and 1992 over 300 hips with osteonecrosis (ON) were treated by the senior author using a modified technique of core decompression with a supplemental loosely fitting cancellous graft. Hips range from Stage I (pre-radiologic) through IV-A (early collapse of the femoral head). Partial weightbearing was maintained for three months. X-rays and Harris scores were obtained preoperatively and at regular intervals postoperatively. Radiographic changes were measured and quantitated by a technique developed at our institution. Results were determined by change in Harris score, radiographic progression, and the need for total hip replacement. These results were compared to 55 hips previously treated by non-operative means and to data obtained from the literature.

Bone Circulation and Vascularization in Normal and Pathological Conditions
Edited by A. Schoutens *et al.*, Plenum Press, New York, 1993

361

RESULTS

Four complications occurred: one non-fatal pulmonary embolism; one femoral thrombophlebitis; one pneumonia; and one subcapital fracture sustained in a fall one month postoperatively.

230 of these hips were followed from two through eleven years postoperatively (mean - 43 months). 28 % (65 hips) required total hip replacement (THR) between 3 months and 60 months following surgery (mean - 22 months). This contrasts with 74 % of control hips with a similar degree of involvement and duration of follow up which required THR (P < 0.005).

Of the hips treated prior to collapse (Stages I and II) 73 % experienced pain relief, 46 % showed overall clinical improvement. 54 % had some degree of radiographic progression as compared to 81 % of controls, and the extent of the progression was less in the treated hips. The results correlated closely with lesion size: THR was required in 7 %, 31 %, and 33 % of small, intermediate, and large lesions respectively. Results in hips treated after collapse (Stage III and IV-A) were not as good as those treated before collapse had occurred but were still better than controls: 43 % required THR as compared to 90 % of controls.

DISCUSSION

The classical core decompression was modified somewhat by making a large central core (8 mm) plus two smaller cores (5 mm) into the lesion. A very loosely fitting cancellous graft was placed in only the central core. No attempt was made to compare results of this technique to core decompression alone without the graft. The advantage of this study over many others is that it is prospective, all procedures were done by a single surgeon, a large number of hips were involved, the duration of follow up is lengthy, and, most important, the results were evaluated by independent observers using objective clinical parameters and a quantitative method of radiographic measurement. It should therefore help dispel much of the contradiction and ambiguity of previous studies.

CONCLUSION

These results clearly indicate that the technique reported has a very low complication rate and is effective in treating avascular necrosis of the femoral head prior to collapse. The results are closely correlated not only with the stage of the lesion at the time of treatment, but also with the size or extent of involvement.

PROXIMAL FEMORAL OSTEOTOMIES IN THE TREATMENT OF IDIOPATHIC AND STEROID-INDUCED OSTEONECROSIS OF THE FEMORAL HEAD

Y. SUGIOKA

Department of Orthopaedic Surgery
Faculty of Medicine, Kyushu University
Fukuoka, Japan

INTRODUCTION

First of all, I would like to mention the principle of the treatment of the osteonecrosis of the femoral head based on the following clinical observations.

In silent hips which show typical osteonecrosis of the femoral head on roentgenogram or MRI without clinical symptoms, the necrotic lesion was not expansive on MRI or roentgenograms by serial investigations. On the other hand, in the cases which showed necrotic lesions in the non-weightbearing area, the necrotic area was reduced in size. These observations indicate that the necrosis can heal when mechanical stress is withdrawn from the reparative area of necrosis. Conversely, if the necrotic lesion is located in the weightbearing area and the collapse occurs at the junction between the intact and necrotic areas, the femoral head tends to subluxate anterolaterally, resulting in further collapse. Once the collapse progresses, the reparative area is exposed to a greater mechanical stress, which causes instability of the lesion, inhibiting the reparative process as seen in pseudoarthrosis of the fracture. This finding supports the argument claiming that withdrawal of mechanical stress from the necrotic lesion is one of the most important factors in successful treatment.

There are two important factors that must be considered in the treatment of the femoral head necrosis. First is the elimination of shear forces from the necrotic focus. This leads to the prevention of progressive collapse, which impedes the reparative process as mentioned previously. The second is joint realignment of the subluxated femoral head which is due to the collapse of the articular surface.

Femoral osteotomy is a very effective method of preventing collapse by diverting the mechanical stress from the lesion.

Osteonecrosis of the femoral head is generally localized in the antero-superior aspect of the head. In contrast, the posterior aspect of the head often retains a normal, smooth contour containing healthy cartilage even in many advanced cases. Considering this

Bone Circulation and Vascularization in Normal and Pathological Conditions
Edited by A. Schoutens *et al.*, Plenum Press, New York, 1993

363

characteristic anatomical localization I devised a new approach to osteotomy, which is a transtrochanteric transposition osteotomy with anterior torsion of the neck of the femur around its longitudinal axis (Sugioka, 1978; 1982; 1984; 1992).

It is important to evaluate the exact location of the intact area of the femoral head on plain A-P and lateral roentgenograms (Sugioka, 1982), tomograms, CT scans, and MRI to determine whether the osteotomy is indicated .

If the intact area is located in the postero-superior or antero-superior portion of the femoral head, anterior rotational or posterior rotational osteotomy is indicated.

Varus osteotomy may be indicated if the intact area occupies a large area in the superolateral portion than the posterosuperior or antero superior portions.

RESULTS OF TRANSTROCHANTERIC CURVED VARUS OSTEOTOMY

From 1969 to 1991, Nishio's transtrochanteric curved varus osteotomy (Sugioka, 1980) was performed on 42 hips in 39 cases with idiopathic and steroid-induced femoral head necrosis. The results of this varus osteotomy in 17 hips were reviewed. Follow-up periods ranged from 3 to 21 years.

All cases were classified into four stages on the basis of the preoperative roentgenographic findings according to the roentgenographic staging system by the Japanese Investigation Committee for Idiopathic Femoral Head Necrosis.

Stage 1: preroentgenographic stage, only detectable by bone scintigraphy, MRI or core biopsy.

Stage 2: early stage, roentgenographic evidence of necrosis with no collapse at all or a collapse of less than 2 mm.

Stage 3: advanced stage, more than 2 mm of collapse.

Stage 4: late stage, advanced stage associated with osteoarthritic changes.

The outcome of the operation was evaluated clinically and roentgenographically (Table 1)

Table 1. Success Rate

(Varus Osteotomy)		
Stage 2	2/2 hips	100 %
Stage 3	10/11 hips	91 %
Stage 4	1/4 hips	25 %
Total	13/17 hips	76 %

In the stage 2 the postoperative clinical and roentgenographic findings were excellent in both hips (Figure 1). Of 11 stage 3 hips, 10 showed excellent results, constituting a success rate of 91 %. However, in stage 4 hips only one out of 4 hips showed excellent results. The total success rate of cases in stage 2 to 4 was 76 %.

Complications of the varus osteotomy were noted in two hips of 47 operated hips. Deep infection was observed in one hip and delayed union of the osteotomy site in the other.

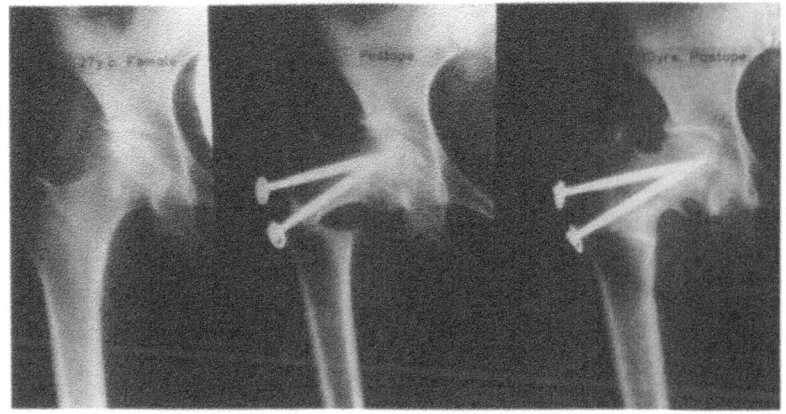

Figure 1. A 27-year-old woman with stage 2 steroid-induced femoral head necrosis was treated with varus osteotomy in 1980. The necrotic lesion healed completely as evident on roentgenograms. Left: preoperative A-P roentgenogram. Middle: postoperative roentgenogram. Right : roentgenogram A-P view, 10 years after osteotomy.

RESULTS OF TRANSTROCHANTERIC ROTATIONAL OSTEOTOMY

From 1972 to 1990, transtrochanteric rotational osteotomy was effectively applied as a new approach to 533 hips in 429 patients with idiopathic and steroid-induced femoral head necrosis.

The results of the operation of 295 hips were reviewed. The longest follow-up was 19 years and the shortest 3 years.

Analysis of the Results Based on Preoperative Staging

All these cases were graded into the previously mentioned four stages on the basis of analysis of the preoperative radiographic findings.

87 out of the 98 hips belonging to stage 2 manifested no pain and no progression to collapse of the newly established weightbearing area. Only twelve hips showed a slight collapse of the necrotic area. Therefore, the success rate was 89 % (Figure 2). In the stage 3 hips a 73 % success rate was obtained. In the stage 4 hips the success rate was 70 %. Success rate of total cases in stage 2 to 4 was 78 % (Table 2).

Table 2. Success Rate

(Transtrochanteric Rotational Osteotomy)

Stage 2	87/98 hips	89 %
Stage 3	98/134 hips	73 %
Stage 4	44/63 hips	70 %
Total	229/295 hips	78 %

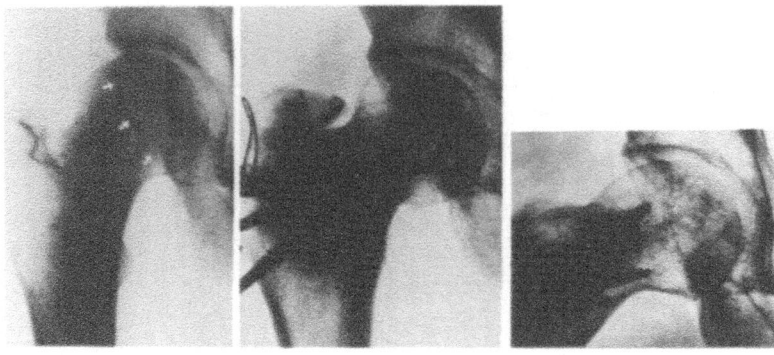

Pre—op   19yrs. after ARO

Figure 2. A 24-year-old man with stage II idiopathic femoral head necrosis was treated with transtrochanteric anterior rotational osteotomy in 1972. Nineteen years after osteotomy excellent clinical and roentgenographic results have been obtained and necrotic lesion has completely healed.
Left: preoperative A-P roentgenogram showed band-like sclerotic shadow in the anter-uperior aspect of the head (white arrows) with slight collapse of articular surface. Middle and right: roentgenogram A-P and lateral view 19 years after osteotomy.

Relationship between Ratio of Transposed Intact Articular Surface to the Acetabular Weightbearing Area and Progression of Collapse

The weightbearing area of the acetabulum was measured from the edge of the acetabular to a point on the acetabular roof determined by lines drawn on the postoperative anteroposterior roentgenogram (Sugioka, 1982; Sugioka, 1992). The ratio of the intact area of the femoral head to the acetabular weightbearing area is then determined. Based on this ratio 295 hips were classified into three groups. Out of 203 hips in which the ratio of the intact area was greater than 36 % only 15 hips showed further collapse which gives a success rate of 93 %.
In the group with intact area being grater than 60 % no progression of collapse was observed.
Fifty-one hips were classified into the second group in which the ratio ranged from 21 to 35 %. Success rate was 65 %.
Forty-one hips were classified into the third group in which the ratio was less than 20 %. Success rate was only 29 %.
Following analysis of the ratio of transposed intact articular surface to the acetabular weight-bearing area after osteotomy, I conclude that the intact area of the posterior part of the femoral head should whenever possible be transposed into the weightbearing area by an adequate rotation and intentional varus position (Sugioka, 1984).

Complications

Early complications were noted in 3 hips and late complications in 11 hips.
Neck fracture was observed in 4 hips. Of these 4 hips 3 hips with extensive lesions were

rotated by 180 degrees posteriorly. In this way the small intact portion of the infero-medial aspect of the head was brought into the weightbearing area.

Osteonecrosis was observed in 2 hips.

Salvage operations were performed on 18 out of 533 hips following rotational osteotomy. Eight out of 18 hips were a challenge to treat, because of the patients were young and the necrosis was advanced with an extensive lesion. The intact area on lateral hip roentgenograms was not well interpreted in two cases. Of the remaining eight hips, two hips had a salvage operation due to advanced osteoarthrosis.

Lastly, I would like to talk on pathological findings of the femoral heads which were removed at the time of a total hip replacement in two patients. They developed osteoarthritic changes several years after anterior rotational osteotomy. Also one autopsy case is included.

The first patient was a forty year old man with bilateral stage 3 hips who had bilateral rotational osteotomy 8 years ago. His left hip did very well but his right hip remained in a valgus position after the osteotomy. This resulted finally in osteoarthritic changes which required total hip replacement. Histologic findings showed complete healing replacement of the necrotic bone by living bone.

The second patient was a forty-nine year old male who developed osteoarthritic changes in his left hip after osteotomy for a stage 4 lesion and he had total hip replacement. In this case also, no necrotic lesion was shown in cross section microscopy. Only osteoarthritic changes were noted.

The third case was a sixty-five year old male who had anterior rotational osteotomy on his left hip 18 years ago and died of heart attack in 1991. A-P view of his left hip showed normal smooth articular surface without development of collapse and sclerotic focus in

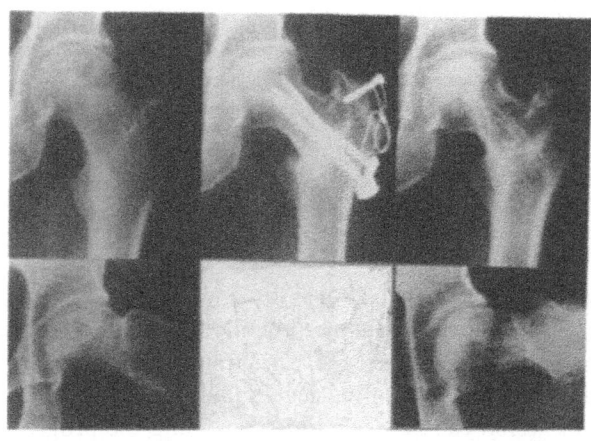

Pre. Ope.   Post.Ope   Post.Ope after 18 yrs.

Figure 3. A 47-year-old man with stage 3 idiopathic femoral head necrosis was treated with transtrochanteric anterior rotational osteotomy in 1972. Eighteen years after osteotomy excellent clinical and roentgenographic results have been obtained. He died of heart attack. Left: preoperative A-P and lateral view of the hip X-ray. Middle: postoperative roentgenogram. Right: A-P and lateral view showed normal smooth articular surface without development of collapse and sclerotic shadow in the antero-medial aspect of the femoral head.

the non-weightbearing area (Figure 3). The removed left femoral head showed a normal contour containing healthy cartilage in the weightbearing area. Cross section of the femoral head also showed normal articular cartilage and healthy subchondral bone. The necrotic bone was replaced by new bone apposition. However, the necrotic trabeculae with calcification was still visible in the small area of the femoral head far away from the joint surface. The reason for the incomplete healing of the necrotic lesion in this case still remains unsolved (Figure 4).

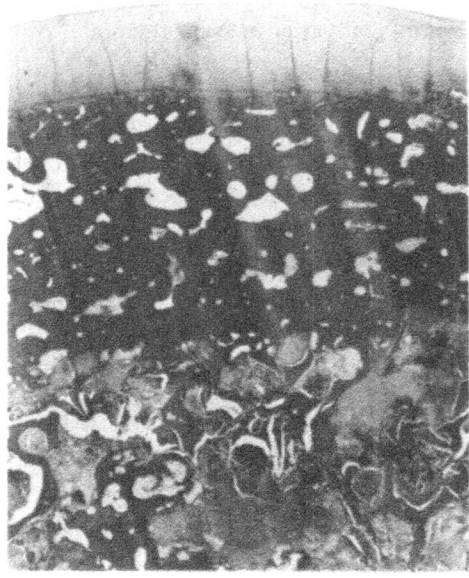

Figure 4. Cross section of the femoral head of the autopsy case of figure 3 shows normal articular cartilage and healthy subchondral bone which was replaced by appositional new bone. However, necrotic trabeculae with calcification still remain in the small area of the femoral head far away from joint surface

Conclusion

The present study showed that about 80 % of the cases with anterior rotational osteotomy achieved excellent results both clinically and radiographically. Progression of collapse was prevented in 93 % of cases in which intact areas of the head occupied more than 36 % of the weightbearing area of the acetabulum after osteotomy.

The fact that necrotic lesion can be healed by preventing the progression of collapse has been well demonstrated roentgenographically and histopathologically in the patients with long follow-up period.

Overall findings have demonstrated that the necrotic lesion can heal if the lesion is free from mechanical stress changing its weightbearing function into a non-weightbearing one by osteotomy. This has further proved my theory that the most important aspect in the treatment of femoral head necrosis is the elimination of shear forces in the necrotic focus which impede reparative process.

INDICATIONS FOR FEMORAL OSTEOTOMY

The indications for femoral osteotomy as clearly suggested by our results, are as follows (Figure 5).

Nonsurgical treatment is recommended for cases where osteonecrosis is in front of the medial third of the acetabular weightbearing surface or non-weightbearing area.

Femoral head necrosis in front of the middle third of the acetabular weightbearing surface necessitates, careful observation and partial weightbearing. If the collapse progresses, varus osteotomy and/or rotational osteotomy should be carried out as soon as possible. In cases where an extensive lesion is positioned in the lateral third of the acetabular weightbearing surface or in the more lateral part of the femoral head rotational osteotomy is absolutely indicated and should be carried out immediately.

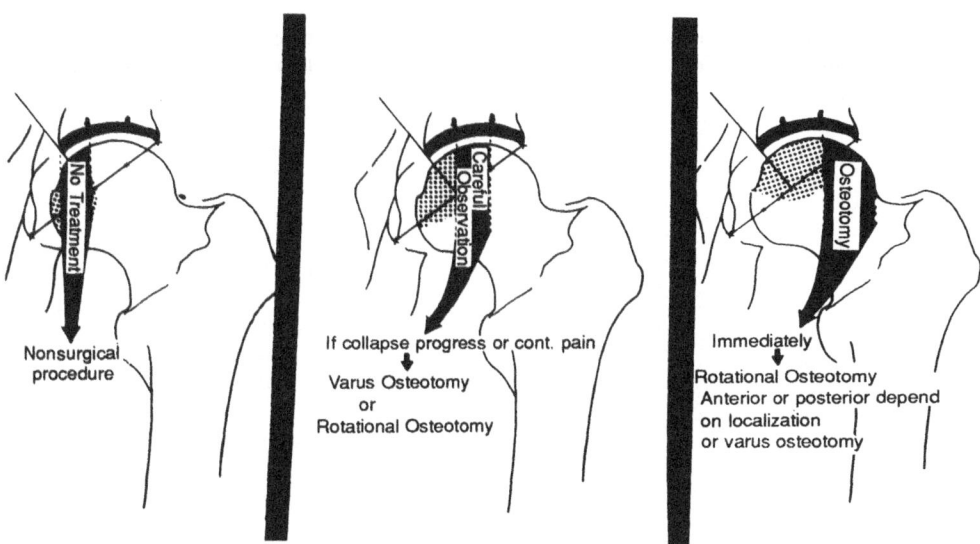

Figure 5. Schematic illustration of the indication for femoral osteotomy in the treatment of osteonecrosis of the femoral head.

ACKNOWLEDGMENT

This work is supported by the grant of Research Committee in the Ministry of Health and Welfare for Avascular Necrosis of the Femoral Head.

REFERENCE

Sugioka, Y.: Transtrochanteric anterior rotational osteotomy of the femoral head in the treatment of osteonecrosis affecting the hip. A new osteotomy. Clin. Orthop. 130: 191-201 (1978).

Sugioka, Y.: Transtrochanteric rotational osteotomy of the femoral head. In: The Hip, 3-23. Excerpta Medica, Amsterdam 1980.

Sugioka, Y., K. Ogata, T. Kitajima, A. Nishio: Transtrochanteric curved varus osteotomy in the treatment of dysplastic hip. Ibid., 227-244 (1980).

Sugioka, Y., I. Katsuki, T. Hotokebuchi: Transtrochanteric rotational osteotomy of the femoral head for the treatment of osteonecrosis. Follow up statistics. Clin. Orthop. 169: 115-126 (1982).

Sugioka, Y.: Osteotomies transtrochantériennes de rotation de la tête fémorale. Rev. Chir. Orthop. 69: 9-22 (1983).

Sugioka, Y.: Transtrochanteric rotational osteotomy in the treatment of idiopathic and steroid-induced femoral head necrosis, Perthes' disease, slipped capital femoral epiphysis and osteoarthritis of the hip. Indications and results. Clin. Orthop. 184:12-23 (1984).

Sugioka, Y.: Transtrochanteric rotational osteotomy in the treatment of the osteonecrosis affecting the hip. In: The hip, 235-238. Excerpta Medica, Amsterdam1984.

Sugioka, Y.: Transtrochanteric rotational osteotomy in the treatment of osteoarthritis of the hip and slipped capital femoral epiphysis. Ibid., 231-234 (1984).

Sugioka, Y.: Résultats et indications de l'ostéotomie de rotation transtrochantérienne. Chirurgie. 114: 617-623 (1987).

Sugioka, Y.: Transtrochanteric rotational osteotomy in the treatment of idiopathic and steroid-induced femoral head necrosis. Microsurgery for Major Limb Reconst., 189-199 (1987).

Sugioka, Y., K. Ogata: Avascular necrosis of the femoral head. Current Opinion in Orthop., 2: 437-441 (1991).

Sugioka, Y., T. Hotokebuchi, H. Tsutsui: Transtrochanteric anterior rotational osteotomy for idiopathic and steroid-induced necrosis of the femoral head. Indications and long-term results. Clin. Orthop. 277: 111-120 (1992).

OSTEONECROSIS
4. Arco Perspective for Staging

METHODOLOGIC PROBLEMS IN STAGING AND EVALUATING OSTEONECROSIS

Bernard N STULBERG[*] and JWM GARDENIERS[**]

[*] The Cleveland Clinic Foundation, Cleveland, USA
[**] Academisch Ziekenhuis, Nijmegen, Netherlands

The early diagnosis of osteonecrosis (ON) is based on the perception that ON is an evolutionary disease whose symptoms may precede any radiographic evidence of the presence of the disease. The presence of high-risk populations, and the general consensus that all treatment modalities show improved success rates when the disease is treated early in its course, reinforce the desire to identify the disease early. These efforts to identify the disease early could result in confusion as to the definition of the disease, particularly as early changes identified with new diagnostic techniques may spontaneously resolve.

It therefore becomes imperative to standardize the methods for establishing diagnosis, for quantitating the extent of the disease process, and for establishing minimum requirements and uniformity in the manner in which the success of intervention is determined .

In the first part of this discussion, we will identify several recent reports that underscore this need to develop a common language of diagnosis, staging, and results reporting. In the second part of this discussion, uniform diagnostic and staging system for ON of the femoral head will be presented. It is anticipated that wider application of such a system will identify areas requiring supplementation or modification. It is presented in earnest, however, as a commonality of language will allow orthopedists, rheumatologists, endocrinologists, and basic sciences, to work together to successfully understand and treat what often has proven in the past to be a disabling condition of the hip joint.

Several studies have dealt with the use of multiple diagnostic tests to identify the disease of osteonecrosis (ON) . Stulberg et al. used a protocol that called for prospective comparison of anterior-posterior and frog lateral radiographs, bone scanning with single photon emission computerized tomography (SPECT) using ^{99m}Tc-diphosphonate, magnetic resonance imaging (MRI), and invasive procedures of the functional exploration of bone and core biopsy. All methods were compared statistically for sensitivity, specificity, predictive value of a positive test, and predictive value of a negative tests. What emerged was an understanding that:

- no single test made the diagnosis of ON 100 % of the time;
- that biopsy was 100 % specific, but not 100 % sensitive for the diagnosis of ON, as sampling error of the biopsy placement and histologic processing artifacts can result in erroneous information from the biopsy;

Bone Circulation and Vascularization in Normal and Pathological Conditions
Edited by A. Schoutens *et al.*, Plenum Press, New York, 1993

373

- even though some tests were initially negative subsequent radiographic evaluation and/or femoral head resection for total hip arthroplasty (THA), would demonstrate changes diagnostic of ON and would, therefore, confirm the diagnosis;
- if patients were treated non-operatively, diagnosis by non-biopsy means could be established.

Perhaps it becomes clear that criteria other than histology need to be accepted as diagnostic for ON if all cases of the disease are to be included. This study led to the development of a diagnostic algorithm which allowed for proper staging of ON while making appropriate use of the diagnostic modalities available. This study, along with those of other investigators confirm the need for a classification system that would allow for the diagnosis of ON in the absence of, or in addition to the presence of histologic evidence of the disease.

The second area of controversy that points to the need for uniformity of results presentation relates to the attempt to compare success, or lack thereof, of the modalities used in treatment of osteonecrosis.

We performed a study which directly compared conservative and operative methods of treatment for ON of the femoral head. While it had long been felt that non-operative therapy with non-weightbearing was of no clinical benefit, we proposed an evaluation in a controlled randomized fashion to directly compare non-weightbearing management with core decompression.

In reporting our results, we attempted to compare our findings with those of others. It was difficult to do so, however, for authors reported their findings in different ways. In addition, patients included in studies by certain diagnostic criteria, would have been excluded in others. This made accurate comparison of different series of treatment regimens very difficult. Attempts to perform these comparisons assume that our diagnostic criteria for the presence or absence of disease and our method of reporting results, was adopted for each of the other ivestigators' work. For what should be clear reasons, it is fundamentally neither fair nor accurate for different investigators to attempt to combine or compare results of comparable or different techniques applied to a disease which is characterized differently. The Association Research Circulation Osseous (ARCO) meeting of 1990 brought together many investigators with similar experiences, who similarly were experiencing difficulties in the interpretation of their experiences in light of others. They believed and still believe that the time had approached to pursue international agreement on a classification, staging, and reporting sytem. The major concerns to be addressed were:
- to develop diagnotic criteria for ON that would allow inclusion of all cases of true ON and eliminate those cases which are not ON;
- to establish diagnostic criteria that will include all histologically positive cases as those where histology is either not available or would perhaps be misleading;
- to provide a classification system that would allow accommodation for new diagnostic procedures;
- to allow a system that would provide quantitation of the extent and location of the disease;
- to establish minimum criteria for the reporting of results of treatment. The ARCO Committee on Terminology and Staging, under the Chairmanship of JWM Gardeniers, was established to address these issues. Results of the efforts of that committee are presented here in a proposed international classification for diagnosis and staging of osteonecrosis.

THE ARCO PERSPECTIVE FOR REACHING ONE UNIFORM STAGING SYSTEM OF OSTEONECROSIS

By the ARCO-Committee on terminology and staging
JWM GARDENIERS, Chairman

At the time of the 2nd General Assembly of the ARCO, which was held on the island of Ischia in Italy on the 29th and 30th of November 1990, the "International ARCO-Committee on Terminology and Staging" was installed.

The assignment of the Committee is to come to an agreement on the terms used in bone necrosis, especially in the histopathological terminology and to establish one uniform classification of Osteonecrosis. The reason for this was that in literature we can find many different classifications on Osteonecrosis based on slightly or completely different definitions, terminologies and diagnostic criteria. These will always cause authors to have differences in opinions and different interpretations of each others work. Therefore the Committee had to be "international" because in many countries many authors have given their best to make a classification for this disease. You have therefore to keep in mind that the proposed classification is a composition of recently published scientific and clinical work and staging systems by many distinguished authors, made by the Committee.

On the 4th of May 1991 the members of the Committee assembled for the first time in Nijmegen, the Netherlands.

The following problems had to be solved first: - Terminology; - Diagnostic Criteria; -An Uniform International Classification (Staging System); - Histological Criteria.

TERMINOLOGY

The many terms used for necrosis of bone are at the least confusing. We all know the many terms used like avascular necrosis, osteonecrosis, infarction of bone and the list is much longer. Everybody uses these terms at random. Within the Committee it was agreed upon that the definition of necrosis of bone had to be the following:

"Bone" in an organ that consists of mineralized and non-mineralized tissues. Bone necrosis is a disease which causes death of bone and is called:

"OSTEONECROSIS."

Bone Circulation and Vascularization in Normal and Pathological Conditions
Edited by A. Schoutens *et al.*, Plenum Press, New York, 1993

375

DIAGNOSTIC CRITERIA

The goal has to be to make to diagnosis "true osteonecrosis" as early as possible in the course of the disease.The diagnostic criteria that are available at this moment are: Radiography, Scintigraphy, Functional Bone Investigation, MRI, Histology.

The problem with radiography is that when the signs are visible on the X-rays the disease has already progressed beyond the initial stage.

This is also true for scintigraphy and the scintigraphic sign "cold in hot", is not always reproducible. The reliability of scintigraphy is low.

"Functional Bone Investigation" as published by J. Arlet, R. Ficat and D. Hungerford is not to be used in a staging system. It is an invasive technique and the derived data, the intraosseous pressure and intraosseous venography are not pathognomonic for Osteonecrosis. You will find them in many other bone disorders too.

MRI is a non-invasive method to visualize early changes in bone. It is both more sensitive and more specific than all other diagnostic measures to diagnose Osteonecrosis.

From 1979 onwards the Steinberg-group, M. Steinberg, G. Haken and D. Steinberg have published many papers about that subject and many authors followed. MRI gives us an image of the existing process in the bone. Classic T1-views in a frontal and axial plane with sections of 5 mm, taken every 2.5 mm are recommended. This is the minimum requirement in Stage 0 and 1 when there are no radiological signs. Every other MRI technique may be included, but again as a minimal addition, a T2-weighted image. MRI gives us an image of a process in the bone seen as a "Band-Like-image of Low Intensity'"on the classic T1-weighted image and a "Double-Line-Sign" on a classic T2-weighted image.

Histology should be done at every possible opportunity. It remains the "Golden Standard", although it is an invasive method. However a descriptive histologic classification of the disease, especially in the early stages is not yet available.

PROPOSED CLASSIFICATION

Staging is a method of defining the development of the disease. Staging has therefore to include the very onset of the disease, Stage 0, and extend till the final end, the complete joint destruction; it must cover the full spectrum of disease.

In this proposal two items are very important which the committee discussed extensively on the 24th September 1992:

1. The "Japanese Orthopaedic Association Committee on Osteonecrosis" pointed out clearly that the position of the Osteonecrosis on the weightbearing dome of the femoral head is very important. A subclassification in a medial, central and lateral located disease has to be made.

2. M. Steinberg and his co-workers were the ones that taught us the importance of quantitation. They recently published a paper in 1991 about this quantitation by means of area involvement (Stage 0 and I), "Extension of the Crescent Sign" (Stage II and III) and the "Extension of the Dome Depression" and/or "Estimated Acetabular Involvement" (Stage IV and V). The Committee has added this as a subclassification to the proposal.

INTERNATIONAL CLASSIFICATION

Stage 0. All present diagnostic techniques are normal or non-diagnostic except histology. Future diagnostic techniques will enable us at some point to make the diagnosis in this Stage 0. Diagnostic techniques are beyond the plain X-rays, the use of scintigraphy and MRI.

Stage 1. Plain X-Ray and CT-scan are normal. Scintigraphy or MRI are abnormal. The classic MRI image is pathognomonic. An open biopsy will confirm the diagnosis.
In this Stage the Subclassification according to location and Femoral Head involvement must be used:
Subclassification: in "medial", "central" and "lateral" located Osteonecrosis and quantitation of Femoral Head involvement in A ="minimal" (area involvement of less than 15 %); B = "moderate" (area involvement between 15 % - 30 %); C = "extensive" (area involvement of more than 30 %).

Stage 2. Radiography shows areas of "abnormalities": mottled aspect, cysts, sclerosis and porosis. However there are no signs of collapse and the femoral head remains spherical on AP and lateral views on X-ray and CT-scan. Again scintigraphy and MRI are positive. Subclassification is important, as described in Stage 1.

Stage 3. The femoral head has begin to fail mechanically. The axial X-Ray shows a fine radiolucent subchondral fracture line, referred to as the "Crescent Sign". However there are no signs of flattening of the femoral dome. The spherical configuration remains intact.
Subclassification according to location remains the same.
Subclassification according to femoral head involvement is now: first it is determined if the Crescent is most prominent on the AP or Lateral view and then the length of the Crescent is expressed as a percentage of the entire articular surface: A = < 15 %; B = 15% - 30 %; C= >30 %.

Stage 4. Radiographs show the articular surface of the femoral head to be flattened, but there is no evidence of joint line narrowing. CT-scanning might be helpfull if there is no evidence of collapse on the plain X-rays.
Subclassification according to the location remains the same, but subclassification according to Femoral Head involvement must be done by "length of crescent" as described in Stage II and/or by the measurement of the "Dome Depression": A = < 2 mm depression of the articular surface; B = 2 mm- 4 mm; C= > 4 mm.

Stage 5. Radiographically the articular surface is flattened and the joint-space starts narrowing. This is often associated with changes on the acetabular side of the joint and with signs of a beginnig osteoarthritis with areas of sclerosis, cysts and marginal osteophytes. Subclassification according to location and Femoral Head involvement remains the same, but in the latter the extension of acetabular involvement can be estimated.

Stage 6. The radiographic examinations show advanced degenerative changes and finally a complete joint destruction.

The ARCO Committee suggests strongly that the minimum requirement for reporting results on scientific and clinical work must include the following:

1: A clinical quantitation by means of any hipscore-rating, e.g. the Harris Hip Score.

2: Staging according to this "ARCO International classification", in which this classification is meant to be a minimum requirement.

3: Radiographical and MRI subclassification is very important.

4: Therapeutic results on femoral head survival.

All other techniques or examinations can be added freely but separately from the above mentioned classification criteria. The above mentioned requirements are as mentioned before a minimum to be able to compare all the different reports on Osteonecrosis by all the different authors.

DISCUSSION

The Committee's effort to try to establish one uniform staging system for Osteonecrosis, the ARCO-classification, has not ended with the report. The discussion will continue.

The ARCO-classification should be a good compromise between the original systems predominating in Europe, USA and Japan respectively, as K. Ono clearly stated. It is not an easy task, but the enthusiasm and extensive ideas from all Committee members makes it a promising one.

The above proposal is an extensive one, because it consists of 7 stages. During the discussion at the Committee meeting on the 24th of September, as well as during the ARCO-conference, it became clear that a simplification is needed.

The system developed by J. Arlet and P. Ficat, as well as the systems made in the USA and Japan, contain fewer stages than the proposed ARCO-classification.

Discussions and recently received comments by mail suggest strongly that the above proposed Stage III and IV should be reestablished in just one new Stage III. In both stages the structural distortion of the integrity of the femoral head is compromised. The "Crescent Sign" is a fracture line and collapse starts and is progressing during Stage III and IV.

The same argument is true for the above mentioned Stage V and VI, but in these two stages the distortion of the integrity expands also to the acetabulum.

Stage V and VI can be compromised in one new Stage IV.

Subclassification for the new Stage III and IV should be proposed above for the old Stage IV and the old Stage V.

The Committee is working on a histological classification to be included in or added to the ARCO-classification.

The problem is that there are not enough histological data available in the very early stages, Stage 0 and I, to make such a classification.

Furthermore synchronisation between the MRI data and histological data is urgently needed. This will finally enable us to establish the early diagnosis in Stage 0 an I and differentiate between these two stages.

Histological data collection and synchronisation with the MRI data is therefore of utmost importance.

Table 1. Considerable heterogeneity exists in intraosseous vascularity, hemodynamics, and bone metabolism. Regional blood flow (RBF), vascular volume (VV), mean blood transit time (TT), tissue hematocrit (tHct), and uptake of bone seeking diphosphonate (^{99m}Tc-DPD) in selected functionally and anatomically different regions of the femur in growing dogs (mean, SEM, n=8)

INTERNATIONAL CLASSIFICATION OF OSTEONECROSIS

STAGE	O	I	II	III	IV	V	VI
FINDINGS	all present techniques normal or non-diagnostic histology only !	X-Ray & CT normal at least one of the below mentioned positive	X-Ray : abnormal the motled aspect sclerosis cysts porosis	X-Ray : crescent sign	X-Ray : flattening of articular surface	stage IV plus narrowing of joint and/or acetabular changes	complete joint destruction
			NO FLATTENING				
TECHNIQUES	X-Ray, CT scintigraph MRI	X-Ray, CT scintigraph MRI QUANTITATE ! only MRI	X-Ray, CT scintigraph MRI QUANTITATE ! MRI & X-Ray	X-Ray, CT ONLY QUANTITATE ! MRI & X-Ray	X-Ray, CT ONLY QUANTITATE ! on X-Ray	X-Ray ONLY	X-Ray ONLY
SUBCLASSIFICATION	NO	QUANTITATION % AREA INVOLVEMENT minimal A < 15% moderate B 15 % - 30 % extensive C > 30 % LOCATION		LENGTH of CRESCENT A < 15 % B 15 % - 30 % C > 30 %	% SURFACE COLLAPSE & DOME DEPRESSION A < 15 % < 2 mm B 15 % - 30 % 2 mm - 4 mm C > 30 % > 4 mm	NO	

medial   central   lateral

379

SIGNATORIES

J.W.M. Gardeniers, chairman - P.C. Acquaviva - J. Arlet - R. Burkhardt

M. Gagne - D. Hungerford - J.P. Jones - B. Mazieres

K. Ohzono - K. Ono - M.F. Steinberg - B. Stulberg

Y. Sugioka

REFERENCES

Arlet, J., P. Ficat: Diagnostic de l'ostéonécrose fémoro-capitale primitive au stade 1. Revue de Chirurgie Orthop. 54: 637-648 (1968).

Enneking, W.F.: The choice of surgical procedures in idiopathic aseptic necrosis. The hip (Proc 7th Mtg of the Hip Soc): 238-243, St. Louis, CV Mosby Co. 1979

Ficat, R.P.: Early diagnosis of osteonecrosis by functional bone investigation. Progress Orthop. Surg. 5: 17-28 (1981).

Hungerford, D.S., Th.M. Zizic: Alcoholism associated ischemic necrosis of the femoral head: Early diagnosis and treatment. Clin. Orthop. Rel. Res. 130: 144-153 (1978).

Marcus, N.D., W.F. Enneking, R.A. Massam: The silent hip in idiopathic aseptic necrosis. J Bone Joint Surg. 55-A: 1351-1366 (1973).

Ono,K.: Criteria for ANFH in Japan, JOA, 1986.

Ohzono, K, M. Saito, K. Takaoka, K. Ono, S. Saito, T. Nishina, T. Kadowaki: Natural history of non-traumatic avascular necrosis of the femoral head. J. Bone Joint Surg. 73-B: 68-72 (1991).

Springfield, D.S., W.F. Enneking: Surgery for aseptic necrosis of the femoral head. Clin. Orthop. Rel. Res. 130: 175-185 (1978).

Steinberg, M.E., D.R. Steinberg: Evaluation and staging of avascular necrosis. Seminars in arthroplasty. 2: 175-181 (1991).

Sugioka, Y.: Pathomechanism and treatment of idiopathic avascular necrosis of the femoral head (in Japanese). J. Jpn. Orthop. Assoc. 50: 1173-1192 (1976).

INDEX

The manufacturer's authorised representative in the EU is Springer
Nature Customer Service Centre GmbH, Europaplatz 3, 69115 Heidelberg,
Germany. If you have any concerns regarding our products, please
contact ProductSafety@springernature.com

Printed and bound by CPI Group (UK) Ltd, Croydon, CR0 4YY
23/04/2026
02095629-0017